The Inductive Brain in Development and Evolution

The Inductive Brain in Development and Evolution

Nelson R. Cabej

Department of Biology, University of Tirana, Tirana, Albania

ACADEMIC PRESS

An imprint of Elsevier

ELSEVIER

Academic Press is an imprint of Elsevier
125 London Wall, London EC2Y 5AS, United Kingdom
525 B Street, Suite 1650, San Diego, CA 92101, United States
50 Hampshire Street, 5th Floor, Cambridge, MA 02139, United States
The Boulevard, Langford Lane, Kidlington, Oxford OX5 1GB, United Kingdom

Notices
Knowledge and best practice in this field are constantly changing. As new research and experience broaden
our understanding, changes in research methods, professional practices, or medical treatment may
become necessary.

Practitioners and researchers must always rely on their own experience and knowledge in evaluating and
using any information, methods, compounds, or experiments described herein. In using such information
or methods they should be mindful of their own safety and the safety of others, including parties for whom
they have a professional responsibility.

To the fullest extent of the law, neither the Publisher nor the authors, contributors, or editors, assume any
liability for any injury and/or damage to persons or property as a matter of products liability, negligence or
otherwise, or rom any use or operation of any methods, products, instructions, or ideas contained in the
material herein.

Library of Congress Cataloging-in-Publication Data
A catalog record for this book is available from the Library of Congress

British Library Cataloguing-in-Publication Data
A catalogue record for this book is available from the British Library

ISBN 978-0-323-85154-1

For information on all Academic Press publications
visit our website at https://www.elsevier.com/books-and-journals

Publisher: Charlotte Cockle
Acquisitions Editor: Anna Valutkevich
Editorial Project Manager: Devlin Person
Production Project Manager: Kiruthika Govindaraju
Cover Designer: Alan Studholme

Typeset by SPi Global, India

Working together
to grow libraries in
developing countries

www.elsevier.com • www.bookaid.org

Contents

Introduction

As material systems, living beings adhere to the second law of thermodynamics by spontaneously losing structural order. The fact that living systems, nevertheless, survive for long periods of time, extending from minutes to thousands of years, is a highly improbable event that the Austrian physicist, Erwin Schrödinger (1887–1961), explained with the fact that these systems compensate for their loss of order by absorbing with food new order or negative entropy in Schrödinger's term (now generally known as negentropy). This is partly true; metazoans disassemble most of their nutritional substances (proteins, carbohydrates, and fats) into their molecular components, amino acids, sugars, fatty acids, and glycerol, to create species-specific order. Thus, what metazoans get with food is matter and energy rather than homospecific order. From this heterospecific order they get with food, metazoans create species-specific order, most importantly proteins, based on the genetic information carried by specific DNA nucleotide sequences.

At the cell and subcellular levels, a metazoan loses matter (chemical components) and energy; at the cell and supracellular level, it loses cells that it must replace to survive and reproduce. Yet, animals maintain to a relative steady state both the body fluid composition and the amount and proportions of different types of cells. This implies that metazoans detect and assess abnormal changes and generate appropriate instructions to restore their normal state.

How do animals identify the changes in the level of numerous body fluid variables and how do they perceive when these limits are exceeded? How do they identify where the cell loss occurred, assess how many cells are lost, and determine the amount and sites of cells to replace and send the relevant instructions to replace them?

It is textbook physiological knowledge that many blood and body fluid variables, including the body temperature, in warm-blooded animals are regulated by the CNS (central nervous system). That all vital functions (respiration, heart work, digestion, excretion, perspiration, blood pressure, etc.) and animal physiology in general are neurally regulated is also common knowledge. The nervous system also determines animal behavior. While animal functions and behavior are controlled by the nervous system, what determines the development and maintenance of the animal structure?

Of course, the development and maintenance of the animal structure and morphology cannot be reasonably derived from "common knowledge"; one cannot extrapolate a similar function of the nervous system in maintaining the animal structure from the role of the nervous system in animal physiology and behavior, but we can rely on the philosophical postulate that the mechanism of regulation and maintenance of the animal body must also reside within the animal body. Otherwise, we land in the realm of teleology. Given that the animal functions and behaviors are under the control of the nervous system and they are inseparable from animal structure (it is wisely said that function and behavior are the structure's raisons d'être), it is tempting to inquire about a possible role of the nervous system in the development and maintenance of the animal structure/anatomy.

1. Levels of organization and the integrated control

Since first postulated almost two centuries ago by Theodor Schwann (1810–1882) and Matthias Schleiden (1804–1881), cells are regarded as the basic units of life (die Elementartheile der Thiere und Pflanzen) (Schwann, 1839), from unicellulars up to the most complex multicellular organisms.

The animal organism is a hierarchically regulated system of cells, not a Virchowian social organization of individual cells with a separate existence (einzelne Existenz) (Virchow, 1858). While the basic structural hierarchy of metazoans consists of two main levels of organization, the cell and organismic levels, intermediate levels of tissues, organs, and organ systems between them lay as parts of the integrated organismal structure rather than independently acting biological entities, operating for the sake of the organism. This suggests that their functions must be under ultimate control at the organismic level.

The structure's hierarchy implies a hierarchical system of control exists, whereby every subsystem level of structural organization possesses its own subsystem of control. All the subsystems in the hierarchy of control constitute the integrated control system (ICS) in which higher levels control and restrict the degrees of freedom of lower levels. Each level "functions according to laws of behavior appropriate to that level" (Ellis, 2012), but the higher level imposes constraints or "sets the context for lower level actions … or … influence what happens at the lower levels, even if the lower levels do the work" (Ellis, 2012). Arguably, no multicellular supracolonial system could arise and exist in the absence of an ICS, capable of integrating and coordinating the activity of all the lower levels of organization, so that they all serve to maintain the function and structure of the whole system (Cabej, 2012, 2019).

The characterization of living systems as ordered structures that resist overall disorder as determined by the law of entropy implies that these systems have evolved the ICS (not necessarily a nervous system) capable of restoring lost order at the molecular, cell, and supracellular levels. To perform this entropy-resisting function, the ICS must

1. receive information on the state of the system and assess deviations from the normal state at the cell, tissue, organ, and systemic levels,
2. possess information about the normal species-specific structure to identify by comparison deviations from the norm,
3. integrate and process incoming information to produce chemical instructions and start signal cascades for restoring the homeostasis and normal structure, and

4. use biochemical pathways to send chemical instructions to the target cells, tissues, and organs and switch on and off specific genes/GRNs in strictly determined spatiotemporal patterns.

2. Where does the ICS reside in the animal body?

Several lines of empirical evidence may help us ascertain what performs these functions of the ICS in animals:

- the flow of developmental information in the process of organogenesis,
- the flow of information for inter- and transgenerational plasticity,
- the evidence on the source of information in the process of regeneration of the lost parts or organs, and
- the evidence on the control of the ordered deposition of parental factors in gametes.

Most of this work is devoted to reviewing and discussing the evidence, which points in the direction of the nervous system as the ICS in eumetazoans.

The metazoan ICS did not arise as the nervous system from the beginning; lower metazoans, placozoa, and sponges do not possess a nervous system. Yet, they have a nonneural control system that regulates their development, homeostasis, and reproduction (Cabej, 2020). How difficult its evolution was is indicated by the extremely long time it required to occur; for almost 3 billion years since its emergence, about three-fourths of its history, life on Earth remained at a basic cell level of organization and complexity. The paramount role of the neural ICS for the evolution of the metazoan life is demonstrated by the fact that most of the progress in the evolution of animal life took place during the last half-billion years since the neural ICS emerged during the Cambrian explosion, especially with the centralization of the nervous system.

3. Nervous system in early embryonic development

The eumetazoan life starts from a unicellular structure, egg or zygote. The genetic machinery at this moment is silenced, and depending on the animal species, it is activated at different embryonic stages, from the 2-cell stage in the sea urchin, but parental cytoplasmic factors continue to be active up to the ~16 thousand-cell stage (14 cleavage cycles) in *Xenopus laevis*. Activation of the embryonic genome during the maternal-to-zygotic transition (MZT) stage is a function of maternal/paternal factors (Winata and Korzh, 2018).

Of course, at this stage, the embryo lacks a nervous system, but there is evidence suggesting that the crucial process of the emplacement of the regulatory parental factors into gametes may be influenced by the nervous system. So, e.g., vitellogenin, one of the main components of insect eggs, is not synthesized by the oocyte but is incorporated into the oocyte via receptor-induced endocytosis in the process of formation of the endocytic complex that involves juvenile hormone (JH), which in turn is induced by brain allatotropins. Neuromodulators serotonin and dopamine and other neuroactive substances play a similar role (Handler and Postlethwait, 1977; Raote et al., 2013).

Vitellogenin deposition in the oocyte may not be considered patterned enough to require specific information for its emplacement into the oocyte, but other transcription factors that are crucially important for early embryonic development in the oocyte are deposited in specific regions of the oocyte. So, e.g., in *Drosophila*, several maternal cytoplasmic factors, including developmentally critical mRNAs such as *bicoid*, *gurken*, *oskar*, *nanos*, etc., are provided to the oocyte by the neighboring nurse cells in a process that starts with the transport of various maternal mRNAs before the nurse cell squeezes its whole contents into the oocyte. The process of forced dumping follows the programmed death of nurse cells induced by ecdysone, whose secretion in turn is centrally regulated by the neuropeptide PTTH (prothoracicotropic hormone) (Soller et al., 1999). These mRNAs are deposited to specific sites in the egg (*oskar* to the posterior end of the egg, *bicoid* to the anterior, *gurken* to the anterior-dorsal, and *nanos* to the posterior), clearly indicating that information is used for the patterned deposition of mRNAs.

There is no direct evidence on the mechanism of the movement of these transcripts to specific sites of the egg. The only thing we know from empirical studies of the process is that mRNAs originating in the nurse cells upon entering the egg are transported along microtubules and the length and direction of microtubules appear to determine their location in the egg. The only mechanism of regulation of the length of microtubules we know of is a neural one. It derives from studies on the regulation of the color change by fish iridophores in which the colors reflected by skin reflectors are regulated by neurally determined changes in the length of microtubules (via their polymerization/depolymerization), which in turn regulate the distance between guanine plates, thus determining the adaptive color reflected by iridophores (Oshima and Fujii, 1987; Mäthger et al., 2003, 2004).

Abundant evidence has been accumulated in the last two decades on the selective transport of various miRNA types into sperm cells and the consequent development of new traits in the carriers of these miRNAs. It is observed that a number of sperm miRNAs induce transgenerational effects in mice offspring (Rassoulzadegan et al., 2006; Grandjean et al., 2009, 2016; Sarker et al., 2019), and it is demonstrated that most of these sperm miRNAs are provided to sperms via epididymosomes released by epididymal epithelial cells while sperms move along the epididymis (Belleannée, 2015; Chen et al., 2016). The process of the incorporation of epididymal miRNAs in sperm appears to be controlled by the local sympathetic innervation, as indicated by the evidence that sympathetic denervation inhibits the development of the embryo (Ricker, 1998; Ricker et al., 1997). A stunning experiment in Australia demonstrated that the male mouse brain transfers in sperms an experimentally injected human RNA, which is also detected in the brain, but not in other organs, of the offspring of injected male mice mated with uninjected females (O'Brien et al., 2020).

4. Self-organization of the nervous system

When the bulk of the parental cytoplasmic factors is exhausted, the CNS is operational as indicated by the start of neural activity. The spontaneous neural activity of neurons instructively determines the specificity of synaptic connections between partner neurons resulting in the formation of neural circuits. The spontaneous activity of neurons in placental animals takes place during the uterine life, without contact with the natural environment, i.e., experience independently, and yet, we have no reasonable or hypothetic explanation on how neurons generate this activity that is responsible for the formation of trillions to quadrillions of specific neuronal connections and generation of corresponding enormous amounts of information. Observations on synaptogenesis in *Drosophila melanogaster* embryos show that all neurons participate in the spontaneous activity that is characterized by brain-wide periodic active and silent phases. How neurons find and connect with the matching partners among thousands to billion/trillions of neurons remains an enigma (Akin et al., 2019). Spontaneous neuronal activity is responsible for the amazing establishment of the myriad of specific synaptic connections between partner neurons. How neurons via axons find their partners is still an unresolved problem. It is said that axons find their way to the matching partner based on the neurons' "best guess" (Katz and Shatz, 1996), but this hardly explains anything.

The idea that neurons make their way to the matching neurons implies a teleonomic purpose and possession of information to find them. Crucial in this case is the origin of the information (=molecular "instructions") that the neuron uses to find its matching partner among the myriad of other neurons. The prevailing idea is that the axon is guided to its target by attractants and chemorepellents (Polleux et al., 2000), implying that the chemoattractants and chemorepellents provide neurons with information necessary to reach their matching partner. To make this hypothesis believable, one must explain why this information is provided only to specific neurons and not to all as if the chemoattractant or chemorepellent is customized just for a particular type of neuron. Are certain neurons selectively "informed" on the "meaning" of chemical signals? There is no genetic answer to this question because all neurons of an animal have the same genotype.

In the same vein, it is sometimes said of the role of leucine-rich repeat (LRR) proteins (Ledda and Paratcha, 2016) and cells (Schuldiner and Yaron, 2015) that provide axon with instructive signals, about "instructive cues" (Wang et al., 2013) that guide axons, or about a "molecular cue that instructs pruning" (Lu and Mizumoto, 2019). Again, such hypotheses imply an unidentified mechanism that customizes "instructions" to make them intelligible to some neurons but not to the rest of them. Certainly, specific neurons use specific cues to find their way to the target cells, but this again implies that these neurons *know* the "meaning" of these cues, which other neurons do not; this is another way of saying that the information is in specific neurons rather than in cues (chemoattractants and chemorepellents).

This indicates that the information guiding neurons to the target neurons, and cells in general, is not provided externally to neurons in any form of instruction. Biochemical mileposts are not information; they are simply cues recognized only by specific neurons to find their way to intended destinations. In an anthropomorphic context, mileposts or traffic signs serve as information only to the humans who know their "meaning," not to animals. Even per se, the neuron's drive to search for its matching partner demonstrates that it intends to reach its destination and "knows" how to discern the partner from the myriad of neurons in the brain.

Hypothetical speculations aside, let us consider reliable experimental evidence on the migration of the neural crest cells from the dorsal neural tube/CNS to their destinations throughout the animal body. Leading investigators in the field suggest that the information these cells use to reach destination sites is inherent and that before leaving the neural tube/CNS neural crest cells are provided with information on where to go and what to do: "the proper program of events governing the migration of crest may need first to be established in the hindbrain, to allow migratory crest cells *to interpret and respond to environmental signals* (emphasis mine) is set up through a series of tissue interactions" (Trainor et al., 2002).

5. Nervous system in organogenesis

Although experiments specially designed to explore the possible role of the nervous system in the induction of organ development in metazoans have not been performed, firm relevant evidence abounds in biological publications. To the best of my knowledge, B.K. Hall was the first to emphasize this role by pointing out that from its inception, the CNS engenders a network of inductions that give rise to the different cells, tissues, and organs (Hall, 1998a) and further embryonic structures "arise in relation to this central axis. This is especially evident in the development of paired elements such as the somites that presage the vertebrae, and paired organ rudiments such as left and right limb buds and the primordia of the gonads, kidney, lung and heart" (Hall, 1998b). Chapter 3 is devoted to presenting evidence on the role of the nervous system in the development of various organs in animals.

6. Nervous system in evolution

At a global evolutionary scale, it is generally observed that great transitions in the animal evolution coincided with dramatic changes in the Earth's environment, such as the drop in the sea level, meteoritic impacts, orogenic and volcanic activity, acidification of the sea, retreat of glaciers, etc., which challenged the survival of species and larger taxonomic groups and often led to substantial extinction rates.

Drastic changes in the environment that cause enduring stressful conditions lead to activation of the neurohormonal stress response mechanism, accompanied with adaptive behavioral and physiological changes as well as morphological changes within the limits of the reaction norm, or cause stress-induced adaptive or maladaptive phenotypic changes.

The observation that the onset of adaptive behavioral changes often coincides with adaptive morphological changes has been emphasized a long time ago (Mayr, 1988), and from a general biological viewpoint, the behavior is the raison d'être of the animal morphology/organ. However, the correlation between the onset of the morphological and behavioral changes does not prove the existence of a causal relationship between them, although it has been proposed that the correlated onset of the behavioral and morphological traits suggests that they are induced by one and the same mechanism (Carvalho and Mirth, 2015).

Both inborn behavior (instincts) and learned behavior are products of the activation of special neural circuits (Sato et al., 2020). It has been argued that the return of ancestral environmental conditions by forcing animals to perform a new learned behavior, after the loss of the organ that used to perform that behavior, may reactivate the relevant circuitry and lead to the reevolution of the lost organ and evolution of the learned behavior into an inborn one. This is impressively indicated by the switching of the aquatic mammals from walking to swimming behavior that was associated with stepwise reevolution of the interdigital webbing, elongation of the body, reduction up to the loss of hind limbs, flattening of the tail, and modification of other traits of mammals that after a long history of completely terrestrial life reevolved the ancestral body form to invade seas and rivers throughout the world (Cabej, 2012).

Empirical evidence from studies on the sympatric speciation accumulated in the last decades unmistakably shows that the neural mechanisms are responsible for reproductive isolation of particular groups of a population of a species and the consequent formation of new species, in a process that precludes geographic isolation and involvement of new or modified relevant genes.

7. Where does the information for evolution of metazoans come from?

Evolutionary change is a change in the hereditary endowment of living beings at the molecular-genetic, cellular, or supracellular level. For the scope of this work, evolutionary change is considered any modified or new discrete, ancestrally not possessed, morphological trait that is transmitted to the progeny "permanently" or for an indefinite number of generations. Most of the knowledge on this issue comes from experimental evidence on inter- and transgenerational plasticity.

The crucial issue in the process of evolving a new, ancestrally not possessed morphological trait is the source of the appropriate information for erecting an amazingly complex structure consisting of thousands to billions of cells of different types arranged in strictly determined and often intricate spatial patterns. The newly emerged structure is put to the test of the natural selection immediately and over generations.

When it comes to the source of information for inherited morphological changes, the first thing that comes to mind is the genetic information, but decade-long studies on inherited plasticity have failed to provide evidence on the involvement of new or changed genes in the process. We still have no formalized hypothesis how changes in a gene or a group of genes might erect a multicellular adaptive morphology.

7.1 On the role of environmental stimuli in emergence of new traits

Emergence of new/modified traits in studied cases of inter- and transgenerational plasticity is an adaptive response of the organism to stressful environmental stimuli. Environmental stimuli are received by sensory organs and transmitted to the CNS via afferent pathways. They represent the first link in the causal chain leading to the development of the new/changed trait and are often considered to provide the organism with the information for the development of the new trait. No effort has been made to elaborate on, or substantiate, the statement. Hence, until proved otherwise, the statement remains unvalidated for the following reasons:

First, if the stimulus would be the source of information, it would not be species-specific, but would induce expression of the trait at least in closely related or sibling species, which it does not.

Second, the emergence of new traits is related to specific changes in the patterns of gene expression, but environmental stimuli per se do not and cannot induce expression of any gene.

Third, an environmental agent or condition is perceived as a stimulus to induce a specific response only when its size/intensity exceeds an upper limit that is neurally determined.

The species-specific threshold is responsible for the fact that what acts as a stimulus in a species does not in another closely related species. Such thresholds are also known as set points and are determined in specific brain areas (Hammel et al., 1963; Boulant, 2000; Nakamura, 2011). Again, were stimuli information or "instructions," we could predict that these "instructions" would induce similar changes in different animals; this is clearly not the case. At best, stimuli are cues, to which different animals assign different "meanings" and respond accordingly. To the brain, the stimulus is a challenge or "problem," whose solution in most cases emerges in the form of a phenotypic adaptation.

While the hypothetical role of the stimuli in providing the organism with information for building new traits is clearly unsubstantiated, there is no doubt that the stimulus is the first link in the causal chain that leads to the emergence of the new/modified trait, and we need to know where in the causal chain the adaptive morphological information is generated in cases of inter- and transgenerational plasticity and by extrapolation to evolutionary changes.

7.2 On the role of the nervous system in emergence of new traits

In a generalized scheme, the input of environmental stimuli is received by neurons of the sensory (visual, olfactory, auditive, and tactile) organs, which encode it into a patterned spike train and in this form transmit it to higher brain centers where it is processed in specific neural circuits. The processing results in the production of a chemical output that starts a signal cascade that, upon reaching the target cells, activates particular transduction pathways, leading to the expression of specific genes and/or GRNs and the development of the new or modified adaptive trait.

Exposing an animal to stressful stimuli or conditions may lead to the appearance of the inter- or transgenerational plasticity in individuals exposed to the stimulus and to their offspring (F1) for one or more generations, or the trait may appear first in the offspring (F1) and persist for one or a number of consecutive generations.

The development of the trait is inherently intended to reduce/avoid harmful effects of the stimulus or adapt the organism to the new environmental conditions. Key in the event is the processing of the stimulus in the neural circuits that translates the unintelligible language of the environmental stimuli or conditions into the genetic-biochemical language intelligible to genes (stimuli per se cannot induce expression/suppression of any genes). Neural processing solves the problem posed by the stressful stimulus by establishing a unique, naturally not existing, relationship between the environmental stimulus and the gene (Cabej, 2019), embodied in a signal cascade that provides the genome with the instructions (=information) for expressing specific genes. The cascade functions as a communication channel for the flow of adaptational information from the brain to the genome. Noteworthy, the neural activity via nerve endings can also act directly on cells to induce gene expression (Hegstrom et al., 1998), regulate gene splicing (Iijima et al., 2016; Hermey et al., 2017; reviewed by Cabej, 2020) and expression of microexons in human brain neurons (Scheckel and Darnell, 2015), etc.

In anticipation of controversies from extrapolating to evolutionary change knowledge from the transgenerational plasticity, two facts are noteworthy. First, both phenomena refer to qualitatively one and the same thing; that is the emergence and inheritance in the progeny of new/modified traits. Second, the quantitative difference between them in the number of generations that inherit the new trait is often blurred and there are known cases when, under laboratory conditions, the transgenerational plasticity switches (transforms) to evolutionary change and reversely, the evolutionary change within a moderate number of generations of returning to the ancestral conditions reverts back to the ancestral state (Teotonio and Rose, 2000, 2001; Teotonio et al., 2002).

Following the flow of information through signal cascades for intergenerational and transgenerational plasticity in all these cases leads to the nervous system as the ultimate source of adaptive information. One of the most stunning facts revealed by these studies is that no changes in relevant genes are involved in their emergence and inheritance of new/modified traits in animals. Moreover, in some cases, no epigenetic modifications (DNA methylation and chromatin remodeling) and in other cases their involvement are identified as correlational rather than causal.

References

Akin, O., Bajar, B.T., Keles, M.F., Frye, M.A., Zipursky, S.L., 2019. Cell-type-specific patterned stimulus-independent neuronal activity in the *Drosophila* visual system during synapse formation. Neuron 101 (5), 894–904.e5.

Belleannée, C., 2015. Extracellular microRNAs from the epididymis as potential mediators of cell-to-cell communication. Asian J. Androl. *17*, 730–736.

Boulant, J.A., 2000. Role of the preoptic-anterior hypothalamus in thermoregulation and fever. Clin. Infect. Dis. *31* (Suppl. 5), S157–S161.

Cabej, N. R. (2020). *Epigenetic Mechanisms of the Cambrian Explosion* (pp. 19–21). London, UK; San Diego; Cambridge, MA; Oxford, UK: Academic Press.

Chen, Q., Yan, M., Cao, Z., Li, X., Zhang, Y., Shi, J., et al., 2016. Sperm tsRNAs contribute to intergenerational inheritance of an acquired metabolic disorder. Science *351* (6271), 397–400. https://doi.org/10.1126/science.aad7977.

Cabej, N.R., 2012. *Epigenetic Principles of Evolution* (pp. 588–594). Elsevier Inc., London, UK/Waltham MA.

Cabej, N.R., 2019. *Epigenetic Principles of Evolution* (2nd ed., p. XXXVIII). MA/Oxford: Academic Press, London/San Diego/Cambridge.

Carvalho, M.J.A., Mirth, C.K., 2015. Coordinating morphology with behavior during development: an integrative approach from a fly perspective. Front. Ecol. Evol. https://doi.org/10.3389/fevo.2015.00005.

Ellis, G.F.R., 2012. Top-down causation and emergence: some comments on mechanisms. Interface Focus *2* (1), 126–140.

Grandjean, V., Gounon, P., Wagner, N., Martin, L., Wagner, K.D., Bernex, F., et al., 2009. The miR-124-Sox9 paramutation: RNA-mediated epigenetic control of embryonic and adult growth. Development *136* (21), 3647–3655. https://doi.org/10.1242/dev.041061.

Grandjean, V., Fourré, S., De Abreu, D., Derieppe, M.A., Remy, J.J., Rassoulzadegan, M., 2016. RNA-mediated paternal heredity of diet-induced obesity and metabolic disorders. Sci. Rep. *5*, 18193. https://doi.org/10.1038/srep18193.

Hall, B.K., 1998a. *Evolutionary Developmental Biology* (2nd ed., p. 134). Chapman & Hall, London.

Hall, B.K., 1998b. *Evolutionary Developmental Biology* (2nd ed., p. 163). Chapman & Hall, London.

Hammel, H.T., Jackson, D.C., Stolwijk, J.A.J., Hardy, J.D., Stroeme, S.B., 1963. Temperature regulation by hypothalamic proportional control with an adjustable set point. J. Appl. Physiol. *18*, 1146–1154.

Handler, A.M., Postlethwait, J.H., 1977. Endocrine control of vitellogenesis in *Drosophila melanogaster*: effects of the brain and corpus allatum. J. Exp. Zool. *202*, 389–402. https://doi.org/10.1002/jez.1402020309.

Hegstrom, C.D., Riddiford, L.M., Truman, J.W., 1998. Spatial restriction of expression during metamorphosis of muscle in the moth, *Manduca sexta*. J. Neurosci. *18*, 1786–1794.

Hermey, G., Blüthgen, N., Kuhl, D., 2017. Neuronal activity-regulated alternative mRNA splicing. Int. J. Biochem. Cell Biol. *91*, 184–193. https://doi.org/10.1016/j.biocel.2017.06.002.

Iijima, T., Hidaka, C., Iijima, Y., 2016. Spatio-temporal regulations and functions of neuronal alternative RNA splicing in developing and adult brains. Neurosci. Res. *109*, 1–8. https://doi.org/10.1016/j.neures.2016.01.010.

Katz, L.C., Shatz, C.J., 1996. Synaptic activity and the construction of cortical circuits. Science *274*, 1133–1138.

Ledda, F., Paratcha, G., 2016. Assembly of neuronal connectivity by neurotrophic factors and leucine-rich repeat proteins. Front. Cell Neurosci. *10*, 199.

Lu, M., Mizumoto, K., 2019. Gradient-independent Wnt signaling instructs asymmetric neurite pruning in *C. elegans*. eLife *8*, e50583.

Mäthger, L.M., Land, M.F., Siebeck, U.E., Marshall, N.J., 2003. Rapid colour changes in multilayer reflecting stripes in the paradise whiptail, *Pentapodus paradiseus*. J. Exp. Biol. *206*, 3607–3613.

Mäthger, L.M., Collins, T.F.T., Lima, P.A., 2004. The role of muscarinic receptors and intracellular Ca^{2+} in the spectral reflectivity changes of squid iridophores. J. Exp. Biol. *207*, 1759–1769.

Mayr, E., 1988. *Toward a New Philosophy of Biology* (p. 408). Harvard University Press, Cambridge, MA.

Nakamura, K., 2011. Central circuitries for body temperature regulation and fever. Am. J. Physiol. Regul. Integr. Comp. Physiol. *301* (5), R1207–R1228.

O'Brien, E.A., Ensbey, K.S., Day, B.W., Baldocket, P.A., Barry, G., 2020. Direct evidence for transport of RNA from the mouse brain to the germline and offspring. BMC Biol. 18 (1), 45. https://doi.org/10.1186/s12915-020-00780-w.

Oshima, N., Fujii, R., 1987. Motile mechanism of blue damselfish (*Chrysiptera cyanea*) iridophores. Cytoskeleton *8*, 85–90.

Polleux, F., Morrow, T., Ghosh, A., 2000. Semaphorin 3A is a chemoattractant for cortical apical dendrites. Nature *404*, 567–573.

Raote, I., Bhattacharyya, S., Panicker, M.M., 2013. Functional selectivity in serotonin receptor 2A (5-HT2A) endocytosis, recycling, and phosphorylation. Mol. Pharmacol. *83*, 42–50.

Rassoulzadegan, M., Grandjean, V., Gounon, P., Vincent, S., Gillot, I., Cuzin, F., 2006. RNA-mediated non-mendelian inheritance of an epigenetic change in the mouse. Nature 441 (7092), 469–474. https://doi.org/10.1038/nature04674.

Ricker, D.D., 1998. The autonomic innervation of the epididymis: its effects on epididymal function and fertility. J. Androl. *19* (1), 1–4.

Ricker, D.D., Crone, J.K., Chamness, S.L., Strader, L.F., Ferrell, J., Goldman, J.M., et al., 1997. Partial sympathetic denervation of the rat epididymis permits fertilization but inhibits embryo development. J. Androl. 18 (2), 131–138. https://doi.org/10.1002/j.1939-4640.1997.tb01893.x.

Sarker, G., Sun, W., Rosenkranz, D., Pelczar, P., Opitz, L., Efthymiou, V., et al. (2019). Maternal overnutrition programs hedonic and metabolic phenotypes across generations through sperm tsRNAs. *Proc. Natl. Acad. Sci. U. S. A.*, *116*(21), 10547–10556. https://doi.org/10.1073/pnas.1820810116.

Sato, K., Tanaka, R., Ishikawa, Y., Yamamoto, D., 2020. Behavioral Evolution of *Drosophila*: Unraveling the Circuit Basis. Genes (Basel). 2020;11(2): 157. https://doi.org/10.3390/genes11020157.

Scheckel, C., Darnell, R.B., 2015. Microexons—tiny but mighty. EMBO J. 34 (3), 273–274. https://doi.org/10.15252/embj.201490651.

Schuldiner, O., Yaron, A., 2015. Mechanisms of developmental neurite pruning. Cell. Mol. Life Sci. 72, 101–119.

Schwann, T. (1839). *Mikroskopische Untersuchungen über die Ueberstimmung in der Struktur und dem Wachsthum der Thiere und Pflanzen* (p. 193). Berlin: Sander'schen Buchhandlung.

Soller, M., Bownes, M., Kubli, E., 1999. Control of oocyte maturation in sexually mature *Drosophila* females. Dev. Biol. *208*, 337–351. https://doi.org/10.1006/dbio.1999.9210.

Teotónio, H., Rose, M.R., 2000. Variation in the reversibility of evolution. Nature *408*, 463–466.

Teotónio, H., Rose, M.R., 2001. Perspective: reverse evolution. Evolution *55*, 653–660.

Teotónio, H., Matos, M., Rose, M.R., 2002. Reverse evolution of fitness in *Drosophila melanogaster*. J. Evol. Biol. *15* (4), 608–617.

Trainor, P.A., Sobieszczuk, D., Wilkinson, D., Krumlauf, R., 2002. Signalling between the hindbrain and paraxial tissues dictates neural crest migration pathways. Development *129*, 433–442.

Virchow, R., 1859. *Die Zellularpathologie in ihre Begründung auf physiologische und patho-logische Gewebelehre* (pp. 12–13). August Hirschwald, Berlin.

Wang, F., Julien, D.P., Sagasti, A., 2013. Journey to the skin—somatosensory peripheral axon guidance and morphogenesis. Cell Adh. Migr. *7* (4), 388–394.

Winata, C.L., Korzh, V., 2018. The translational regulation of maternal mRNAs in time and space. FEBS Lett. *592*, 3007–3023.

Preneural stage of development

1 Parental cytoplasmic factors control the earliest stage of the embryonic development

The early embryonic development, before the beginning of the zygotic gene transcription at the midblastula transition, is under the control of parental (maternal + paternal) cytoplasmic factors (Winata and Korzh, 2018), hence the necessity to take a closer look at their deposition in the germ cells and their function in the early development.

In many organisms, including vertebrates, oocytes produce maternal cytoplasmic factors (maternal effect genes) in the form of mRNAs, miRNAs (Soni et al., 2013; Tóth et al., 2016), secreted proteins, hormones, neurotransmitters/neuromodulators, nutrients, etc., but in organisms such as *Drosophila*, most maternal cytoplasmic factors are supplied to the oocyte primarily by the supporting nurse cells. Most of the cytoplasmic factors in the oocyte are "subcellularly localized" in discrete regions rather than dispersed randomly in the cytoplasm (Lécuyer et al., 2007).

In the oocyte and the sperm cell, cytoplasmic mRNAs are in a state of repression, preventing their translation into proteins, but beginning from the first zygotic division, they are stepwise translated at appropriate points in time when their proteins are needed. It is observed that mRNA poly(A) tails (long stretches of adenine nucleotides) added to mRNA transcripts by the enzyme polyadenylate polymerase are shortened directly after their transport into the cytoplasm, a fact that led to the suggestion that deadenylation may be the cause of the translational inactivation of mRNAs in oocytes (Huarte et al., 1992), while elongation of the poly(A) tails facilitates mRNA activation. In *C. elegans* oocytes, maternal mRNAs form cytoplasmic granules where the sequestered and stabilized mRNAs are repressed (Winata and Korzh, 2018).

The first zygotic divisions and, depending on species, the early embryonic development proceed under the control of maternal cytoplasmic factors followed by a short period of the binary maternal (parental) and zygotic control until around the phylotypic stage, but in oviparous animals not earlier than the formation of the operational central nervous system (CNS), when the reserve of the parental cytoplasmic factors is exhausted.

Parental cytoplasmic factors in the egg/sperm cell and zygote of fish and amphibians are localized at specific regions rather than distributed randomly (Fig. 1.1). An analysis of 3370 genes in *Drosophila* showed that 71% of those expressed mRNAs

The Inductive Brain in Development and Evolution. https://doi.org/10.1016/B978-0-323-85154-1.00005-9

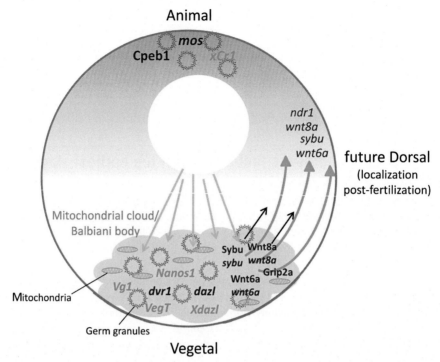

FIG. 1.1

Distribution of key maternal factors in the zebrafish oocyte. Studies in different organisms have shown that maternal mRNAs are organized in cytoplasmic granules together with several regulatory proteins responsible for their posttranscriptional processing and thus translational regulation. In fish and amphibians, a large structure known as the mitochondrial cloud or Balbiani body is present at the vegetal pole of the oocyte. This structure consists of a large accumulation of mitochondria and cytoplasmic granules (specifically termed germ granules) containing silenced mRNAs. The mitochondrial cloud serves as a vehicle for transporting and localizing maternal factors to the vegetal cortex during oogenesis by means of the microtubule network and motor proteins (*yellow arrows*). At egg activation and fertilization, *Sybu* and *Wnt8* are translocated to the future dorsal axis through microtubule-mediated transport (*blue arrows*).

From Winata, C.L., Korzh, V., 2018. The translational regulation of maternal mRNAs in time and space. FEBS Lett. 592, 3007–3023.

were in specific localized sites and this localization was tightly correlated with their translated proteins in embryonic cells during early development (Lécuyer et al., 2007). The specific localization of parental cytoplasmic factors is necessary for their local confined activity and the establishment of the embryonic body axes (Bastock and St. Johnston, 2008). Localization of a number of mRNAs in the zebrafish and *Xenopus* is shown, respectively, in Figs. 1.1 and 1.2.

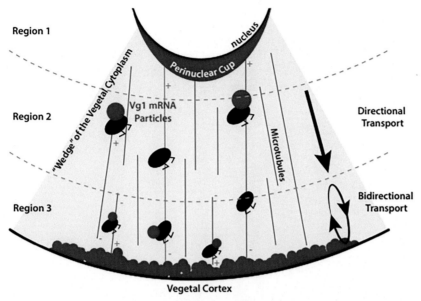

FIG. 1.2

Schematic of mRNA localization in the vegetal cytoplasm of the *Xenopus* oocyte. mRNA cargoes (*red*) are transported by motor proteins (*black*) along polarized microtubule filaments (*blue lines*) toward the vegetal cortex. The vegetal cytoplasm is schematically divided into three regions where transport parameters have been found to differ.

From Ciocanel, M.-V., Sandstede, B., Jeschonek, S.P., Mowry, K.L., 2018. Modeling microtubule-based transport and anchoring of mRNA. SIAM J. Appl. Dyn. Syst. 17, 2855–2881.

The number of maternal cytoplasmic factors identified in mammals is relatively small, apparently because of the direct maternal control of the embryo during the intrauterine life. In mice, until 2010 were identified only 30 maternal cytoplasmic factors, but investigators believed that "this number is likely to increase dramatically in the near future" (Li et al., 2010).

2 Ordered deposition of parental factors in gametes

Drosophila melanogaster is one of the organisms where the ordered deposition of maternal cytoplasmic factors is studied best. Maternal factors that determine the establishment of body axes primarily come from the neighboring nurse cells through ring canals (specific cytoplasmic bridges) before, and during, the process of the squeezing of the nurse cell cytoplasm into the oocyte. Among the most important mRNAs provided to the oocyte in this process are *bcd* (*bicoid*), *grk* (*gurken*), *osk* (*oskar*), and *nos* (*nanos*) mRNAs (Yamashita, 2018).

bicoid mRNA is one of the first maternal cytoplasmic factors to enter the oocyte from nurse cells at stage 10B. In the form of the *bicoid* mRNA-Exu complex (Cha et al., 2001), it is transported to the cortical anterior end of the oocyte through ring canals via both microtubules (Mineyuki and Furuya, 1986) and the actomyosin cyto-skeleton (Nebenführ et al., 1999; Shimmen, 2007). Later, the cytoplasmic streaming refines its position at the center of the anterior end (Fig. 1.3).

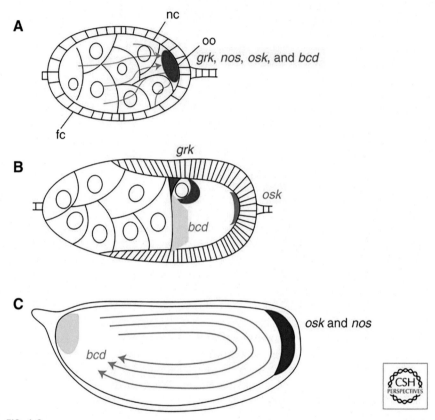

FIG. 1.3

Localization of patterning mRNAs in *Drosophila* oogenesis. (A) In early oogenesis, several mRNAs, including *grk, nos, osk,* and *bcd,* are transported from the nurse cells through cytoplasmic bridges called ring canals into the oocyte. This involves minus-end-directed transport along microtubules (*blue arrows*) mediated by the dynein motor complex. *nc,* nurse cells; *fc,* follicle cells; *oo,* oocyte. (B) In mid-oogenesis, *osk* mRNA localizes to the posterior of the oocyte, *grk* mRNA localizes to the anterodorsal corner in close association with the oocyte nucleus, and *bcd* mRNA localizes to the anterior pole. (C) In late oogenesis, centrifugal cytoplasmic streaming (delineated by *arrows*) coupled with posterior anchoring brings about a further posterior enrichment of *osk* mRNA as well as posterior enrichment of *nos* mRNA. The distribution of *bcd* mRNA at the anterior pole is further refined.

From Lasko, P., 2012. mRNA localization and translational control in Drosophila oogenesis. Cold Spring Harb. Perspect. Biol. 4, a012294.

Two factors are involved in the anterior localization of bicoid mRNA in the oocyte, the maternal protein Staufen (a principal component of various ribonucleoproteins involved in the localization and trafficking of cellular mRNAs) and the ESCRT-II (Vps22, Vps25, and Vps36) mRNA-binding complex, but recently it is demonstrated that the Golgi system is also involved in the process (Cai et al., 2019).

In *Xenopus*, Staufen and kinesin, as well as vegetally localized mRNAs, *Vg1* and *VegT*, form a ribonucleoprotein complex, which is responsible for the transport with the help of a kinesin (a motor protein) of *oskar* mRNA to the posterior end (Micklem et al., 2000; Yoon and Mowry, 2004). The opposite localization of the *bicoid* mRNA and *oskar* mRNA, respectively, at the anterior and posterior ends of the oocyte (Fig. 1.3) determines the establishment of the embryonic anteroposterior (AP) axis.

The nurse cell (or oocyte-derived) *gurken* mRNA, via microtubules, is initially transported to the posterior end of the oocyte and then to the dorsoanterior corner of the oocyte where it is translated under the influence of vasa (Tomancak et al., 1998). The breakage of the radial symmetry cannot be explained with centrosomal microtubules alone but also involves centrosome-independent microtubules (Hayashi et al., 2014). *nanos* is also deposited to the posterior end of the oocyte (Lehmann and Nüsslein-Volhard, 1991). From the total number of the known *Drosophila* maternal mRNAs, 106 are identified in the anterior side of the oocyte and 119 in the posterior side (Jambor et al., 2015).

Not only is localization in the oocyte relatively consistent, but so is the total number of various types of mRNAs. A Drosophila oocyte comprises of nearly 1 million bicoid (bcd) mRNA molecules, varying in less than 10% between individual Drosophila embryos, and the number of these molecules is determined by signals that are external to the oocyte or "generated by perfectly linear feedforward processes" (Petkova et al., 2014).

The role of *bicoid* as a morphogen was explained with a diffusion model (synthesis, diffusion, and uniform degradation, or SDD model). According to the model, the morphogen forms a gradient that determines the expression of gap genes, which in turn regulate the expression of pair-ruled genes and the latter induce the expression of segmentation genes. The model is incompatible with the latter evidence that egg activation leads to movement of *bcd* mRNA away from the cortex (Weil et al., 2008) and the formation of the *bcd* gradient requires its active or passive movement (Little et al., 2011), and especially with the fact that *bcd* moves not in front to form a gradient as the SDD would predict, but it moves along the oocyte cortex (Cai et al., 2019).

3 Maternal factors are deposited to specific regions of the oocyte via microtubules

In the oocyte cytoplasm, mRNAs are transported in the form of RNA granules, large structures containing RNAs and RNA-binding proteins (RBPs) (Martin and Ephrussi, 2009). The deposition of maternal factors in particular regions of the oocyte requires their active transport along microtubules and the association of RBPs with motor proteins. Microtubules are also known to control and regulate cell shape, division, motility, and differentiation (Eliscovich et al., 2013; Gadadhar et al., 2017). Two hormones, hGH (human growth hormone) (Goh et al., 1997) and ecdysone (Soller et al., 1999), regulated, respectively, by neuropeptides GHRH (growth hormone-releasing hormone) and insect PTTH (prothoracicotropic hormone), are known to influence the length of microtubules. Among the known factors influencing the length of microtubules are MAPs (microtubule-associated proteins), which tend to bind to microtubules. The most important among them are the destabilizing (MAP-depolymerizing) protein XKCM1 and stabilizing (MAP-polymerizing) protein MAP215 that stabilizes the structure of the microtubule (Tournebize et al., 2000). Generally, MAPs shift the microtubule dynamics toward assembly and stability (Dubey et al., 2015).

Experimental observations demonstrate that microtubules direct the movement of maternal factors to the oocyte cortex (Theurkauf and Hazelrigg, 1998). The ordered emplacement of mRNAs in the oocyte implies the use of information in the process and the lack of any evidence that microtubules "know" where the oocyte-specific mRNAs must go, suggest that we have to look for other sources of information for the directed movement and regionalization of mRNAs in the oocyte via microtubules. There is empirical evidence suggesting that the nervous system is involved in adjusting the length of microtubules to the destination distance.

4 On the possible role of the nervous system in the ordered deposition of parental cytoplasmic factors in gametes

As already mentioned, the transport of maternal factors in specific regions of the oocyte takes place along microtubules with the help of motor proteins, dynein and kinesin. In most eukaryotes, dynein is the only microtubule minus-end-directed motor that by binding microtubules moves mRNA cargoes away from the nucleus (Kardon and Vale, 2009) toward the periphery of the cell, while kinesin moves them in the opposite direction. The fact that most of the maternal cytoplasmic factors are deposited in specific regions of the cell rather than randomly distributed throughout the oocyte implies that microtubules transporting them somehow adjust their length to reach the cargo's destination.

What is the mechanism that adjusts the length of the microtubule to the distance from the departing site to the destination point and where does the relevant information come from? The only known form of the control and regulation of the microtubule length, to the best of my knowledge, is a neural one. There is a considerable number of studies on the proximate mechanisms of the controlled deposition of maternal and paternal cytoplasmic factors in gametes in various animals, especially in *Drosophila*, but no studies aimed at revealing the origin and the nature of the information for ordered asymmetric deposition of these factors in the oocyte and sperm have been conducted so far. Hence, our discussion on the topic will mainly be based on the evidence obtained from other studies on the neural mechanisms of the regulation length of microtubules and partly on the direct evidence of the involvement of the nervous system in emplacement of parental cytoplasmic factors in gametes.

Muscarinic acetylcholine receptors (mAchRs) are identified in many animal species, including *Xenopus laevis* (Miledi et al., 1982) and humans (Eusebi et al., 1984). It is observed that the neurotransmitter acetylcholine is involved in oocyte maturation by downregulating cAMP rather than changing levels of Ca^{2+} (Yoon et al., 2011). Although there is no evidence yet on the physical contact of the ovarian innervation or the ovarian cholinergic neurons with the oocyte, Ach may reach the oocyte to bind its receptors AchR via blood and other body fluids like reproductive hormones (FSH and LH) do. That the AchR expression may be induced by the presence of the neurotransmitter in the oocyte is suggested by the experimental demonstration that extracts of mRNAs from the electric organ of *Torpedo torpedo* fish (Barnard et al., 1982) and from chicken brains (Sumikawa et al., 1984) injected in *Xenopus* oocytes induced expression of neurotransmitter receptors and voltage-operated channels in the oocyte membrane. While the above evidence strongly suggests the presence of neurotransmitters within the oocyte, it does not give any hint whether, and how, the local neurons and innervation can "at will" determine the length of oocyte microtubules and the regionalization or polarization of maternal factors in the oocyte.

In recent years, several reports were published on the direct involvement of the nervous system in ordered deposition of parental factors in gametes (Dias and Ressler, 2014; Toker et al., 2020). Other reports indicate that during the transit in the epididymal lumen, mammalian sperm is provided with developmentally crucial miRNAs (Belleannée, 2015; Eaton et al., 2015; Gapp and Bohacek, 2018; Zhou et al., 2019) that are also involved in the transgenerational plasticity of various traits. The release of these miRNAs is regulated by local innervation as indicated by the fact that denervation of the epididymis inhibits development of eggs fertilized by sperms from denervated epididymis (Ricker et al., 1997; Ricker, 1998). In a tour de force of experimental ingenuity, an Australian research team demonstrated that injection of human pre-MIR94-1 in the mouse brain led to the appearance of human pre-MIR94-1 in the contralateral side of the brain, lymph nodes, and the epididymis, but not in the blood or liver. Human pre-MIR94-1 was also detected in the same organs in the offspring of the injected male mice mated with naive mice, indicating that the brain-derived miRNA was emplaced in the sperm and incorporated into the zygote (O'Brien et al., 2020).

Regarding the regulation of the transport of molecular cargoes on microtubules, it is demonstrated that neuronal electric activity regulates motor-driven AMPAR (α-amino-3-hydroxy-5-methyl-4-isoxazolepropionic acid receptor) transport by ATP-powered kinesin motors moving along the length of microtubules of neuronal processes (Hoerndli et al., 2015). Neural activity is also involved in polyglutamylation (Maas et al., 2009). In turn, polyglutamylation regulates the binding of MAPs to microtubules (Bonnet et al., 2001) and serves as a permissive signal for spastin-mediated microtubule severing (Lacroix et al., 2010), leading to an increase in the number of microtubules. Activation of the neurotransmitter serotonin 1A

(5-HT1A) receptor alters the association of MAP2 with microtubules and induces depolymerization (shortening) of microtubules (Yuen et al., 2005).

Another line of evidence on the role of the nervous system in regulating the length of the microtubules comes from studies on two known forms of body coloration in animals: dynamic structural coloration and pigmentary coloration. The first results from nanostructures deposited in parallel layers, whose reflected light may interfere, thus creating a sensation of different colors when seen from different angles or when the angle of illumination changes. In both cases, the visible change in color, iridescence, results from a change in the spacings between stacks of nanostructure layers that change the angle of sight. Examples are feather colors in some birds, scales in butterfly wings, etc. Pigmentary coloration results from the presence in cells of pigment-containing granules, of various color, and the intensity of the colors they reflect depends on whether the pigment is dispersed throughout the chromatophore, producing intense color, or concentrated in its center, generating lighter color; as a result, polymerization or depolarization of microtubules on which the transport of pigment-containing granules, from the center to the periphery and the reverse, within chromatophores takes place (McNiven et al., 1984). In the cases of pigmentary coloration, the color looks the same from all viewing angles.

The direct evidence on the neural regulation of the length of cell microtubules is impressive. It comes from the study of iridescence in fish and cephalopods. Various squid species (*L. pealeii, L. vulgaris, Lolliguncula brevis*, and *Alloteuthis subulata*) and teleost fish (blue damselfish, neon tetra, etc.), as well as a polymorphic lizard, possess iridophores, pigmentless light-reflecting cells of the skin containing one or several stacks of light-reflective purine (guanine) platelets (Oshima and Kasai, 2002; Wardill et al., 2012; Lewis et al., 2017). The guanine stacks form multilayer reflectors with refractive indices higher than spaces separating them and the neighboring tissues. In some iridophore types, the distance between the adjacent platelets is held constant by intervening microtubules, while in others, the distance adaptively changes as a result of the adaptive changes in the length of microtubules that hold platelets apart, hence the name dynamic structural coloration. The agents of the adaptive color change are neurotransmitters epinephrine, acetylcholine (ACh), and the cholinomimetic drug carbachol, whose experimental administration produces the color change within 15 s (Mäthger et al., 2004; Chiou et al., 2007). Now, it is known that the skin dynamic structural coloration in squids can be controlled by neurons descending from the CNS reaching within the bounds of iridophores (Wardill et al., 2012). In squids, e.g., Ach (acetylcholine) release by nerves beneath the iridophore layer induces changes in the structural coloration and reflectance peak of iridophores as a result of changes in patterns of phosphorylation of the four different reflectins. Similar changes induced the electrical stimulation of these nerves (Izumi et al., 2010; Wardill et al., 2012; Levenson et al., 2015). In the absence of neural signals in the case of denervation, iridophores remain transparent. In squids, iridophores and chromatophores are innervated by different motor neurons originating in the CNS. Iridophores are innervated through the stellate ganglion and chromatophores by the fin nerve (Gonzalez-Bellido et al., 2014).

In support of the neural regulation of iridophore adaptive coloration comes also the fact that stress induces a rapid color change related to changes in platelet spacing in the male tawny dragon lizards, *Ctenophorus decresii* (Lewis et al., 2017).

5 The maternal-zygotic transition and the complementary roles of the maternal and zygotic expression in early development

The existence of maternal cytoplasmic factors in insect eggs was discovered 40 years ago (Nüsslein-Volhard et al., 1980), and maternal cytoplasmic factors in mammalian species were discovered two decades ago (Tong et al., 2000; Christians et al., 2000). The parental (maternal and paternal) cytoplasmic factors (mRNAs, miRNAs, secreted proteins, hormones, neuromodulators, etc.) deposited in gametes in the process of spermatogenesis and oogenesis control the early embryonic development. Their action, depending on species, lasts a varying number of cell divisions.

In all studied cases, the zygotic genome is initially repressed, and, as a rule, the first cell divisions are under the control of parental cytoplasmic factors until they begin to induce the expression of zygotic genes. This moment is the so-called MZT (maternal-to-zygotic transition), which, depending on species, takes place in different stages of early embryonic development. So, in *C. elegans* it takes place at the 100-cell stage (Edgar et al., 1994), and in *Drosophila* after the 8192-cell stage (Yartseva and Giraldez, 2015). In vertebrates such as zebrafish, the onset of the zygotic gene expression starts at the midblastula, i.e., at the 512-cell stage, in the round goby (*Neogobius melanostomus*) transcription of the zygotic genes starts at the 32-cell stage (Adrian-Kalchhauser et al., 2018), and in the African clawed frog *X. laevis* in 4000–5000 cell embryos (Yang et al., 2015). Many maternal factors are translated beyond the midblastula, and the specification of dorsal identity requires maternal involvement in the Wnt/β-catenin pathway (Wagner et al., 2004). A great number of parental mRNAs are polyadenylated (multiple adenine bases are added to the 3′ end of the mature mRNAs), i.e., derepressed, right before the MZT (Aanes et al., 2011) when a massive degradation of maternal mRNAs begins (Fig. 1.4).

In placental mammalians, however, the MZT takes place very early. For example, in mice, it begins at the 2- to 4-cell stage (Piko and Clegg, 1982; Wang et al., 2004), in humans and pigs in the 4- to 8-cell stage (Braude et al., 1988), in rabbits transcription of the zygotic genes starts from the 8- to 16-cell stage (Manes, 1973), and in sheep embryos it begins in the second cleavage and is fully activated in the 4th cell cycle (Crosby et al., 1988), while in cattle the MZT takes place between the 8- and 16-cell stage (Frei et al., 1989). It is clear that in the course of evolution, in placental mammalians, part of the role of parental cytoplasmic factors is taken over by the maternal organism that can exert directly its influence owing to the intrauterine contact with the embryo.

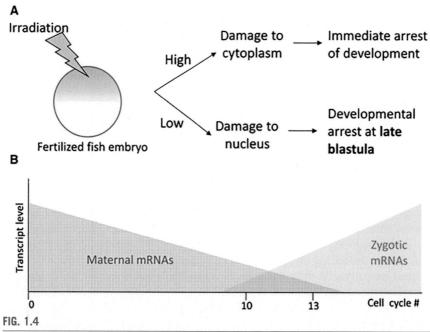

FIG. 1.4

Maternal mRNAs control development up to the point when the zygotic genome is activated during the MZT occurring at late blastula in fishes and amphibians. (A) Neyfakh's experiment with irradiated loach embryos shows different phenotypes depending on the doses of radiation, which affects either the cytoplasm or the nucleus. The delayed developmental arrest resulting from nuclear damage gave rise to the understanding of the morphogenetic function of the nuclei, which is responsible for development from late blastula onward.
(B) Current understanding at the molecular level established that maternal mRNAs present at high levels prior to MZT drive development up to this point before it is taken over by zygotic genes expressed from here onward.

From Winata, C.L., Korzh, V., 2018. The translational regulation of maternal mRNAs in time and space. FEBS Lett. 592, 3007–3023.

    The MZT is preceded by the beginning of the process of degradation of the maternal cytoplasmic factors by a particular subset of maternal cytoplasmic factors (and later by a group of zygotic genes) (Tadros and Lipshitz, 2009). The clearance of maternal factors, thus, is based on the activation of two independent mechanisms: the maternal and zygotic mechanism. The maternal mechanism involves poly(A) tail-microRNAs shortening for mRNA elimination, which is independent of the zygotic mechanism. The activation of the zygotic clearance of maternal RNAs, which intensifies with the MZT, is accomplished by zygotic miRNAs and is intended to take over the control of further development (Yartseva and Giraldez, 2015). The proportion of mRNAs eliminated by the maternal and zygotic mechanisms varies among different species.

The expression of zygotic genes begins from the MBT - or more generally from the MZT. Many maternal/paternal factors, depending on species, continue to be translated beyond the MBT/MZT, until the phylotypic stage, when the embryo has developed a functioning CNS, and beyond as in the case of the purple sea urchin *S. purpuratus*, where some maternal mRNAs, such as *SoxB1*, *hnf6*, *β-catenin*, suppressor of hairless *su(H)*, *lef1*, and *tbr* continue to be expressed in later stages of development (Wei et al., 2006). In zebrafish, maternal *Nanog*, *Pou5f1*, and *SoxB1* mRNAs regulate zygotic gene expression, whereas their loss causes failure to transcript 75% of zygotic genes and the arrest of development before gastrulation (Lee et al., 2013). Screening of 500 paternal and 600 maternal zebrafish genomes identified at least 5 paternal cytoplasmic factors and 63 maternal cytoplasmic factors that continued being translated after the onset of the MZT and were necessary for the later embryonic development. Mutations in these maternal/parental genes caused defects in the Bauplan and other morphological developmental defects (Wagner et al., 2004).

The number of maternal cytoplasmic factor types identified in mammals is relatively small because of the direct maternal transplacental control of the embryo in this class. So, e.g., the number of the known maternal cytoplasmic types investigated in mice until 2010 was less than 30 (Li et al., 2010). In mice is identified a considerable number of maternally inherited miRNAs, including let-7 family miRNAs that are active at least until the eight-cell stage and are critical for the early development in mammalians (Tang et al., 2007). However, even in mammals, miRNAs from both the egg and sperm are required for embryonic development (McJunkin, 2018). Sperm of mice deficient in particular miRNAs can fertilize the oocyte but leads to impaired development, which is rescued by injecting wild-type total RNAs or only miRNAs (Yuan et al., 2016).

The crucial role of maternal cytoplasmic factors in development is also supported by irradiation experiments conducted in the early 1950s on the fertilized eggs of the loach (*Misgurnus fossilis*), which showed that lower doses of irradiation damaged only the nucleus and led to the arrest of the development at the late blastula stage, while higher X-ray irradiation led to cytoplasmic damage and the immediate developmental arrest (Winata and Korzh, 2018), demonstrating the important role of the parental cytoplasmic factors as well as the overlapping period of translation of the maternal/paternal factors and expression of zygotic genes during the embryonic development in fish and amphibians (Fig. 1.4).

As mentioned earlier, the onset of the zygotic gene expression marks neither the exhaustion nor the end of the role of the parental cytoplasmic factors in embryonic development. There is evidence that in zebrafish the maternal chromatin-modifying methyl CpG binding protein (MeCP), which is necessary for neuronal differentiation, is present at least up to the pharyngula stage, i.e., 24 hpf (hours postfertilization) (Gao et al., 2015), whereas maternally deposited cortisol in the zebrafish oocyte is essential for the functioning of the HPI (hypothalamic-pituitary-interrenal) axis, is present in the embryo at least until 48 hpf, and is involved in "programming stress axis development and function in zebrafish" during the embryonic and posthatch stages (Nesan and Vijayan, 2016). Evidence from a variety of species shows that parental cytoplasmic factors persist, continue to be expressed, and actively induce developmental processes along with the expression of zygotic genes (Hrabé de Angelis et al., 1995). So, e.g., despite the onset of the zygotic gene expression, the maternal FGF2 in the uterine secretion of the day-6 rabbit blastocyst promotes gastrulation both in vivo and in vitro (Fig. 1.5).

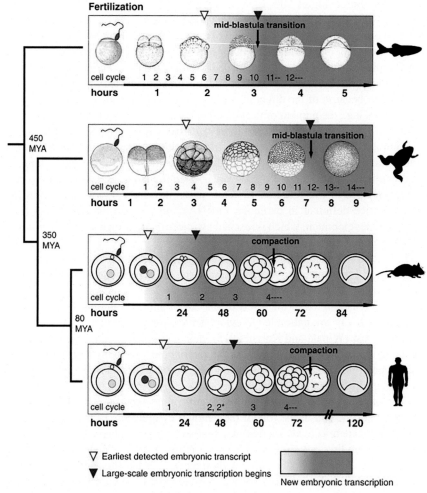

FIG. 1.5

Timing of transcriptional onset and early embryonic events varies in vertebrates.

From Jukam, D., Shariati, S.A.M., Skotheim, J.M., 2017. Zygotic genome activation in vertebrates. Dev. Cell 42, 316–332.

The study of divergence and heterochronies in the development of the blind cave and eyed surface morphotypes of *Astyanax mexicanus* during gastrulation showed that these phenomena were associated with the action of the maternal factors (Torres-Paz et al., 2019).

A comparative study between the cave and surface morphotypes of the fish *A. mexicanus* reports that maternal determinants are fully penetrant up to the end of gastrulation and they are active up to the phylotypic stage, as proved by the fact that the larval lens apoptosis and eye regression in the cave morphotype appear to be under maternal control (Ma et al., 2018).

How extensive may be the influence of parental factors in modifying expression patterns in animals is eloquently demonstrated by the experimental fact that even as early as the two-cell stage of the surface and cave morphotypes of *A. mexicanus*, 32% of the 20,730 genes examined were differentially expressed in the two morphs (Torres-Paz et al., 2019). Striking increases in the contents of several maternal cytoplasmic factors (*axin1*, *pou2f1b*, *runx2b* mRNAs) and a decrease in the quantity of *Shh* mRNA are observed in the cave morph compared with the surface morph of *A. mexicanus*. Although the influence of maternal factors is weakened after the gastrulation, surprisingly, the maternal factors in the cave morph persisted way beyond this stage, influencing the development of the CNS and even the latter process of lens apoptosis in the cave morph (Ma et al., 2018). Maternal components had the strongest impact on the number and patterns of as well as the differentiation of hypothalamic neurons of the fish (Torres-Paz et al., 2019).

The onset of the MZT paves the way for gastrulation, during which the germ layers of the embryo endoderm, mesoderm, and ectoderm form through a combination of cell migration, ingression, and invagination (Tadros and Lipshitz, 2009) (Fig. 1.6).

6 Parental mRNAs regulate the formation of germ layers

Three maternal and zygotic POU-V mRNAs, *Oct60*, *Oct25*, and *Oct91*, are localized in the animal half of the embryo and maintain the undifferentiated state by restricting mesendoderm-inducing signals and promoting the formation of the neuroectoderm (Cao, 2013) (Fig. 1.7). *Oct60* that is maternally transcribed and *Oct25* that is maternally and zygotically transcribed overlap with *VegT* in the vegetal-equatorial region, whereas *Oct91* is embryonically little expressed in the vegetal half (Cao et al., 2007). *Oct25* blocks the activities of maternal *VegT* and β-catenin that induce signaling cascades for mesendoderm formation. The appropriate balance between activities of these maternal and embryonic mRNAs is required for correct patterning of the embryonic germ layers that arise in the process of gastrulation.

The ordered deposition of maternal mRNAs in the egg cell determines at a large extent the cell fate during cell divisions, the establishment of the embryonic dorsoventral and anteroposterior axes, and the formation of germ layers (the endoderm, ectoderm, and mesoderm) during the process of gastrulation. The vegetally localized

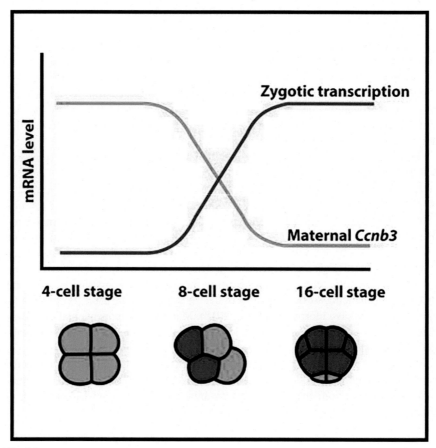

FIG. 1.6

Diagrammatic representation of the simultaneous decline of the maternal *Ccnb3* and beginning of the zygotic gene expression in *C. intestinalis*.

From Treen, N., Heist, T., Wang, W., Levine, M., 2018. Depletion of maternal cyclin B3 contributes to zygotic genome activation in the ciona embryo [published correction appears in Curr. Biol. 28:1330-1331]. Curr. Biol. 28, 1150–1156.

maternal *nanos* mRNA regulates the formation of the primordial germ cells (Lai and King, 2013). Depletion of maternal mRNAs suppresses the formation of germ layers (Sheets et al., 2017). Vegetally localized maternal mRNAs induce the formation of the endoderm, whereas depletion of the vegetally localized *Vg1* mRNA can only develop the endoderm and mesoderm of reduced size (Birsoy et al., 2006). Depletion of maternal *VegT* mRNA suppresses the formation of the endoderm; depleted maternal *Vg1* leads to the formation of reduced mesoderm, while depletion of maternal *Wnt11* inhibits the formation of the Spemann organizer (Sheets et al., 2017). Depleted maternal *FGFR* mRNA induces the formation of defective gastrulation

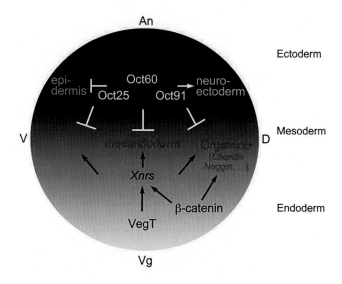

FIG. 1.7

Correlation between pluripotency factors and differentiation signals during *Xenopus* germ layer formation. Germ layers are formed in strict spatiotemporal patterns. Either POU-V factors or germ layer-inducing signals have specific localizations and exhibit different levels of activities during the process of germ layer formation of *Xenopus* embryos. POU-V factors antagonize the differentiation signals so as to ensure that germ layer formation can occur in the correct spatiotemporal patterns. *An*, animal pole; *D*, dorsal side; *V*, ventral side; *Vg*, vegetal pole.

Reproduced from [42] (Cao, Y., Siegel, D., Donow, C., Knöchel, S., Yuan, L., Knöchel, W., 2007. POU-V factors antagonize maternal VegT activity and β-Catenin signaling in Xenopus embryos. EMBO J. 26(12), 2942–2954) with modifications from Cao, Y., 2013. Regulation of germ layer formation by pluripotency factors during embryogenesis. Cell Biosci. 3 (1), 15. https://doi.org/10.1186/2045-3701-3-15.

(Yokota et al., 2003). Maternal and paternal miRNAs also play important roles in embryonic development. They are required for embryonic development (McJunkin, 2018). In mice, the sperm from mice deficient in particular maternal miRNA can fertilize the oocyte, but the early development is impaired, whereas injection of total mRNAs or only miRNAs from the wild type sperm rescues the defect (Yuan et al., 2016). The nematode-specific *mir-35* family is involved in multiple regulatory networks responsible for embryonic development in *C. elegans* and is required to prevent the premature expression of sex-specific genes for normal sex determination in the worm (McJunkin and Ambros, 2017). Just two miRNAs, one *miR-35* and one *miR-51* family member, out of 150 canonical miRNAs are sufficient for embryonic patterning in *C. elegans* (Dexheimer et al., 2020). In zebrafish, maternal *miR-430* starts clearance of maternal mRNAs to promote the onset of zygotic gene expression (Giraldez et al., 2006).

References

Aanes, H., Winata, C. L., Lin, C. H., Chen, J. P., Srinivasan, K. G., Lee, S. G., et al. (2011). Zebrafish mRNA sequencing deciphers novelties in transcriptome dynamics during maternal to zygotic transition. *Genome Res.*, *21*, 1328–1338.

Adrian-Kalchhauser, I., Walser, J.-C., Schwaiger, M., & Burkhardt-Holm, P. (2018). RNA sequencing of early round goby embryos reveals that maternal experiences can shape the maternal RNA contribution in a wild vertebrate. *BMC Evol. Biol.*, *18*, 34.

Barnard, E. A., Miledi, R., & Sumikawa, K. (1982). Transplantation of exogenous messenger RNA coding for nicotinic acetylcholine receptors produces functional receptors in *Xenopus* oocytes. *Proc. R. Soc. Lond. B Biol. Sci.*, *215*, 241–246.

Bastock, R., & St. Johnston, D. (2008). mRNA localization and translational control in *Drosophila* oogenesis. *Curr. Biol.*, *18*, R1082–R1087.

Belleannée, C. (2015). Extracellular microRNAs from the epididymis as potential mediators of cell-to-cell communication. *Asian J. Androl.*, *17*, 730–736.

Birsoy, B., Kofron, M., Schaible, K., Wylie, C., & Heasman, J. (2006). Vg 1 is an essential signaling molecule in *Xenopus* development. *Development*, *133*(1), 15–20. https://doi.org/10.1242/dev.02144.

Bonnet, C., Boucher, D., Lazereg, S., Pedrotti, B., Islam, K., Denoulet, P., et al. (2001). Differential binding regulation of microtubule-associated proteins MAP1A, MAP1B, and MAP2 by tubulin polyglutamylation. *J. Biol. Chem.*, *276*, 12839–12848.

Braude, P., Bolton, V., & Moore, S. (1988). Human gene expression first occurs between the four- and eight-cell stages of preimplantation development. *Nature*, *332*, 459–461.

Cai, X., Fahmy, K., & Baumgartner, S. (2019). *bicoid* RNA localization requires the *trans*-Golgi network. *Hereditas*, *156*, 30.

Cao, Y. (2013). Regulation of germ layer formation by pluripotency factors during embryogenesis. *Cell Biosci.*, *3*(1), 15. https://doi.org/10.1186/2045-3701-3-15.

Cao, Y., Siegel, D., Donow, C., Knöchel, S., Yuan, L., & Knöchel, W. (2007). POU-V factors antagonize maternal VegT activity and β-Catenin signaling in *Xenopus* embryos. *EMBO J.*, *26*, 2942–2954. https://doi.org/10.1038/sj.emboj.7601736.

Cha, B. J., Kopetsch, B. S., & Theurkauf, W. E. (2001). *In vivo* analysis of *Drosophila bicoid* mRNA localization reveals a novel microtubule-dependent axis specification pathway. *Cell*, *106*, 35–46.

Chiou, T.-H., Mäthger, L. M., Hanlon, R. T., & Cronin, T. W. (2007). Spectral and spatial properties of polarized light reflections from the arms of squid (*Loligo pealeii*) and cuttlefish (*Sepia officinalis* L.). *J. Exp. Biol.*, *210*, 3624–3635.

Christians, E., Davis, A. A., Thomas, S. D., & Benjamin, I. J. (2000). Maternal effect of Hsf1 on reproductive success. *Nature*, *407*, 693–694.

Crosby, I. M., Gandolfi, F., & Moor, R. M. (1988). Control of protein synthesis during early cleavage of sheep embryos. *J. Reprod. Fertil.*, *82*, 769–775.

Dexheimer, P. J., Wang, J., & Cochella, L. (2020). Two microRNAs are sufficient for embryonic patterning in *C. elegans*. *Curr. Biol.*, *30*(24). https://doi.org/10.1016/j.cub.2020.09.066. 5058–5065.e5.

Dias, B. G., & Ressler, K. J. (2014). Parental olfactory experience influences behavior and neural structure in subsequent generations. *Nat. Neurosci.*, *17*(1), 89–96.

Dubey, J., Ratnakaran, N., & Koushika, S. P. (2015). Neurodegeneration and microtubule dynamics: death by a thousand cuts. *Front. Cell. Neurosci.*, *9*, 343.

Eaton, S. A., Jayasooriah, N., Buckland, M. E., Martin, D. I., Cropley, J. E., & Suter, C. M. (2015). Roll over Weismann: extracellular vesicles in the transgenerational transmission of environmental effects. *Epigenomics*, *7*(7), 1165–1171.

Edgar, L. G., Wolf, N., & Wood, W. B. (1994). Early transcription in Caenorhabditis elegans embryos. *Development*, *120*, 443–451.

Eliscovich, C., Buxbaum, A. R., Katz, Z. B., & Singer, R. H. (2013). mRNA on the move: the road to its biological destiny. *J. Biol. Chem.*, *288*, 20361–20368.

Eusebi, F., Pasetto, N., & Siracusa, G. (1984). Acetylcholine receptors in human oocytes. *J. Physiol.*, *346*, 321–330.

Frei, R. E., Schultz, G. A., & Church, R. B. (1989). Qualitative and quantitative changes in protein synthesis occur at the 8-16-cell stage of embryogenesis in the cow. *J. Reprod. Fertil.*, *86*, 637–641.

Gadadhar, S., Bodakuntla, S., Natarajan, K., & Janke, C. (2017). The tubulin code at a glance. *J. Cell Sci.*, *130*, 1347–1353.

Gao, H., Bu, Y., Wu, Q., Wang, X., Chang, N., Lei, L., et al. (2015). Mecp2 regulates neural cell differentiation by suppressing the Id1 to Her2 axis in zebrafish. *J. Cell Sci.*, *128*, 2340–2350.

Gapp, K., & Bohacek, J. (2018). Epigenetic germline inheritance in mammals: looking to the past to understand the future. *Genes Brain Behav.*, *17*(3), e12407. https://doi.org/10.1111/gbb.12407.

Giraldez, A. J., Mishima, Y., Rihel, J., Grocock, R. J., Van Dongen, S., Inoue, K., et al. (2006). Zebrafish MiR-430 promotes deadenylation and clearance of maternal mRNAs. *Science*, *312*(5770), 75–79. https://doi.org/10.1126/science.1122689.

Goh, E. L., Pircher, T. J., Wood, T. J., Norstedt, G., Graichen, R., Lobie, P. E., et al. (1997). Growth hormone promotion of tubulin polymerization stabilizes the microtubule network and protects against colchicine-induced apoptosis. *Endocrinology*, *139*, 4364–4372.

Gonzalez-Bellido, P. T., Wardill, T. J., Buresch, K. C., Ulmer, K. M., & Hanlon, R. T. (2014). Expression of squid iridescence depends on environmental luminance and peripheral 2 ganglion control. *J. Exp. Biol.*, *217*, 850–858.

Hayashi, R., Wainwright, S. M., Liddell, S. J., Pinchin, S. M., Horswell, S., & Ish-Horowicz, D. (2014). A genetic screen based on in vivo RNA imaging reveals centrosome-independent mechanisms for localizing *gurken* transcripts in *Drosophila*. *G3 (Bethesda)*, *4*, 749–760.

Hoerndli, F. J., Wang, R., Mellem, J. E., Kallarackal, A., Brockie, P. J., Thacker, C., et al. (2015). Neuronal activity and CaMKII regulate kinesin-mediated transport of synaptic AMPARs. *Neuron*, *86*, 457–474.

Hrabé de Angelis, M., Gründker, C., Herrmann, B. G., Kispert, A., & Kirchner, C. (1995). Promotion of gastrulation by maternal growth factor in cultured rabbit blastocysts. *Cell Tissue Res.*, *282*, 147–154.

Huarte, J., Stutz, A., O'Connell, M. L., Gubler, P., Belin, D., et al. (1992). Transient translational silencing by reversible mRNA deadenylation. *Cell*, *69*, 1021–1030.

Izumi, M., Sweeney, A. M., Demartini, D., Weaver, J. C., Powers, M. L., Tao, A., et al. (2010). Changes in reflectin protein phosphorylation are associated with dynamic iridescence in squid. *J. R. Soc. Interface*, *7*, 549–560.

Jambor, H., Surendranath, V., Kalinka, A. T., Mejstrik, P., Saalfeld, S., & Tomancak, P. (2015). Systematic imaging reveals features and changing localization of mRNAs in *Drosophila* development. *Elife*, *4*, e05003.

Kardon, J., & Vale, R. (2009). Regulators of the cytoplasmic dynein motor. *Nat. Rev. Mol. Cell Biol., 10*, 854–865.

Lacroix, B., van Dijk, J., Gold, N., Guizetti, L., Aldrian, G., Rogowski, K., et al. (2010). Tubulin polyglutamylation stimulates spastin-mediated microtubule severing. *J. Cell Biol., 189*, 945–954.

Lai, F., & King, M. L. (2013). Repressive translational control in germ cells. *Mol. Reprod. Dev., 80*(8), 665–676. http://www.ncbi.nlm.nih.gov/pubmed/23408501.

Lécuyer, E., Yoshida, H., Parthasarathy, N., Alm, C., Babak, T., Cerovina, T., et al. (2007). Global analysis of mRNA localization reveals a prominent role in organizing cellular architecture and function. *Cell, 131*, 174–187.

Lee, M. T., Bonneau, A. R., Takacs, C. M., Bazzini, A. A., DiVito, K. R., Fleming, E. S., et al. (2013). Nanog, Pou5f1 and SoxB1 activate zygotic gene expression during the maternal-to-zygotic transition. *Nature, 503*, 360–364.

Lehmann, R., & Nüsslein-Volhard, C. (1991). The maternal gene *nanos* has a central role in posterior pattern formation of the *Drosophila* embryo. *Development, 112*, 679–691.

Levenson, R., Bracken, C., Bush, N., & Morse, D. E. (2015). Cyclable condensation and hierarchical assembly of metastable reflectin proteins, the drivers of tunable biophotonics. *J. Biol. Chem., 291*, 4058–4068.

Lewis, A. C., Rankin, K. J., Pask, A. J., & Stuart-Fox, D. (2017). Stress-induced changes in color expression mediated by iridophores in a polymorphic lizard. *Ecol. Evol., 7*, 8262–8272.

Li, L., Zheng, P., & Dean, J. (2010). Maternal control of early mouse development. *Development, 137*, 859–870. https://doi.org/10.1242/dev.039487.

Little, S. C., Tkačik, G., Kneeland, T. B., Wieschaus, E. F., & Gregor, T. (2011). The formation of the bicoid morphogen gradient requires protein movement from anteriorly localized mRNA. *PLoS Biol., 9*(3), e1000596.

Ma, L., Strickler, A. G., Parkhurst, A., Yoshizawa, M., Shi, J., & Jeffery, W. R. (2018). Maternal genetic effects in Astyanax cavefish development. *Dev. Biol., 441*, 209–220.

Maas, C., Belgardt, D., Lee, H. K., Heisler, F. F., Lappe-Siefke, C., Magiera, M. M., et al. (2009). Synaptic activation modifies microtubules underlying transport of postsynaptic cargo. *Proc. Natl. Acad. Sci. U. S. A., 106*, 8731–8736.

Manes, C. (1973). The participation of the embryonic genome during early cleavage in the rabbit. *Dev. Biol., 32*, 453–459.

Martin, K. C., & Ephrussi, A. (2009). mRNA localization: gene expression in the spatial dimension. *Cell, 136*, P719–P730.

Mäthger, L. M., Collins, T. F. T., & Lima, P. A. (2004). The role of muscarinic receptors and intracellular Ca^{2+} in the spectral reflectivity changes of squid iridophores. *J. Exp. Biol., 207*, 1759–1769.

McJunkin, K. (2018). Maternal effects of microRNAs in early embryogenesis. *RNA Biol., 15*(2), 165–169. https://doi.org/10.1080/15476286.2017.1402999.

McJunkin, K., & Ambros, V. (2017). A microRNA family exerts maternal control on sex determination in *C. elegans*. *Genes Dev., 31*(4), 422–437. https://doi.org/10.1101/gad.290155.

McNiven, M. A., Wang, M., & Porter, K. R. (1984). Microtubule polarity and the direction of pigment transport reverse simultaneously in surgically severed melanophore arms. *Cell, 37*, P753–P765.

Micklem, D. R., Adams, J., Gruenert, S., & Johnston, D. S. (2000). Distinct roles of two conserved Staufen domains in oskar mRNA localization and translation. *EMBO J., 19*, 1366–1377.

Miledi, R., Parker, I., & Sumikawa, K. (1982). Properties of acetylcholine receptors translated by cat muscle mRNA in *Xenopus* oocytes. *EMBO J.*, *1*, 1307–1312.

Mineyuki, Y., & Furuya, M. (1986). Involvement of colchicine-sensitive cytoplasmic element in premitotic nuclear positioning of *Adiantum protonemata*. *Protoplasma*, *130*, 83–90.

Nebenführ, A., Gallagher, L. A., Dunahay, T. G., Frohlick, J. A., Mazurkiewicz, A. M., Meehl, J. B., et al. (1999). Stop-and-go movements of plant Golgi stacks are mediated by the actomyosin system. *Plant Physiol.*, *121*, 1127–1142.

Nesan, D., & Vijayan, M. (2016). Maternal cortisol mediates hypothalamus-pituitary-interrenal axis development in zebrafish. *Sci. Rep.*, *6*, 22582.

Nüsslein-Volhard, C., Lohs-Schardin, M., Sander, K., & Cremer, C. (1980). A dorso-ventral shift of embryonic primordia in a new maternal-effect mutant of *Drosophila*. *Nature*, *283*, 474–476.

O'Brien, E. A., Ensbey, K. S., Day, B. W., Baldock, P. A., & Barry, G. (2020). Direct evidence for transport of RNA from the mouse brain to the germline and offspring. *BMC Biol*, *18* (1), 45.

Oshima, N., & Kasai, A. (2002). Iridophores involved in generation of skin color in the zebrafish, *Brachydanio rerio*. *Forma*, *17*, 91–101.

Petkova, M. D., Little, S. C., Liu, F., & Gregor, T. (2014). Maternal origins of developmental reproducibility. *Curr. Biol.*, *24*, 1283–1288.

Piko, I., & Clegg, K. B. (1982). Quantitative changes in total RNA, total poly(A), and ribosomes in early mouse embryos. *Dev. Biol.*, *89*, 362–378.

Ricker, D. D. (1998). The autonomic innervation of the epididymis: its effects on epididymal function and fertility. *J. Androl.*, *19*(1), 1–4. 9537285.

Ricker, D. D., Crone, J. K., Chamness, S. L., Klinefelter, G. R., & Chang, T. S. (1997). Partial sympathetic denervation of the rat epididymis permits fertilization but inhibits embryo development. *J. Androl.*, *18*(2), 131–138. https://doi.org/10.1002/j.1939-4640.1997.tb01893.x.

Sheets, M. D., Fox, C. A., Dowdle, M. E., Blaser, S. I., Chung, A., & Park, S. (2017). Controlling the messenger: regulated translation of maternal mRNAs in *Xenopus laevis* development. *Adv. Exp. Med. Biol.*, *953*, 49–82. https://doi.org/10.1007/978-3-319-46095-6_2.

Shimmen, T. (2007). The sliding theory of cytoplasmic streaming: fifty years of progress. *J. Plant Res.*, *120*, 31–43.

Soller, M., Bownes, M., & Kubli, E. (1999). Control of oocyte maturation in sexually mature *Drosophila* females. *Dev. Biol.*, *208*, 337–351.

Soni, K., Choudhary, A., Patowary, A., Singh, A. R., Bhatia, S., Sivasubbu, S., et al. (2013). miR-34 is maternally inherited in *Drosophila melanogaster* and *Danio rerio*. *Nucleic Acids Res.*, *41*(8), 4470–4480. https://doi.org/10.1093/nar/gkt139.

Sumikawa, K., Parker, I., & Miledi, R. (1984). Partial purification and functional expression of brain mRNAs coding for neurotransmitter receptors and voltage-operated channels. *Proc. Natl. Acad. Sci. USA*, *81*(24), 7994–7998.

Tadros, W., & Lipshitz, H. D. (2009). The maternal-to-zygotic transition: a play in two acts. *Development*, *136*, 3033–3042.

Tang, F., Kaneda, M., O'Carroll, D., Hajkova, P., Barton, S. C., Sun, Y. A., et al. (2007). Maternal microRNAs are essential for mouse zygotic development. *Genes Dev.*, *21*, 644–648.

Theurkauf, W. E., & Hazelrigg, T. I. (1998). In vivo analyses of cytoplasmic transport and cytoskeletal organization during *Drosophila* oogenesis: characterization of a multi-step anterior localization pathway. *Development*, *125*, 3655–3666.

Toker, I. A., Lev, I., Mor, Y., Gurevich, Y., Fisher, D., Houri-Zeevi, L., et al. (2020). *Transgenerational regulation of sexual attractiveness in C. elegans nematodes*. bioRxiv. https://doi.org/10.1101/2020.11.18.389387. 2020.11.18.389387.

Tomancak, P., Guichet, A., Zavorsky, P., & Ephrussi, A. (1998). Oocyte polarity depends on regulation of *gurken* by vasa. *Development*, *125*, 1723–1733.

Tong, Z. B., Gold, L., Pfeifer, K. E., Dorward, H., Lee, E., Bondy, C. A., et al. (2000). Mater, a maternal effect gene required for early embryonic development in mice. *Nat. Genet.*, *26*, 267–268.

Torres-Paz, J., Leclercq, J., & Rétaux, S. (2019). Maternally-regulated gastrulation as a source of variation contributing to cavefish forebrain evolution. *Elife*. https://doi.org/10.7554/eLife.50160.

Tóth, K. F., Pezic, D., Stuwe, E., & Webster, A. (2016). The piRNA pathway guards the germline genome against transposable elements. *Adv. Exp. Med. Biol.*, *886*, 51–77. https://doi.org/10.1007/978-94-017-7417-8_4.

Tournebize, R., Popov, A., Kinoshita, K., Ashford, A. J., Rybina, S., Pozniakovsky, A., et al. (2000). Control of microtubule dynamics by the antagonistic activities of XMAP215 and XKCM1 in *Xenopus* egg extracts. *Nat. Cell Biol.*, *2*, 13–19.

Wagner, D. S., Dosch, R., Mintzer, K. A., Wiemelt, A. P., & Mullins, M. C. (2004). Maternal control of development at the midblastula transition and beyond: mutants from the zebrafish II. *Dev. Cell.*, *6*, 781–790.

Wang, Q. T., Piotrowska, K., Ciemerych, M. A., Milenkovic, L., Scott, M. P., Davis, R. W., et al. (2004). A genome-wide study of gene activity reveals developmental signaling pathways in the preimplantation mouse embryo. *Dev. Cell*, *6*, 133–144.

Wardill, T. J., Gonzalez-Bellido, P. T., Crook, R. J., & Hanlon, R. T. (2012). Neural control of tuneable skin iridescence in squid. *Proc. Biol. Sci.*, *279*, 4243–4252.

Wei, Z., Angerer, R. C., & Angerer, L. M. (2006). A database of mRNA expression patterns for the sea urchin embryo. *Dev. Biol.*, *300*, 476–484.

Weil, T. T., Parton, R., Davis, I., & Gavis, E. R. (2008). Changes in *bicoid* mRNA anchoring highlight conserved mechanisms during the oocyte-to-embryo transition. *Curr. Biol.*, *18*, 1055–1061.

Winata, C. L., & Korzh, V. (2018). The translational regulation of maternal mRNAs in time and space. *FEBS Lett.*, *592*, 3007–3023.

Yamashita, Y. M. (2018). Subcellular specialization and organelle behavior in germ cells. *Genetics*, *208*, 19–51.

Yang, J., Aguero, T., & King, M. L. (2015). The Xenopus maternal-to-zygotic transition from the perspective of the germline. *Curr. Top. Dev. Biol.*, *113*, 271–303.

Yartseva, V., & Giraldez, A. J. (2015). The maternal-to-zygotic transition during vertebrate development: a model for reprogramming. *Curr. Top. Dev. Biol.*, *113*, 191–232.

Yokota, C., Kofron, M., Zuck, M., Houston, D. W., Isaacs, H., Asashima, M., et al. (2003). A novel role for a nodal-related protein; Xnr3 regulates convergent extension movements via the FGF receptor. *Development*, *130*(10), 2199–2212.

Yoon, Y. J., & Mowry, K. L. (2004). *Xenopus* Staufen is a component of a ribonucleoprotein complex containing Vg1 RNA and kinesin. *Development*, *131*, 3035–3045.

Yoon, S.-. Y., Choe, C., Kim, E.-. J., Kim, C.-. W., Han, J., & Kang, D. (2011). Acetylcholine controls mouse oocyte maturation via downregulation of cAMP. *Clin. Exp. Pharmacol. Physiol.*, *38*, 435–437.

Yuan, S., Schuster, A., Tang, C., Yu, T., Ortogero, N., Bao, J., et al. (2016). Sperm-borne miR-NAs and endo-siRNAs are important for fertilization and preimplantation embryonic development. *Development*, *143*(4), 635–647.

Yuen, E. Y., Jiang, Q., Chen, P., Gu, Z., Feng, J., & Yan, Z. (2005). Serotonin 5-HT1A receptors regulate NMDA receptor channels through a microtubule-dependent mechanism. *J. Neurosci.*, *25*, 5488–5501.

Zhou, W., Stanger, S. J., Anderson, A. L., Bernstein, I. R., De Iuliis, G. N., McCluskey, A., et al. (2019). Mechanisms of tethering and cargo transfer during epididymosome-sperm interactions. *BMC Biol.*, *17*, 35. https://doi.org/10.1186/s12915-019-0653-5.

Development of the central nervous system

1 Exhaustion of the parental epigenetic information, neurulation, and the phylotypic stage

The development of the nervous system is a primarily epigenetic phenomenon (Tierney, 1996). Depending on the species, but more in oviparous animals and less in viviparous animals, the early development is determined by maternal cytoplasmic factors before the onset of the MZT (maternal-to-zygotic transition); parental factors continue to influence the early development until the gastrulation stage and even beyond (Torres-Paz et al., 2019).

After the diversifying stage of the early development, embryos of most different species of a phylum eventually converge with respect to both the morphology and regulatory gene expression, This is the phylotypic stage, the earliest stage where the basic Bauplan of the phylum, including the formation of the neural tube, can be recognized. This is the stage of the greatest similarity among species of a phylum. Development after the phylotypic stage is characterized by morphological and physiological divergence that leads to the adult morphological and physiological diversity of species. The phylotypic stage represents a period rather than a moment of individual development.

This stage, initially named the *pharyngula stage* (Ballard, 1981), the *early somite segmentation stage* (Wolpert, 1991), and the *tailbud stage* (Slack et al., 1993), is generally known as the *phylotypic stage* in a term coined by Sander (Sander, 1983). The phylotypic stage of vertebrates seems to be homologous to the insect "extended germband" stage, while in nematodes it corresponds to stage 7 of development.

The phylotypic stage begins with the process of neurulation (Galis and Metz, 2001), the transformation of the neural plate into the neural tube and the formation of the central nervous system.

2 Formation of the neural tube

During gastrulation three embrynic layers (endoderm, mesoderm, and ectoderm) are differentiated and the AP (antero-posterior) axis is established as a result of the invgination of the early embryonic structure. This is followed by formation along the midline of a transitional mesodermal cylindrical structure, the notochord, which functions as a source of signals that transform the adjacent overlying ectoderm into the neural plate as opposed to the more distant ectodermal layers that develop into the

The Inductive Brain in Development and Evolution. https://doi.org/10.1016/B978-0-323-85154-1.00003-5

epidermis or skin because of high *BMP* (a group of TGF-β family protein) expression.

Ectodermal cells are inherently biased to differentiate into neuronal cells, but this tendency is suppressed by *BMP* expression in the nonneural ectoderm and by mesodermal signals. This is indicated by the fact that when embryonic ectodermal cells are cultured at low density in the absence of mesodermal cells, they undergo neural differentiation (express neural genes), suggesting that neural differentiation is the default fate of ectodermal cells. In explant cultures (which allow direct cell–cell interactions), the same cells differentiate into the epidermis. This is due to the action of BMP4 that induces ectodermal cultures to differentiate into the epidermis (Chang and Harland, 2007).

The neural plate forms during gastrulation from the cranial portion of the primitive streak over and around Hensen's node (Bellairs and Osmond, 2014). Its formation represents the stage of the *primary neurulation*. In zebrafish, a maternal cytoplasmic factor *mib* is involved in the initiation of neurulation by inducing proneural genes that encode basic helix–loop–helix (bHLH) transcription factors. These, in turn, induce expression of Delta and its receptor Notch (Itoh et al., 2003), while the Delta-Notch pathway plays a key role in the development of the neural tube. Then, the two edges of the neural plate bend and gradually elevate to form the neural groove, whose deepening, along with the coalescence of the neural folds, leads to the formation of a tubular structure, the neural tube, which is considered the beginning of the *secondary neurulation* at the so-called neurula stage that follows the gastrula stage in vertebrates. While the primary neurulation takes place in the trunk segment of the embryo, the secondary neurulation takes place in the tail segment of the vertebrate body, where it forms the medullary cord. It consists in a process known as a mesenchymal-to-epithelial transition (Shimokita and Takahashi, 2011) that leads to the formation of the neural tube, a neuroepithelial layer enclosing a hollow space (Fig. 2.1).

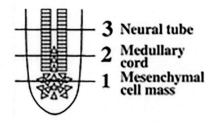

FIG. 2.1

Progressive epithelialization during the secondary neurulation (SN). Three different levels along the anteroposterior axis. Level 1, cells are still mesenchymal; level 2, a transitory zone (medullary cord) where epithelialization proceeds in a dorsoventral direction; level 3, tubulogenesis completed.

From Shimokita, E., Takahashi, Y., 2011. Secondary neurulation: fate-mapping and gene manipulation of the neural tube in tail bud. Dev. Growth Differ. 53, 401–410.

FGF3 signals from the mesoderm are necessary for the closure of the neural tube; they stimulate the extension of the neural tube and downregulate BMP expression. In the absence of FGF3, BMP secretion increases, leading to the delay of the neural tube closure and premature specification of the neural crest (Anderson et al., 2016). In *Ciona*, suppression of the *FGF* expression leads to the development of a truncated anterior brain (Wagner and Levine, 2012).

3 Neuronal specification and proliferation

According to a widely held hypothesis, in vertebrates many embryonic lineages have a "default" neuronal program, which is under inhibitory control, but they switch to the ectodermal fate under the influence of extracellular BMP signals, which in turn are suppressed by organizer signals (Wilson and Hemmati-Brivanlou, 1997), a combination of multiple molecules, including noggin (Smith and Harland, 1992), a BMP antagonist, chordin (Sasai et al., 1994; Karzbrun and Reiner, 2019), and follistatin (Hemmati-Brivanlou et al., 1994).

Shh secreted by the notochord and the floor plate/ventral neural tube (Placzek, 1995) determines the development of interneurons (Martí et al., 1995) and motor neurons in the ventral side of the neural tube (Martí et al., 1995; Ericson et al., 1996). The role of Shh as an inducer of neuronal differentiation in the ventral neural plate/tube is confirmed by the fact that antibodies against Shh suppress the differentiation of motor neurons (Martí et al., 1995).

Neuronal progenitors or neural stem cells (NSCs), via gap junctions, form networks, which via electrical depolarization induce calcium activity to stimulate their proliferation (Malmersjö et al., 2013). Proliferation of neurons results from asymmetric divisions of neural progenitors in a process regulated by electrical signaling and neurotransmitter release (Spitzer, 2006). The process of proliferation of neural progenitors here will be exemplified by knowledge derived mainly from studies on the model organism *Drosophila melanogaster*, not only because the basic processes of neurogenesis are conserved across the animal world (Homem and Knoblich, 2012) but also because the relative simplicity of its nervous system enables a better study of neurogenesis in this fly.

Neuroblasts (NBs) or neuron progenitors represent neuronal stem cells. In *Drosophila*, they start differentiating during embryonic stages 9–11 (Homem and Knoblich, 2012), i.e., approximately 3:40 to 5:20h AEL (after egg laying) starting during the transient segmentation of the mesodermal layer, and begin dividing by the end of this period to produce neurons. After an 8–10h quiescence period (Fig. 2.2), neuroblasts begin a new wave of neurogenesis, leading to the formation of about 90% of neurons of the fly's CNS (Homem and Knoblich, 2012), after which the remaining neuroblasts enter apoptosis and disappear.

In *Drosophila*, NBs are differentiated from the neuroepithelium. Two basic NB types, type I and type II, are identified in the fly. They differ from each other in the type of nuclear transcription factor they express and in the cell types resulting from

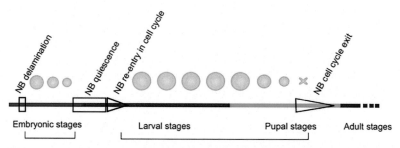

FIG. 2.2

Neurogenesis in *Drosophila* embryos and larvae. Timeline of the two waves of neurogenesis occurring during *Drosophila* development. Neuroblasts (NBs; beige) and their sizes are depicted throughout the timeline. NBs are generated during embryonic stages by delamination from the neuroectoderm. Embryonic NBs do not regrow after each division. They become quiescent during late embryogenesis but reenter the cell cycle to start the second wave of neurogenesis in larvae. Larval NBs regrow after each cell division and therefore can divide more often. During pupal stages, NBs disappear, and this ends the second wave of neurogenesis. Different NBs exit the cell cycle at different time points; the cartoon depicts the cell-cycle exit mechanism described for thoracic NBs, which reduce their size until they undergo a sizewise symmetric division and differentiate.

From Homem, C.C.F., Knoblich, J.A., 2012. Drosophila *neuroblasts: a model for stem cell biology. Development 139, 4297–4310.*

their asymmetric divisions. Type I NB divides to reproduce itself and to produce a GMC (ganglion mother cell) (Wodarz and Huttner, 2003), while type II divides asymmetrically, resulting in the production of daughter cells with two distinct cell fates, an intermediate neural progenitor (INP) that after several rounds of asymmetric divisions of self-renewal generates GMCs, with each of the latter dividing to produce two neurons. Each *Drosophila* embryonic brain lobe contains about 90 type I NBs and 8 type II NBs with the latter located in the posterior side of the brain (Homem et al., 2015). However, in the embryonic central brain area of *Drosophila*, by virtue of the INP (intermediate neural progenitor) transit-amplifying population, the number of type II NB-derived neurons grows much faster than that of the type I NB origin. The driver of the process of proliferation of type II NBs is *bantam* miRNA with *brat* and *prospero* downstream targets (Weng and Cohen, 2015).

Spatiotemporal patterns of calcium transients determined by neuronal ion channels activate or repress enzymes that transduce ionic signals into biochemical products, thus deciding the neurotransmitter phenotype; by regulating the neuronal cytoskeleton, they determine axon and dendrite morphology (Rosenberg and Spitzer, 2011) (Fig. 2.3).

In rodents, and maybe in mammalians in general, neural stem cells may also undergo symmetric and asymmetric divisions. Neural tube cells are neuroepithelial cells or neural stem cells that initially undergo symmetric self-renewing divisions, leading to the emergence of two daughter cells and to the proliferation of neural tube

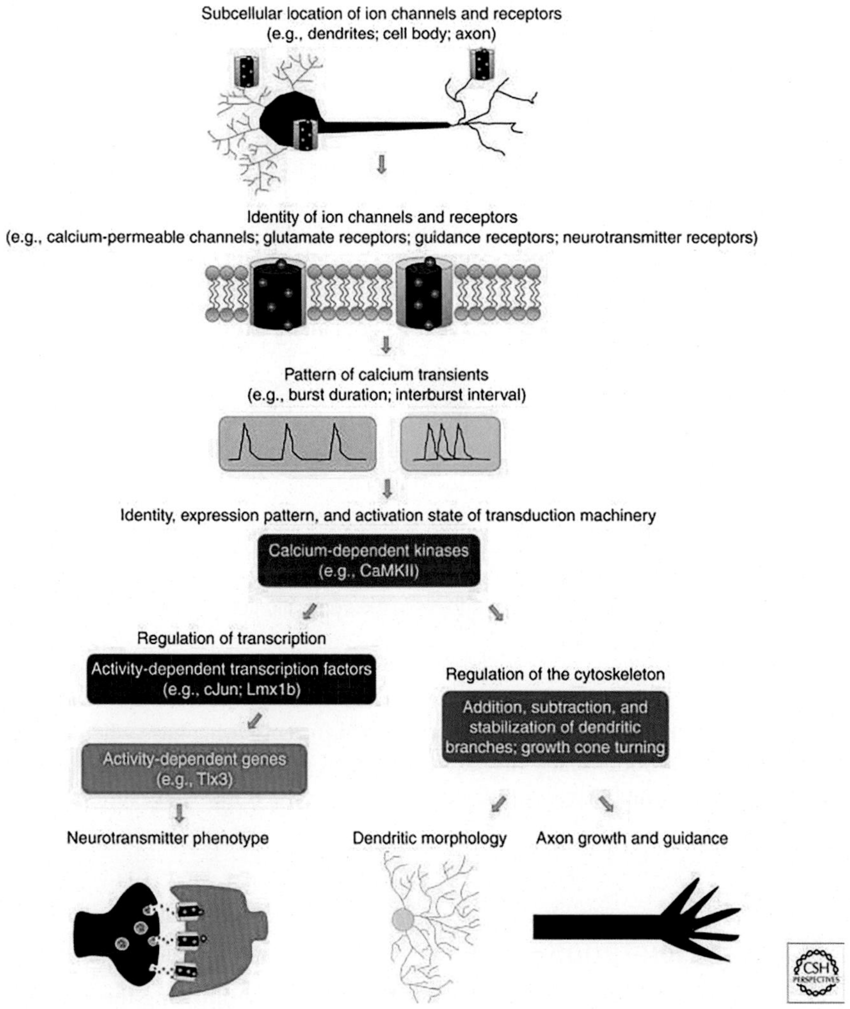

FIG. 2.3

Providing specificity for calcium signaling in neuronal differentiation. Calcium transients direct neuronal differentiation by regulating neurotransmitter phenotype, dendritic morphology, and axonal growth and guidance. Factors dictating intracellular calcium dynamics include the subcellular location of ion channels within the neuron and the neuron-specific constellation of ion channels and receptors expressed by individual cells. The location and identity of these channels and receptors influence the timing and frequency of calcium transients and determine whether the changes in calcium concentration occur in a global or localized fashion. Spatiotemporal patterns of calcium transients select the downstream mechanisms involved in neuronal differentiation. Calcium transients activate enzymes that transduce ionic signals into biochemical ones. These enzymes impact differentiation either through transcriptional mechanisms or by the regulation of cytoskeletal dynamics. Activation or repression of transcription factors controls neurotransmitter expression, whereas cytoskeletal remodeling regulates axon and dendrite morphogenesis.

From Rosenberg, S.S., Spitzer, N.C., 2011. Calcium signaling in neuronal development. Cold Spring Harb. Perspect. Biol. 3 (10), a004259. https://doi.org/10.1101/cshperspect.a004259.

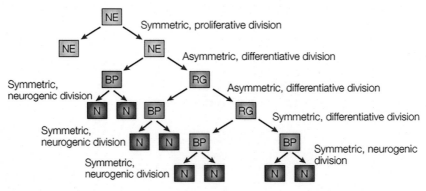

FIG. 2.4

Lineage tree of neurogenesis. The lineage tree shown provides a simplified view of the relationship between neuroepithelial cells (NE), radial glial cells (RG), and neurons (N) with basal progenitors (BP) as cellular intermediates in the generation of neurons. They also show the types of cell division involved.

From Götz M., Huttner, W.B., 2005. The cell biology of neurogenesis. Nat. Rev. Mol. Cell Biol. 6, 777–788.

cells (proliferation phase), which is followed by a differentiation phase whereby each neural stem cell asymmetrically divides into a radial glial cell (RG) and a basal progenitor cell (BP), which divide to produce only neurons (Götz and Huttner, 2005) (Fig. 2.4).

4 Neuronal migration

Neuronal migration is an essential stage of neurogenesis, following the proliferation stage (Fig. 2.5). In their journey from the birthplace to their permanent position, neurons are guided by attractive and repellant signals. Initially, neurons project filopodia and lamellipodia to explore the microenvironment, and based on the integration and processing of the attractive and repulsive findings, neurons form the leading edge of the cell body under the action of forces exerted by microfilaments, while microtubules push the nucleus toward the leading edge (de Rouvroit and Goffinet, 2001).

The pathfinding is a probabilistic rather than a random statistical process, meaning that it is a predictive forward-looking process that includes some uncertainty (in the case of the establishment of the neuronal connections that are eliminated through connection pruning). Both processes of prediction and pruning may be positively selected in the course of the evolution of eumetazoans.

The spontaneous neuronal activity is necessary for axonal pathfinding (Hanson and Landmesser, 2004) and axonal branching (Mizuno et al., 2007; Kalil and Dent, 2014) in vertebrates. The fact that the spontaneous neural activity is necessary for axonal pathfinding indicates that the neuron, after leaving the neural tube, is provided with information on the nature of the molecules it has to be guided by or avoid. Indeed, different types of neurons are attracted to different molecules that they identify as guidance cues.

According to the mode of their movement in the central nervous system, migration of neurons is described to occur in two alternative forms: radial migration and tangential migration (Fig. 2.5). About 80%–90% of all mammalian neocortical neurons (Ayala et al., 2007) differentiate from neural stem cells, proliferating in the ventricular zone (VZ) or subventricular zone of the developing neocortex (Reiner, 2013). They begin migration from the VZ toward the pia mater radially by passing through the neocortical layers or tangentially, i.e., in parallel with the layers (Rao and Wu, 2001). Approaching the pia mater, neurons settle away from their VZ birthplace (Fig. 2.5). Similarly to the human cerebral cortex, the outermost layer of the brain that is composed of six neuronal layers, the layer and columnar structure of the *Drosophila* optic lobe, receiving and processing visual inputs from the retina, consists of four layers (lamina, medulla, lobula, and lobula plate) (Suzuki and Sato, 2014) of tens of different types of neurons.

In humans, neurons begin migration from VZ in the direction of pia mater by the GW (gestational week) 13–14, but most neuronal migration takes place during GW

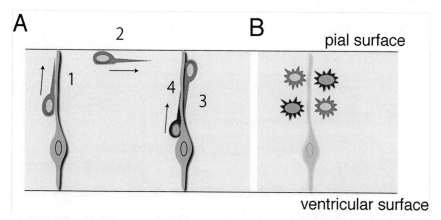

FIG. 2.5

A schematic illustration showing intermixing of neurons from different sites of origin as a result of switching the migratory mode. (A) One neuron first migrates radially toward the pial surface (1), then tangentially (2), and finally radially back toward the ventricular surface (3), a distance away from the site of origin. Another neuron simply migrates toward the pial surface without much tangential displacement (4). (B) Intermixing of different types of neurons as a result of tangential migration. Progenitors in different locations give rise to different types of neurons. Tangential migration of neurons, as shown in (A), allows neurons (*blue*) to move into a region where different types of neurons (*red*) are located.

From Hatanaka, Y., Zhu, Y., Torigoe, M., Kita, Y., Murakami, F., 2016. From migration to settlement: the pathways, migration modes and dynamics of neurons in the developing brain. Proc. Jpn. Acad. Ser. B Phys. Biol. Sci. 92, 1–19.

12–20, with the radial migration almost completed by GW 26 and nonradial migration disappearing by GW 34 (Wilkinson et al., 2017).

---

5 Neuronal maturation

With the termination of migration and settling into the mammalian neocortex, neurons start their maturation, leave their leading process in the rear, and while migrating along the radial glia fibers, form new radially leading processes, one of whom becomes axon (Hatanaka et al., 2016), while on the opposite side of the cell they develop dendrites (Fig. 2.6).

When the neurite fields of neurons begin to overlap, they form homogeneous or clustered networks of functionally similar local interconnected neurons. Migration and clustering increase connectivity and accelerate the onset of electrical activity. At one point, migration and growth of neurites inhibit further clustering (Okujeni and Egert, 2019). Formation of clusters of functionally connected neurons via synapses increases the processing capability of individual neurons, and clusters of at least 30 neurons display network-level electrical activity or firing (Shaw et al., 1982; Idelson et al., 2010).

Generally, neurons producing the same neurotransmitter form separate neuronal aggregations, but their collective behavior in the course of migration is also

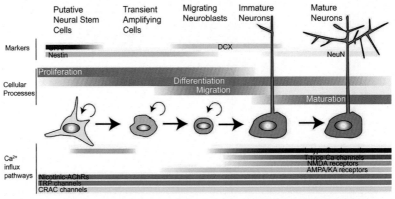

FIG. 2.6

Schematic of the different stages of neurogenesis in the developing brain. Neurogenesis proceeds through several overlapping stages: proliferation, differentiation, migration, and maturation. The Ca^{2+} influx pathways expressed during these stages are indicated. Commonly used immunohistological markers for staging neurogenesis are also shown above. Abbreviations: CRAC, Ca^{2+} release-activated Ca^{2+}; NMDAR, NMDA receptors; TRP channels, Transient receptor potential channels.

From Toth, A.B., Shum, A.K., Prakriya, M., 2016. Regulation of neurogenesis by calcium signaling. Cell Calcium 59(2–3), 124–134. https://doi.org/10.1016/j.ceca.2016.02.011.

geometry-dependent, leading to aggregation of neurons at the center of the migrating collective of neurons and forming monoclusters (Sun et al., 2011). Clusters of functionally connected neurons that respond to specific stimuli by performing a specific function are known as neuronal circuits.

Abundant evidence in these last decades indicates that neuronal matching is regulated by neuronal electrical activity. It would be plausible that in embryos of anamniotes, such as fish and amphibians, whose visual circuit develops by the early visual contact with the environment, axons find their matching partners as a result of the neuronal activity arising in response to the visual stimuli. However, this explanation is not valid for amniotes developing in the absence of the visual information within thick eggs and in utero (Kutsarova et al., 2017). In the absence of the visual stimuli, this group of animals generates neuronal activity to establish the visual circuit. This indicates visual information may not be necessary for the process of neuronal matching and, most importantly, the formation of visual circuits in the absence of visual stimulation is an inherited character. The latter is the essential, however incomprehensible, fact.

In the Drosophila, the nerve cord contributes not only global positional cues, but also partner-derived cues and synaptic activity to the assemblage of the neuronal circuits (Valdes-Aleman et al., 2019). However, the role of electrical activity is supported by most studies. In establishing correct connections with muscles, motoneurons have to distinguish between "a global chemorepellant signal from all muscles, and a local chemoattractive signal from the target cell," which is believed to be determined by Ca^{2+} oscillations (Vonhoff and Keshishian, 2017).

A generalized schematic of stages of neurogenesis is presented in Fig. 2.5.

6 Formation of neuronal connections: Synaptogenesis

The first step in establishing of neural circuits is the extension of special neuronal projections, the neurites. With the formation of neurites begins the neuronal wiring. Like many other cells, neurons begin their life as spheroid structures that soon extend neurites, neuron projections from which axons and dendrites develop that form synapses for transmission and reception of impulses with each other. Neurite initiation results from a coordinated interaction of microtubules (MTs) and actins. Bundled microtubules composed of 13 protofilaments of α- and β-tubulins, (α β) tubulin heterodimers, represent the structural core of the neurite (Flynn, 2013), whose polymerization lengthens neurites and depolymerization has the opposite effect. A number of proteins, such as MAPs, associated with microtubules, promote stabilization of the microtubule structure, whereas other proteins are known to destabilize it.

The tip of the neurite is an actin-rich growth cone (Flynn et al., 2012). During early development, neurites are used for migration of neuroblasts to their destination, where they differentiate into mature neurons. Formation of neurites depends on the presence of two protein types, ADF (actin-depolymerizing factor) and/or cofilin, which induce MT bundling and neurite initiation. Cofilin is responsible for actin assembly and disassembly; it is more efficient in actin polymerization than ADF, which is more potent as an actin depolymerization/disassembling agent (Bernstein and Bamburg, 2010).

More than half a century ago, Roger W. Sperry came out with the chemoaffinity hypothesis: "the cells and fibers of the brain and cord must carry some kind of individual identification tags, presumably cytochemical in nature, by which they are distinguished one from another … the growing fibers are extremely particular when it comes to establishing synaptic connections, each axon linking only with certain neurons to which it becomes selectively attached by specific chemical affinities" (Sperry, 1963). Accordingly, neurons seek out and link with partners that express specific matching molecules. The first axonal tags identified were Eph (erythropoietin-producing human hepatocellular) membrane receptors that bind proteins of the Eph family ligands (Drescher et al., 1997). However, while recognition of molecules expressed by neurons predicted by this "lock-and-key" hypothesis is important for linking them together, the hypothesis cannot explain how this exact matching arises to form specific neuronal circuits involving several to hundreds of neurons. This difficulty seems to have been overcome by another hypothesis, which posits that neurons search to reach specific sites rather than link with specific neurons; the axon extension and branching are regulated by the neural activity (Fig. 2.7). Both hypotheses, however, share, the idea that neurons "search" for their target, implying that they "know" "which" is the target and how to reach it by recognizing cytochemical tags or specific sites. Neural activity determines types of receptor molecules expressed by the presynaptic neuron and the expression of activity-dependent molecules, axon guidance molecules in the postsynaptic neuron. Or, alternatively, it is plausible that the formation of connections between matching neurons results from the combined neural activity of both partners rather than the unilateral exploratory activity of only one of them. Synaptic connections are only established with specific neurons and axons often must travel long distances to the matching neuron. This suggests that the neural activity of the postsynaptic neurons by expressing, or not, a specific ligand for the axonal receptor of the presynaptic neuron somehow determines whether the synapse will be established or not.

Formation of connections is an activity-dependent process. Neural activity can refine the patterns of connections through Hebbian and/or homeostatic plasticity mechanisms. In early embryonic life, the formation of connections proceeds without sensory input (Yamamoto and López-Bendito, 2012), without any sensory experience, but the correct wiring of the neuronal connections before the embryo becomes receptive to sensory information is still puzzling.

After neurons settle in their destinations, they form synapses, communication nodes that connect them to each other. Unlike the formation of neurites, axons, and dendrites, which develop in an activity-independent manner, the formation of synapses is a function of the neural activity (Fig. 2.8). Formation of connections is an

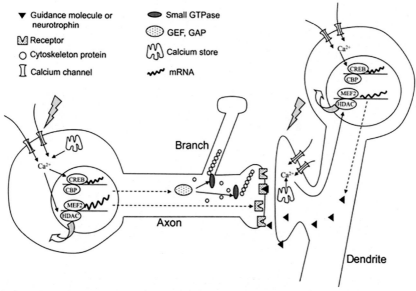

FIG. 2.7

Possible mechanisms of activity-dependent gene expression and axonal remodeling. Neuronal activity induces nucleocytoplasmic translocation of class II HDACs, whereupon particular transcription factors such as myocyte-specific enhancer factor 2 can bind to their target sequence and upregulate downstream gene expression. Simultaneously, HAT activity is increased, and this promotes the activity of transcription factors such as CREB. In postsynaptic target cells, an activity-dependent regulation may alter the expression of guidance molecules or neurotrophins. In presynaptic growing axons, such a regulatory mechanism may affect the expression of their receptors. Cytoskeleton protein polymerization or depolymerization may also be affected by altered expression of regulatory proteins such as GEFs or GAPs. Abbreviations: HDAC, histone deacetylase; HAT, histone acetyltransferase; CREB, cAMP response element-binding protein; GEF, guanine nucleotide-exchange factors; GAPs, GTPase-accelerating proteins.

From Yamamoto, N., López-Bendito, G., 2012. Shaping brain connections through spontaneous neural activity. Eur. J. Neurosci. 35, 1595–1604.

activity-dependent process that in early embryonic life proceeds without external sensory input (Yamamoto and López-Bendito, 2012), without any sensory experience.

Synapses belong to two main types, chemical synapses and electrical synapses, which differ from each other in the ways the neural impulse is transmitted from the presynaptic to the postsynaptic neuron. While in the chemical synapse, the neural signal is transmitted via neurotransmitters released by the presynaptic synapse, in electrical synapses, the smaller gap of the synaptic cleft allows transmission of electrical signals from the presynaptic to the postsynaptic neuron via ions passing through the channel proteins.

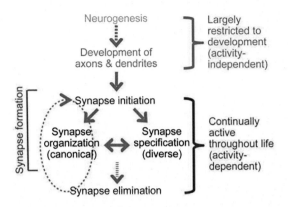

FIG. 2.8

Synapses construct neural circuits. Schematic of neural circuit development. Neurogenesis is followed by neural migration (not shown) and elaboration of axons and dendrites, including extension of especially axons, often over long distances (axon pathfinding). Guided axon-dendrite contacts then form synapses, with three proposed components of synapse formation: target recognition that causes synapse initiation, organization of the canonical components of synapses such as synaptic vesicles and active zones, and specification of synapse properties such as transmitter identity, release probability, or competence for long-term plasticity. Synapse formation is often followed by synapse elimination, resulting in a continuous turnover of some synapses.

From Südhof, T.C., 2017. Synaptic neurexin complexes: a molecular code for the logic of neural circuits. Cell 171, 745–769.

Mature neurons form a great number of synapses between their axons and dendrites of other neurons. Each neuron establishes between one to several thousand connections with other specific neurons, transforming the CNS into an enormous network of the order of quadrillion interneuronal connections and leading to the formation of millions of neural circuits performing the most diverse functions.

Formation of a synapse requires various protein types, including neurexins, neurotrophins, and cell adhesion molecules (Fig. 2.9). The assembly of neural circuits is a result of the interaction of numerous ligands and receptors of neurons participating in the assembly of the neural circuit.

Neurexins are presynaptic membrane cell adhesion molecules produced by alternative splicing of three genes (Aoto et al., 2013), leading to the generation of tens of thousands of neurexin isoforms from these genes. Neurexins act both as membrane adhesion molecules and receptor molecules and serve as anchors for at least seven postsynaptic protein families (Südhof, 2017). Release of neurexins by the presynaptic neurons is the initial step in the formation of synaptic connections. Neural activity, by inducing accumulation of H3K9me3 in chromatin, causes a shift in neurexin (Nrxn1) splice isoforms (Ding et al., 2017).

Neurotrophins initially were identified as secreted proteins involved predominantly in the development of the nervous system, but now it is known that at least

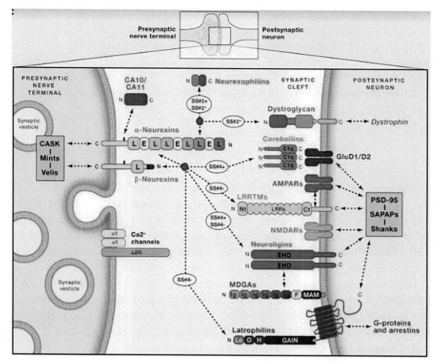

FIG. 2.9

Schematic of the interactions of α- and β-neurexins with selected ligands in the context of the synapse. Requirements for neurexin splice variants are indicated; possible competition between ligands is indicated by junctions marked with a circle. Proteins are not drawn to scale. Abbreviations: E, EGF-like domain; EHD, esterase homology domain; L, LNS domain; LRRs, leucine-rich repeats; Nt and Ct, N- and C-terminal sequences surrounding LRRs; Ig, Ig domain; F, fibronectin III domain; MAM, MAM domain; Lc, lectin domain; O, olfactomedin-like domain; H, hormone-binding domain; GAIN, GAIN domain.

From Südhof, T.C., 2017. Synaptic neurexin complexes: a molecular code for the logic of neural circuits. Cell 171, 745–769.

BDNF (brain-derived growth factor) localized in the DCV (dense core vesicles) is essential in the adult nervous system as a regulator of synaptic connections, neurotransmitter release, synapse formation, and synaptic plasticity (Song et al., 2017). Among neurotrophins, NGF (nerve growth factor), BDNF (brain-derived neurotrophic factor), neurotrophin-3, and neurotrophin-4 are derived from a common ancestral gene (Hallböök, 1999).

Among cell adhesion molecules involved in synaptogenesis, neuroligins are localized in postsynaptic membranes and serve as ligands for α- and β-neurexins. They are encoded by four genes and are crucial for synapse maturation, and the functional impairment of neuroligins is a probable cause of autism

(Varoqueaux et al., 2006). Another important group of postsynaptic cell adhesion proteins that bind neurexins to form a complex transsynaptic signaling system are members of the LRRTM (leucine-rich repeat transmembrane protein) family of postsynaptic cell adhesion molecules (Dagar and Gottmann, 2019) as well as latrophilins, dystroglycans, etc.

All the molecules involved in the formation of synapses are expressed long before the beginning of synaptogenesis (Daly and Ziff, 1997).

7 Neural activity and establishment of neural circuits

Spontaneous activity is necessary for the expression of the axon guidance receptor EphA4. It is involved in the fine-tuning of the axonal arbor (Yamamoto and López-Bendito, 2012), and its experimental alteration leads to axon guidance errors and erratic pathfinding behavior (Hanson and Landmesser, 2004; Valdes-Aleman et al., 2019). Thus, the spontaneous neuronal activity determines the neuron's target, i.e., the matching neuron it is looking for.

Spontaneous activity in different animals emerges at different stages of the early development. In zebrafish, spontaneous activity of individual neurons is observed first at 17.5 hpf (hours postfertilization). That is, during the segmentation stage (Kimmel et al., 1995), at 7 h before the beginning of the phylotypic stage. This corresponds to the 20-somite stage. The patterned neural activity of small groups of neurons appears by 19 hpf, but only by 22 hpf does it rise to the level of a globally synchronized ensemble (Wan et al., 2019). In zebrafish, the spontaneous activity and establishment of the neuronal circuits in the neural tube emerge first during the 20-somite stage at 17.5 hpf (Wan et al., 2019), implying that at the phylotypic stage the CNS is operational.

Correlated neural activity is thought to be related to Hebbian mechanisms, in which neurons with similar activity patterns are likely to synapse together (Torborg and Feller, 2005) or strengthen the synapse. The reverse occurs when their activity is not coordinated. This leads to the assembly of the neurons in specific circuits according to the hypothesis "neurons that fire together wire together." The coordination of the activity of the pre- and postsynaptic neurons is essential for the establishment of neural circuits.

Specific patterned spontaneous neuronal activity instructively determines the development of neural circuits and neuronal connectivity, and although the Hebbian mechanisms are involved in the establishment of synaptic connections and function, they do not have any critical role in neural circuit formation (Xu et al., 2011). Neuronal circuits respond to developmental changes in embryonic structure by changing their connection patterns, morphology, and behavior, including the developmental output by generating inductive signals.

Neurons integrate various incoming signals to generate spike patterns that are translated into output signals that differ not only between different neurons but also between different synapses of the same neuron (Südhof, 2017). Depending on the neural activity of the presynaptic neuron or on the influence of other neurons, transmission of chemical and electrical signals may be improved (facilitation) or impaired (inhibition).

The spontaneous neural activity and later the sensory (e.g., visual) experience also regulate the elaboration of dendritic arbors and dendrite size (Yuan et al., 2011) and synaptogenesis (Sheng et al., 2019). Sensory experience also regulates the physiology of neural circuits (Kaneko et al., 2017). Before the onset of the competence to respond to light, and in the absence of visual input, retinas exhibit the so-called retinal waves consisting in spontaneous bursts of action potentials that starting from particular locations extend to areas comprising up to a hundred neurons (Feller, 2009). A common characteristic of these bursts is that they correlate across tens to hundreds of neurons and reoccur in a matter of minutes (Feller, 1999). Retinal waves play an instructive role in the development of the visual map in mice (Xu et al., 2011) in an experience-independent way. In primates, waves occur only prenatally, whereas in other mammalians such as mice and ferrets, they occur both prenatally and postnatally (Huberman et al., 2008). Adequate evidence shows that patterns of neural activity in the form of action potentials are essential in the development, maintenance, and refinement of adult neural circuits (Bleckert and Wong, 2011; Kirkby et al., 2013).

In *D. melanogaster*, axons of the photoreceptor neurons, R1–R6, select the correct target neurons in the lamina, forming a characteristic topographical map in the fly's brain, but this map is defective in flies that lack or have lost the function of cadherin Flamingo (Fmi) and N-cad (N-cadherin), leading to the idea that the *Drosophila* visual system fine-scale topography does not depend on neural activity (Lee et al., 2003; Hiesinger et al., 2006). However, it is demonstrated that changes in the patterns of the sensory input induce specific changes in the cortical maps (Feller, 2009) and the monocular deprivation of vision leads to specific changes in neural circuits during critical periods of development in diverse mammals, from rodents to humans (Morishita and Hensch, 2008). The establishment of the fine-scale topography system of C4da neurons in *Drosophila* is regulated by the level of neural activity via the brain-specific ligase Trim9 (Yang et al., 2014; Li et al., 2019).

According to the Hebbian principle, the coordination of the activity of the pre- and postsynaptic neurons is a condition of the formation of stable synaptic connections. However, recent evidence from Nakashima et al. (2019) on the activity-dependent formation of the olfactory map in mice refutes the principle. Sensory olfactory neurons producing the same OR (olfactory receptor) project axons to the same glomeruli in the OB (olfactory bulb), but contrary to the Hebbian plasticity rule, even in the absence of synaptic partners, mutant mice form glomerular-like structures in the OB. Their experiments demonstrated that temporal patterns of spontaneous neural activity of the olfactory neurons determine both the olfactory receptor type they produce and the types of axon-sorting molecules, thus regulating glomerular segregation (Nakashima et al., 2019) (Fig. 2.10).

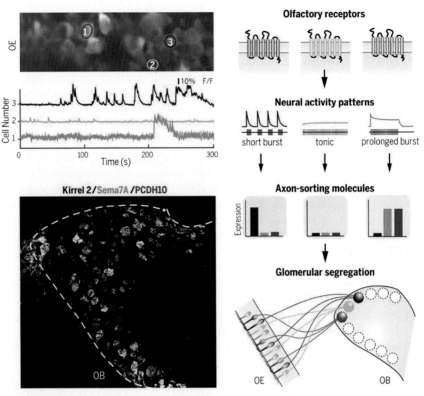

FIG. 2.10

Firing pattern-dependent olfactory map formation. (*Top left*) Diverse patterns of spontaneous neural activity in olfactory sensory neurons. (*Bottom left*) A combinatorial expression pattern of axon-sorting molecules at axon termini of olfactory sensory neurons. (*Right*) A model for activity-dependent olfactory map formation. OE, olfactory epithelium; OB, olfactory bulb. Axon-sorting molecules for glomerular segregation: red, Kirrel2 (kin of IRRE-like protein 2); green, Sema7A (semaphorin 7A); blue, PCDH10 (protocadherin 10).

From Nakashima, A., Ihara, N., Shigeta, M., Kiyonari, H., Ikegaya, Y., Takeuchi, H., 2019. Structured spike series specify gene expression patterns for olfactory circuit formation Science 365, 6448, eaaw5030.

8 Activity-dependent respecification of neurotransmitter phenotype

Contrary to the earlier concept of the neurotransmitter phenotype of neurons as a fixed character of neuronal identity, now we know that during a brief time window of development, even after formation of synapses, neurons can switch from one transmitter phenotype to another as a result of changes in neuronal activity and calcium-dependent activity that are transduced into biochemical signals acting on different transcription factors, leading to the synthesis of new enzymes determining neurotransmitter respecification (Spitzer, 2012). So, e.g., suppression of Ca^{2+} spike activity increases expression of excitatory neurotransmitters ACh and glutamate but decreases expression of inhibitory neurotransmitters, GABA and glycine, and the reverse occurs when the Ca^{2+} spike activity is enhanced (Borodinsky et al., 2004; Wolfram and Baines, 2013) (Fig. 2.11).

9 Establishment of phenotypic set points

Homeostasis, a relatively steady state of the internal environment, is crucial for the life and evolution of animals. The maintenance of homeostasis is a function of the central nervous system (Pacak and Palkovits, 2001).

In the course of evolution, animals have evolved special mechanisms for maintaining, within relatively narrow limits, many phenotypic, morphological, morphometric, and physiological characters, such as the amount/concentration of numerous substances and other parameters in body fluids and interstitial liquid, including maintenance of body size, temperature, and water content. In other words, animals have evolved the ability to maintain these parameters within species-specific set points. The set points are determined at various times of individual development, extending from the early development to adulthood. It is adequately demonstrated that maintenance of relatively steady levels of these phenotypic characters is a function of

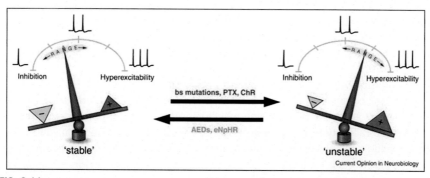

FIG. 2.11

Activity-dependent switching of neurotransmitter phenotypes. Studies in *Xenopus* demonstrate that enhanced (shown on the right) as well as reduced (shown on the left) neuronal activity is sufficient to induce a respecification of the neurotransmitter phenotype in neurons of the spinal cord to maintain an appropriate excitation-inhibition balance. Decreased activity favors an increase in neurons expressing excitatory neurotransmitters (acetylcholine [Ach] and glutamate, *orange circles*), whereas an increase in activity leads to increased numbers of GABA-expressing neurons (*blue circles*).

From Wolfram, V., Baines, R.A., 2013. Blurring the boundaries: developmental and activity-dependent determinants of neural circuits. Trends Neurosci. 36, 610–619.

specific neural circuits, which adaptively respond to the afferent input on the deviations from physiological norms.

Just for illustration, let us consider the set point on body temperature, body mass, and blood glucose level.

9.1 Set points for normal activity levels of neuronal circuits

To maintain the "normal" physiological levels of vital body parameters, animals needed first to evolve mechanisms form maintainingthe stability and strength of the synaptic connections at life-compatible levels. Destabilization of the synaptic strength and connections in neural circuits would be associated with adverse consequences for the homeostasis and the life of the embryo. Nevertheless, under common circumstances, this does not occur, implying that neural circuits themselves possess mechanisms that can stabilize and restore the perturbed neuronal activity within the normal limits. This idea is verified during the last two decades, and the mechanism of the restoration and stabilization of neural activity is known as "homeostatic plasticity" (Turrigiano, 2011).

During embryogenesis, based on the level of synaptic activity, animals encode homeostatic set points of neuronal circuits, but it has been demonstrated that these set points can be changed by experimentally changing the electrical activity (Giachello and Baines, 2017; Kaneko et al., 2017; Tien and Kerschensteiner, 2018). How these set points are established is not known. It is suggested that they are established during the critical developmental periods, depending on the level of the neural activity they are exposed to at the time. It seems plausible that the set points may be determined in an experience-dependent way, at a narrow range between the lower and higher levels that proved to be life-compatible. In other words, by the exclusion principle, the relevant neural circuit adopts as a set point the existing variable when the afferent input indicates that embryonic development proceeds undisturbed.

Studies in *Drosophila* show that this critical time may be at 17–19 h AEL (after egg laying), coinciding with the onset of action potentials in motoneurons and neurons. This is a time window during which neurons integrate into circuitry and use endogenous activity to refine their properties. Hyperactivity of neurons during this time window displaces the homeostatic set point beyond the dynamic range (Fig. 2.12), leading to the onset of permanent seizure-like behavior, which can be rescued by the administration of AED (antiepileptic drugs), restoring the neural activity to physiological levels (Giachello and Baines, 2015; Truszkowski and Aizenman, 2015; Giachello and Baines, 2017). The time window usually coincides with the time of the establishment of the respective neural circuits.

9.2 Set point for body temperature

Thermoregulation in warmblooded animals is an evolutionarily evolved and physiologically vital mechanism under the control of a central neural circuit positioned predominantly in the hypothalamus (Boulant, 2000; Hammel et al., 1963). A continual input on T_{core} (core body temperature) and the brain temperature from skin and

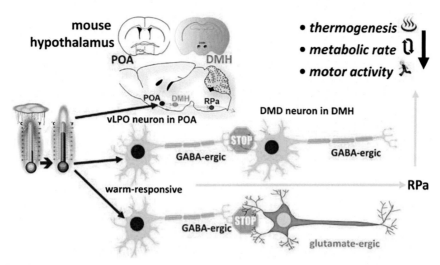

FIG. 2.12

Neurons set homeostatic set points based on synaptic excitation during early development. Developing neurons encode homeostatic set points, which limit their activity range, early in development based on exposure to synaptic excitation. Increased levels of synaptic excitation, due to genetic (mutations), chemical (picrotoxin, PTX), or optogenetic (ChR) modifications, change the set point and allow activity limits, leaving circuits prone to instability (i.e., seizure). Capping activity in these backgrounds, using either AEDs or optogenetics (eNpHR), allows neurons to encode appropriate set points, thus restoring neural activity to a physiologically appropriate range. Abbreviations: AEDs, antiepileptic drugs; ChR, channelrhodopsin; eNpHR, halorhodopsin.

From Giachello, C.N., Baines, R.A., 2017. Regulation of motoneuron excitability and the setting of homeostatic limits. Curr. Opin. Neurobiol. 43, 1–6.

central and visceral thermoreceptors reaches the hypothalamus and other areas of the brain, where it is compared with the species-specific T_{core} set point and, in response, activates mechanisms to restore/maintain the T_{core} set point (Boulant, 2000). In viviparous animals, the mechanism is operational at birth, and in oviparous animals, the mechanism is operational before hatching. The basic mechanisms of thermoregulation during both hyperthermia and hypothermia are as follows:

1. Thermogenesis, production of heat as a result of metabolic processes and motor activity. When T_{core} drops under the set point, signals originating in the hypothalamus increase thermogenesis by inducing secretion of thyroid hormones, neural (adrenergic) induction of constriction of skin blood vessels, and neural stimulation of shivering and movement toward warmer places.

2. Reduction of thermogenesis and the loss of body heat. This is the case when the environment and body temperature rise above T_{core}. In this case, hypothalamic signals inhibit secretion of thyroid hormones and increase heat loss via neural

cholinergic stimulation of vasodilatation and sweating, as well as panting and abandonment of the warm environment. Similar effects have experimental cooling of the hypothalamic POA (preoptic area).

The central circuit receiving information on T_{core} and the brain temperature comprises POA, DMH (dorsomedial hypothalamic nucleus), and the dorsal raphe. The temperature regulating central circuit can be made visible in the case of the elevated environmental temperature. In this case, the environmental heat stimulates glutamatergic neurons in POA, which activate a group of GABAergic neurons in the vLPO (ventral part of the lateral preoptic nucleus), which in turn inhibit thermogenic DMD (dorsal part of dorsomedial hypothalamus) neurons, inducing suppression of EE (energy expenditure) and leading to the declining of T_{core}. In the case of the low environmental temperature the GABAergic and glutamatergic neurons of the DMD are activated, stimulating thermogenesis, metabolic rate, and motor activity, resulting in T_{core} restoration (Zhao et al., 2017) (Fig. 2.13).

Regulation of body temperature in vertebrates, from fish to mammals, is essentially similar (Crawshaw et al., 1985).

Hypothalamic set points are neither inalterable nor permanent. They can change not only in the course of species evolution but, by experimental manipulation, also within the lifetime of an animal; animals that underwent acclimation to lower temperatures established lower T_{core} set points compared with the control ones (Tzschenke and Nichelmann, 1997), and eggs incubated at lower and higher than normal temperatures hatched ducklings that had respectively higher and lower warm sensitive neurons in the hypothalamus than the control ducklings (Tzschenke and Basta, 2002).

9.3 Set points for body mass

In most animals, the body size is species-specific within certain limits. Generally, the species-specific body mass is determined in adulthood, coinciding with the time when the animal stops growing. As a typical example, the tobacco hornworm *Manduca sexta* stops growing when the hemolymph, after suppression of the JH secretion, is cleared of JH (juvenile hormone), and at the very first photoperiodic gate, the insect's brain starts secreting neurohormone PTTH (prothoracicotropic hormone), which stimulates the prothoracic gland to produce hormone ecdysone, inducing growth suppression, ceasing feeding, purging the gut, and initiating metamorphosis (Nijhout and Williams, 1974; Nagamine et al., 2016).

What happens is that stretch sensors send to the brain the input on the degree of stretch caused by body growth: "The assessment of this critical body size is made in the brain and is based on processing of signals sent by the insect's stretch proprioceptive neurons that receive mechanical stimuli of increasing stretch as a result of increased body size" (Gorbman and Davey, 1991). When the proprioceptive input reaches the weight threshold of about 5 g, brain responds by sending efferent signals

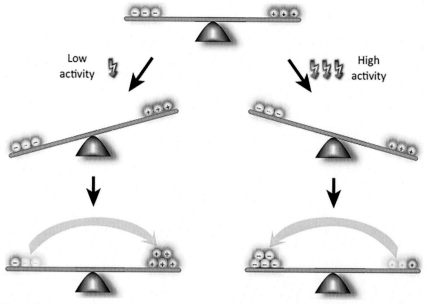

FIG. 2.13

Schematic shows a rise in temperature activating the warm-responsive, GABAergic neurons in the vLPO subnucleus of the preoptic area (POA), illustrated for schematic orientation in the mouse brain. These GABAergic neurons relay synaptically to GABAergic and also to glutamatergic neurons in the DMD subnucleus of the dorsomedial hypothalamic (DMH) nucleus, which is also schematically shown. GABAergic relay (i.e., inhibitory neurotransmission) to both types of DMD neurons attenuates thermogenesis, metabolic rate, and also behavioral motor activity, presumably via efferences to the rostral pallidum (RPa) in the brainstem. Note that GABAergic vLPO neurons could be intrinsically warm-sensitive, as well as receiving warmth-evoked peripheral afferences. Abbreviations: DMD, dorsal part of dorsomedial hypothalamus; DMH, dorsomedial hypothalamus; RPa, rostral pallidum; vLPO, ventral part of the lateral preoptic nucleus.

From Liedtke, W.B., 2017. Deconstructing mammalian thermoregulation. Proc. Natl. Acad. Sci. USA 114, 1765–1767.

(neuropeptides of the family of allatostatins) to corpora allata for suppressing the synthesis of JH and, consequently, stopping growth.

In insects of the genus *Rhodnius*, it is a blood meal that "stimulates stretch receptor nerves in the abdomen, which in turn stimulate specialized cells in the brain to secrete prothoracicotropic hormone, which is released into the blood and stimulates the prothoracic gland to secrete the molting hormone, ecdysone, which stimulates molting" (West-Eberhard, 2003). There is also evidence on the existence of a minimal larval body weight of ca. 600 mg before the brain of the yellow-spotted longicorn beetle, *Psacothea hilaris*, sends signals for ending growth and entering the diapause (Munyiri et al., 2004; Munyiri and Ishikawa, 2005). On reaching that set point, two hormonal processes occur sequentially, repression of JH secretion and its clearing from the hemolymph and the beginning of ecdysone synthesis, both under cerebral control via secretion of neuropeptides allatostatins and PTTH (prothoracicotropic hormone), respectively (Zitnan et al., 1993).

Regeneration experiments in planarians have shown that these very simple animals have a precise accounting mechanism that controls the body mass by determining "both the absolute and relative numbers of different cell types in complex organs such as the brain during cell turnover, starvation, and regeneration" (Takeda et al., 2009).

Set points for body mass and morphological/morphometrical characters in vertebrates are established upon reaching adulthood. Experiments on the deer mouse, *Peromyscus maniculatus*, showed that intraperitoneal implants of inert masses cause a compensatory quasi-equivalent loss of body weight that animals regain after removal of the implant. In investigators' interpretation, this points to the existence of a body mass set point in the animal's brain. They suggested that mechanoreceptors located within muscles and tendons that have afferent pathways to the cerebral cortex provide to the CNS the input on the body mass (Adams et al., 2001). Other experiments suggested that the center that regulates vertebrate body weight and food intake is in the hypothalamus (Baeckberg et al., 2003).

Similar results were obtained in analogous experiments in mice implanted with 10% of their biological body, but with a clear difference that the compensatory decrease of body was half of the implant weight and affected male mice alone. The study points out: "the conclusions drawn from our study could also be applied to the study by Adams et al. (2001)," but in their case "mass-specific set-point regulation of body mass is not accurate or might be impaired by still unknown competing mechanisms" (Wiedmer et al., 2004). Curiously enough, no significant compensatory decrease occurred in implanted female mice. This sexual dimorphism of body weight regulation was explained with an evolutionary pressure on female mice to maintain adequate energy stores during gestation that is more important than maintaining the body weight set point, which would lead to diminution of energy stores (Wiedmer et al., 2004).

Set points for body weight may change by administration of specific substances such as the recombinant B protein that lowers to a sustained 12% loss of the body mass set point in mutant obese (ob/ob) C57BL/6J mice (Pelleymounter et al., 1995; Halaas et al., 1995).

Body weight and glucose homeostasis are under a coordinated neural control by many hypothalamic nuclei, especially by neurons of the arcuate nucleus that secrete proopiomelacortin (POMC), the neuropeptide Y, GABA, and the brain-specific agouti-related peptide (AgRP) (Xu et al., 2017) (Fig. 2.14).

Body weight in vertebrates is maintained at a constant level by a balance between food intake regulated by neural mechanisms and energy expenditure (via the basal metabolic rate, mainly under the control of the hypothalamus–pituitary–thyroid axis), physical and mental activity, and adaptive thermogenesis via the hypothalamus–pituitary–thyroid axis. Generally speaking, the hypothalamus is the part of

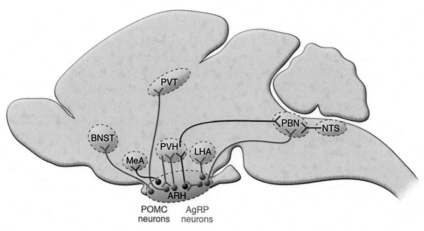

FIG. 2.14

The complex brain neural circuits regulating body weight balance. *Red lines* represent anorexigenic circuits, and *blue lines* represent orexigenic circuits. Abbreviations: ARH, arcuate nucleus of the hypothalamus; BNST, bed nucleus of stria terminalis; LHA, lateral hypothalamic area; MeA, medial amygdala; PVT, paraventricular nucleus of the thalamus; NTS, nucleus tractus solitarius; PBN, parabrachial nucleus; POMC, proopiomelanocortin; PVH, paraventricular nucleus of the hypothalamus.
From Xu, Y., O'Malley, B.W., Elmquist, J.K., 2017. Brain nuclear receptors and body weight regulation. J. Clin. Invest. 127, 1172–1180.

the brain where long-term adiposity signals (leptin and insulin) are integrated and the long-term energy homeostasis and body weight set point are regulated (Rui, 2013).

9.4 Glucose set point

Related to the maintenance of the normal body weight is the regulation of the glucose level as an indicator of the nutritional status of the organism. In mammals, the brain regulates glucose levels in body fluids in two main ways: via the sympathetic innervation it adaptively shuttles glucose between the liver stores and plasma and via the parasympathetic innervation it induces the pancreas to secrete insulin (Kosse et al., 2015) that decreases blood glucose levels by facilitating the intake of glucose by cells. The brain also regulates the glucose levels reactively, post factum, and additionally it controls and regulates glucose level predictively or anticipatorily by preventing anticipated changes in the glucose level.

The role of the CNS in determining the level of glucose is quite unambiguously demonstrated in *Drosophila*. The fly has a pair of glucose-sensing neurons in the brain, which project axons to induce secretion of the insulin-like peptide (dilp2) by insulin-like producing (IPC) neurons and the adipokinetic neurohormone (AKH), a neuropeptide analog of the mammalian glucagon. Experimental suppression of glucose-sensing neurons and dilp2 secretion derepresses secretion of AKH (adipokinetic hormone) in corpora allata, leading to hyperglycemia, abnormally elevated levels of glucose in hemolymph, a condition similar to that of diabetes mellitus. Based on these results, it was concluded that "these glucose-sensing neurons maintain glucose homeostasis by promoting the secretion of dilp2 and suppressing the release of AKH when hemolymph glucose levels are high" (Oh et al., 2019).

In vertebrates, a drop of the glucose blood level, hypoglycemia, is sensed in a number of brain areas, especially in the hypothalamic ventromedial nucleus (VMN). The brain receives additional information on the glucose level in plasma from the portal vein sensory neurons. It is true that the pancreas β-cells themselves sense the rise in the glucose level and, through a feedback mechanism, secrete insulin, which increases utilization of glucose. The role of the CNS in in regulating of insulin secretion is demonstrated in a number of experiments, including those with suppression of the parasympathetic activity by inactivating glucose transporter 2 (Glut2) in NgKO mice that reduce insulin secretion by the pancreas as a result of reduction by 30% of the β-cell mass (Tarussio et al., 2014).

Upon sensing the drop of blood glucose level, the brain triggers appetite by generating an emotional state of "liking" and motivational drive of "wanting." Then the satiety sensory input from the gastrointestinal tract serves as a feedback signal to the hindbrain homeostatic circuit, inducing suppression of feeding. Adiposity hormone and insulin also act as suppressors of feeding via the hypothalamic homeostatic circuits (Rui, 2013). For regulation of glucose levels, see also section 1.1.2 in Chapter 5.

10 On the nature and source of the spontaneous activity

Spontaneous activity is generation by neurons of electrical spike trains not related to environmental stimuli or sensory experience. Spontaneous activity begins in the neural tube before the nervous system starts receiving sensory inputs from the external environment. Earlier it was believed that spontaneous neuronal activity was just "noise," but studies on the development of all the sensory systems, especially the visual and auditory system, indicate that it is essential to the wiring of the sensory systems, clearly implying that it is not random but instructive. It is possible that the neuronal activity during the early stages of development, beginning from its outset around the phylotypic stage, arises not in the absence of any sensory input but, contrary to what is commonly believed, is an adaptive and instructive response to the input of a continuous stream of afferent input on changes taking place in the developing embryonic structure (Cabej, 2019b).

In regard to the assumption that spontaneous activity arises in the absence of sensory input, let us remember that during the early development, "in the absence of external environmental stimuli," changes in the developing embryonic structure are extrinsic to the nervous system and may well serve as environmental stimuli, i.e., as somatosensory stimuli to which the nervous system may respond appropriately and (as later will be shown) often in goal-oriented ways or instructively as it is the case with the development, maintenance, and refining of the sensory systems (Giachello and Baines, 2017; Kaneko et al., 2017; Tien and Kerschensteiner, 2018) considered earlier. As it will be demonstrated in the next chapter, the nervous system, presumably in response to sequential changes taking place in the developing embryonic structure, sends inductive signals for the development of various tissues and organs. It sends signals for the expression of various genes and activation of GRNs (gene regulatory networks) by creating otherwise not existing causal relationships between the stimuli and the specific genes and GRNs. According to this hypothesis, spontaneous activity or spikes of electrical trains produced by neurons represent the output of the processing of the somatosensory input of stimuli (changes in the developing embryonic structure).

Spontaneous activity is necessary for the expression of the axon guidance receptor EphA4, and, as mentioned earlier, its experimental alteration leads to axon guidance errors and erratic pathfinding behavior (Hanson and Landmesser, 2004; Valdes-Aleman et al., 2019). It is involved in the fine-tuning of the axonal arbor (Yamamoto and López-Bendito, 2012). Recently it has been demonstrated that the spontaneous release of the neurotransmitter glutamate by young neurons acts as a long-range signal to NMDA (*N*-methyl-D-aspartate) receptors in distant dendrites and experimental blockade of glutamate release leads to the reduction of the growth and complexity of the dendritic arbor (Andreae and Burrone, 2015). Sensory experience also regulates the physiology of neural circuits (Kaneko et al., 2017).

The prevailing view is that molecular guidance cues are involved in the initial crude connection patterning, but certainly the spontaneous neural activity, which is not related to sensory information, plays the determining role (Kutsarova et al., 2017) in establishing the huge amount of neuronal connections in the brain. Molecular guidance cues provide information for coarse axonal targeting, whereas the neuronal activity fine-tunes these connections by pruning the wrong connections to maintain only the right ones in the development of the neural circuits (Crair, 1999).

The spontaneous neuronal activity "acts as carrier of information" (Eggermont, 2015). According to another hypothesis, neural activity extends in a broad spectrum between two extremes: from a passively permissive role to an instructive function in determining the establishment of neural circuits (Crair, 1999). Firing activity can modify the expression of receptor molecules, signaling pathways, and gene expression (Hanson and Landmesser, 2004) as well as neurotransmitter phenotype (Hanson and Landmesser, 2004; Borodinsky et al., 2004), and experimental alteration of neuronal activity or the expression of the receptor may change the axonal pathfinding (Hanson and Landmesser, 2004; Yamamoto and López-Bendito, 2012). Spontaneous activity is robust, and it will persist even when an element of the circuit is disrupted (Blankenship and Feller, 2009).

11 The neural crest

The neural crest is a vertebrate structure discovered in chick embryos about 150 years ago by Wilhelm His who described it as an intermediate furrow containing, what he named, the intermediate cord (Zwischenstrang) (His, 1868). By the middle of the last century, in neural crest transplantation experiments, it was demonstrated that the neural crest was involved in the development of various vertebrate structures, including the peripheral nervous system and facial skeleton (Bronner and LeDouarin, 2012). Later (1879), A. Marshall coined the name "neural crest" to define this neural tube-related structure.

11.1 Development of the neural crest

Both the incipient neural tissue and the neural crest originate from the same cytological precursors, and the neural crest emerges as part of the dorsal neural tube (Luo et al., 2003) (see Fig. 2.16). In the differentiation of the neural crest from the neural plate folds/dorsal neural tube are involved signals from the neural plate/neural tube, the adjacent nonneural ectoderm, and the underlying mesoderm (BMP, Wnt, Notch, and FGF) (Prasad et al., 2019) where the BMP gradient is moderate, between the high ectodermal level and very low levels of the neural plate BMP (Huang and Saint-Jeannet, 2004). Essential for the development of the neural crest is a specific GRN (gene regulatory network) (Meulemans and Bronner-Fraser, 2004; Simoes-Costa and Bronner, 2015) (Fig. 2.15).

The primordial neural crest population forms after the completion of the gastrulation at the border between the neural plate and the nonneural ectoderm while the neural plate "rolls up" to form the dorsal neural tube where the neural crest develops. Its development is induced by signals from the neural plate, nonneural ectoderm, and the underlying mesoderm. Based on the location of its development in the roof plate of the neural tube, the neural crest may be considered a derivative of the dorsal neural tube (Fig. 2.16).

According to Gerhard Schlosser's model, the neural ectoderm's competence is restricted to the formation of the neural crest, while the nonneural ectoderm's competence is restricted to only generate the preplacodal ectoderm (Schlosser, 2008).

11.2 Migration of neural crest cells

Before beginning to migrate, neural crest cells undergo the processes of delamination, separation from the neural plate/neural tube, and the EMT (epithelial-to-mesenchymal transition). Generally, there is no close correlation between the timing of delamination and the stage of neurulation. Depending on species, delamination takes place during the neural plate stage, before the closure of the neural tube (e.g., mouse and *Xenopus*), or after the formation of the neural tube (e.g., in birds). Delamination and EMT are triggered by different signal cascades.

Being specified as a separate cell type in the dorsal neural tube (in the zebrafish, it coincides with the beginning of segmentation, around 11 hpf), they undergo sequential processes of delamination, epithelial-to-mesenchymal transition (EMT), and migration. In the beginning, there are two distinct NCC subpopulations in mice: one pre-EMT population that has not yet started delaminating and the delaminating

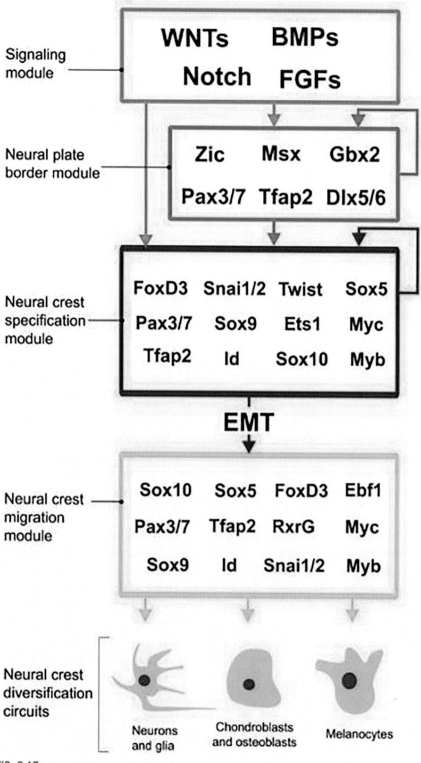

FIG. 2.15

See figure legend on next page

one. EMT proceeds as a gradual process involving changes in the morphology and the transcriptional states of NCCs (Fig. 2.17).

The cranial NCCs (neural crest cells) leave the neural tube in groups as opposed to the trunk NCCs that migrate one by one, but mostly after the closure of the neural tube (Theveneau and Mayor, 2012).

11.3 How do NCCs know where to go and what to do in their destination sites?

The migration behavioral pattern of neural crest cells indicates that they are biased (in B.K. Hall's expression) or programmed to migrate to specific sites and differentiate to cell types that are specific for tissues where they migrate (Fig. 2.18) and that the program is coded in the neural tube/primordial brain.

It was demonstrated for the first time in 1996 that early-emigrating neural crest cells from the posterior midbrain form the dental mesenchyme of the mandibular molar teeth in rat embryos (Imai et al., 1996), and earlier Lumsden (1988) had shown that premigratory NCCs from the neural folds and various regions of the neural crest induced teeth formation when brought in contact with mandibular arch epithelium, indicating that "mammalian neural crest has an odontogenic potential" (Lumsden, 1988). In quail embryos, "[a]lmost half of the neural crest population that first emerges from the explanted neural tube consists of fate-restricted precursors that give rise to a single cell type" (Henion and Weston, 1997), and in chick embryos, virtually all trunk NCCs are developmentally restricted before starting to migrate (Henion and Weston, 1997). In zebrafish, many premigratory NCCs are type restricted before reaching their final locations: "in zebrafish, premigratory trunk crest cells are not specified until after they undergo a restrictive cell division. For most cells, the restrictive division occurs before they begin to migrate" (Raible and Eisen, 1994).

Later, based on the results of interspecies (mouse-chick) neural crest grafting experiments, it was concluded that "[t]he proper program of events governing the migration of crest may need first to be established in the hindbrain, to allow

FIG. 2.15,   CONT'D

A GRN controls neural crest formation. Outline of the GRN controlling neural crest development. Different inductive signals pattern the embryonic ectoderm and induce the expression of neural plate border specifier genes, which define the neural plate border territory. These genes engage in mutual positive regulation and also drive neural crest specification by activating the neural crest specifier genes. The neural crest specification program results in the activation of the EMT machinery that allows the neural crest cells to become migratory. The migratory neural crest cells express a set of regulators that endow them with motility and the ability to initiate different differentiation programs.

From Simoes-Costa, M., Bronner, M.E., 2015. Establishing neural crest identity: a gene regulatory recipe. Development 142, 242–257.

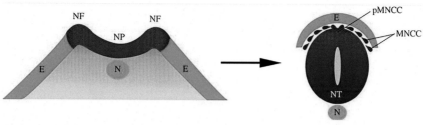

FIG. 2.16

General model of neural crest development in a vertebrate embryo. Shown on the left is a cross section of an open neural plate-stage embryo with the neural plate (NP, central nervous system primordium) in the middle flanked bilaterally by the neural plate borders, which elevate as neural folds (NF) and ventrally by the notochord (N). The epidermal ectoderm (E), presumptive skin, is shown extending ventral and lateral to the NPB (neural plate border). Shown on the right is a cross section of the neural tube (NT) with premigratory neural crest cells (pMNCC) dorsally and migratory neural crest cells (MNCC) exiting this region. Dorsal is up and ventral is down.

From York, J.R., McCauley, D.W., 2020. The origin and evolution of vertebrate neural crest cells. Open Biol. 10, 190285.

migratory crest cells to interpret and respond to environmental signals set up through a series of tissue interactions" (Trainor et al., 2002). Homotopic transplantation of quail (*Coturnix japonica*) neural crest cells to ducks (*Anas platyrhynchos*) leads to the appearance of quail beak morphology in ducks, clearly indicating that neural crest cells are programmed for producing donor morphology before being involved in the development in the recipient beak morphogenesis (Schneider and Helms, 2003).

Transplantation experiments of the neural tube between embryos of different species contributed to understanding the relative morphogenetic role of the neural crest in their destination. When neural crests from the dorsal tube of duck embryos were grafted homotopically in the Japanese quail neural tube, as well as in reverse experiments, it was observed that the shape and the size of the emerging skeletal elements resembled that of the donor rather than the receiver, clearly indicating that the donor provided the patterning information. NCCs in quails (Erickson and Goins, 1995) and chicks (Krispin et al., 2010; Le Douarin and Dupin, 2003; Thomas and Erickson, 2009) are specified to melanoblasts before leaving the neural tube or at the beginning of NCC migration.

CNCCs (cranial neural crest cells) from the hindbrain (rhombomeres 1 and 2) and midbrain (rhombomere 4) migrate to the presumed middle ear to serve as their building blocks of the middle ear ossicles, incus, and malleus from rhombomeres 1 and 2 and stapes from rhombomere 4. Empirical evidence showed that the patterning cues for shape and size of the bones "reside in the neural crest before migration" (Schneider and Helms, 2003; Tucker and Lumsden, 2004).

Despite the large body of solid evidence on the fate restriction of many NCCs before or at the time of emigration, the evidence on the multipotency of migrating NCCs is not lacking. In 2015, it was demonstrated that most murine NC cells in vivo are multipotent,

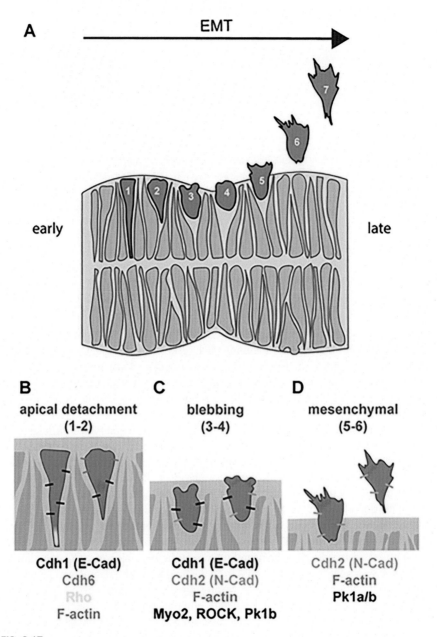

FIG. 2.17

A zebrafish-specific neural crest GRN. This simplified gene regulatory network is built exclusively from zebrafish data. Direct interactions are depicted with *solid lines*, whereas *dashed lines* show interactions inferred from loss-of-function studies. Abbreviations: EMT, epithelial-to-mesenchymal transition; GRN, gene regulatory network.

From Rocha, M., Singh, N., Ahsan K., Beiriger, A., Prince, V.E., 2019. Neural crest development: insights from the zebrafish. Dev. Dyn. 249, 1–24.

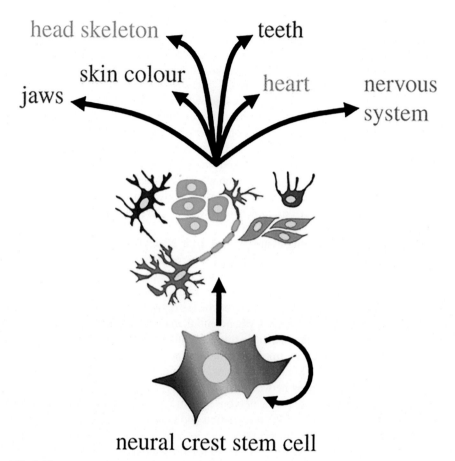

FIG. 2.18

Cartoon schematic of a neural crest cell. Cartoon schematic of a neural crest cell that has stem cell properties (capable of self-renewal, *circular arrow*) and is multipotent by generating diverse cell types that make up numerous vertebrate structures and tissues.

From York, J.R., McCauley, D.W., 2020. The origin and evolution of vertebrate neural crest cells. Open Biol. 10, 190285.

with an undetermined minority of premigratory and migratory NC cells being fate restricted (Baggiolini et al., 2015). Most recently, it is reported that in mice the premigratory stage progenitor NCCs gradually start coactivating competing fate-specific gene expression patterns, but at the bifurcation point, one of them gains the upper hand, while the other vanishes, thus determining the type of the NCC-derived cell (Soldatov et al., 2019). These reports are hardly compatible with most of the relevant evidence. What one should keep in mind, however, is that the first report admits that a small proportion of the murine NCCs investigated were fate restricted before migration and in the second report a premigratory bias or fate restriction is admitted when stated that "acquisition of a fate likely requires appropriate cell-intrinsic state and extrinsic signaling during delamination" and "[c]onsistently, we find that mesenchymal transcriptional bias emerges during delamination" (Soldatov et al., 2019).

Under similar culture conditions, cranial (mesencephalic) and trunk (sacral) neural crest cells display intrinsic differences: growth factors, FGF2 and FGF8, which promote survival, proliferation, and differentiation of cranial NCCs. The absence of such a role in trunk NCCs clearly suggests different biases (transcriptional states) of NCCs from different neural tube regions. Other factors such as BMP2/4 and TGFβ1 involved, respectively, in neurogenesis and smooth muscle formation play opposite roles (inhibitory and stimulatory) in the cranial and trunk neural crest. In vivo, cranial NCCs form osteoblasts whereas trunk NCCs do not, implying that "there is an inherent restriction on the potential of trunk neural crest, which prevents it from becoming skeletal tissue" (Abzhanov et al., 2003).

That the fate restriction biases are decided in the NCCs before they leave the neural tube to begin migration is also demonstrated in recent experiments by Simoes and Bronner. They have identified a cranial-specific regulatory circuit and observed that transfection in the trunk neural crest of the late cranial-specific factors (Sox8, Tfap2b and Ets1) elevated expression of chondrocytic genes *Runx2* and *Alx1* in the trunk NCCs and augmented their chondrocytic potential, inducing a shift from trunk to cranial crest identity. Investigators believe this shows that the cranial crest circuit "conveys regulatory information from the anterior neural plate border to the late migratory neural crest" (Simoes-Costa and Bronner, 2016), implying not only regional intrinsic differences between neural crest types but most importantly that the incipient nervous system/central nervous system "conveys information" to neural crest cells at an early stage of the neurulation and neural tube development.

It is also noteworthy that the zebrafish eye, as an extension of the forebrain, regulates the migration of NCCs from the brain (diencephalon and mesencephalon) to the eye and orbit and eyeless mutants *chokh/rx3* do not develop the dorsal orbit and most of the ethmoid plate. According to Langenberg et al. (2008), the eye is the only known guidance for proper migration of diencephalic and mesencephalic NCCs.

12 The brain-neural crest synergy

Even after migrating from the neural tube to destination sites, many NCCs cooperate with or are regulated by neural signals (Fig. 2.19). NCCs contribute to the formation of teeth (Bhatt et al., 2013) as well as the eye and surrounding structures (Williams and Bohnsack, 2015), but the formation of these structures depends on the release by the peripheral sensory nerves of Shh in their stem cell niches (Adameyko and Fried, 2016).

Although NCCs are necessary for the development of the craniofacial skeleton and musculature, expression of Shh in the FEZ (frontonasal ectodermal zone), the signaling center regulating the patterned growth of the upper jaw, is regulated by Shh (Sonic hedgehog) signals from the forebrain. Blocking Shh signaling in the brain

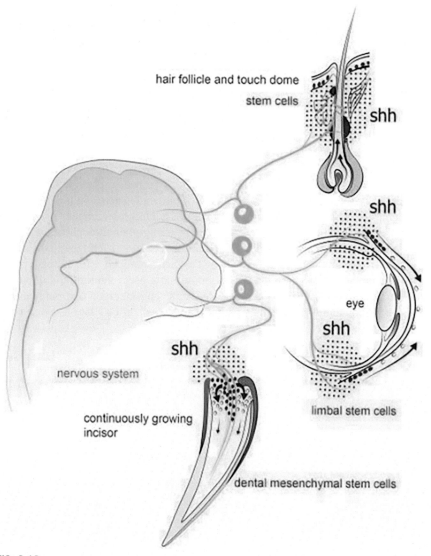

hair follicle and touch dome
stem cells

shh

shh

eye

shh

nervous system

shh

limbal stem cells

continuously growing
incisor

dental mesenchymal stem cells

FIG. 2.19

Peripheral sensory nerves inject Shh into a variety of structures in the head and tune the activity of stem cell compartments. For instance, nerve-derived Shh plays an important role in controlling stem cell niches in continuously growing incisor, hair follicle, touch dome, and limbal region of the cornea in the eye.

From Adameyko, I., Fried, K., 2016. The nervous system orchestrates and integrates craniofacial development: a review. Front. Physiol. 7, 49.

prevents expression of Shh in FEZ, leading to severe malformations in the face (Marcucio et al., 2005; Marcucio et al., 2011; Hu et al., 2015) (Fig. 2.20), which could be rescued by early application of Shh (Chong et al., 2012), while perturbing Shh expression within the forebrain disrupts patterning and growth of the facial bones without affecting the general facial skeleton patterning (Marcucio et al., 2005). Ectopic expression of Shh in the brain leads to the development of chick embryonic faces resembling embryonic mice face traits (Hu and Marcucio, 2009).

As already mentioned, in zebrafish, signals from the optic vesicle, an outgrowth of the neural tube/brain, control and regulate the migration of NCCs from the mesencephalon and diencephalon that populate the eye, orbit, and frontal nasal process (Langenberg et al., 2008).

On the way to their destination, NCCs produce glial cells (default fate of neuroepithelial cells) along the nascent nerves. Thus, embryonic nerves built of axons and associating Schwann cell progenitors establish early links between the CNS and the rest of the body (Jessen and Mirsky, 2005). In their migration throughout the animal body, many NCCs settle along nerves and are induced to differentiate into nerve-associated Schwann cells, known as SCPs (Schwann cell progenitors), by a NRG1 (neuregulin-1) signal secreted by motor and sensory neurons and deposited into the axonal membranes. This signal determines proliferation and maintenance of the SCP population around nerves and their Schwann cell fate as opposed to

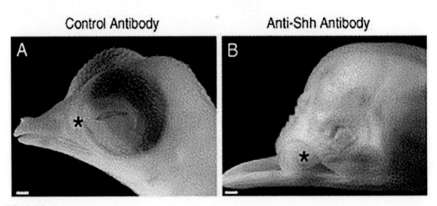

FIG. 2.20

The inhibition of Shh signaling in the neuroectoderm. The inhibition of Shh signaling in the neuroectoderm, beginning at St. 10, results in a truncation of the upper beak by St. 40. (A) Injection of hybridoma cells that express an anti-β-galactosidase antibody into the neural tube does not alter the morphology of the face by 15 days of development (approximately St. 41). (B) In contrast, the injection of hybridoma cells expressing the anti-Shh antibody leads to anophthalmia and truncations of the distal part of the upper beak. The nasal capsule appears relatively unaffected (*asterisk*).

From Marcucio, R.S., Cordero, D.R., Hu, D., Helms, J.A., 2005. Molecular interactions coordinating the development of the forebrain and face. Dev. Biol. 284, 48–61.

the melanoblast fate that it adopts in the absence of axonal contact (Komiyama and Suzuki, 1992), although NCCs may also be differentiated into melanocytes. Growing nerves throughout the body are considered to be a "stem/progenitor niche containing SCPs" (Adameyko et al., 2009) (Fig. 2.21).

Just like NCCs, SCPs are transient cell types destined to differentiate to other cell types. The latter are nerve-associated crest cells that gradually adapt themselves to life at the nerve surface in more advanced stages of embryonic development to differentiate through a transitory immature state into myelinating and nonmyelinating (Remak) Schwann cells as well as terminal Schwann cells and other subtypes of specialized glial cells in sensory skin end organs (Kaukua et al., 2014).

Another case of the controlling role of the nervous system on SCPs is observed in the process of mammalian regeneration. Mammals are able to regenerate fingertips, and the regeneration begins with regeneration of axons, dedifferentiation of the Schwann cells, and their penetration into blastema where mesenchymal (dermal, endothelial, and osteocyte) precursors secrete OSM (oncostatin M), PDGF-AA

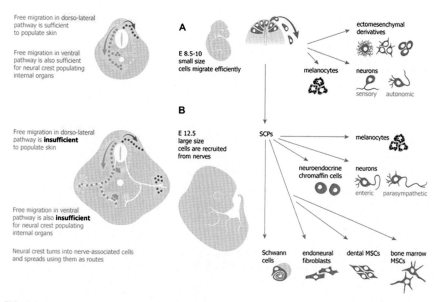

FIG. 2.21

Migratory neural crest and nerve-associated Schwann cell precursors represent a long-lasting source of multipotent progenitors available in any body location due to diversification of dissemination strategies. Outline of dissemination routes in the mouse embryo: E8.5-E10 NCCs (A) and E12.5-onward nerve-dependent SCPs (B). NCCs and SCPs form most of the peripheral nervous system elements and other nonneuronal cell types (shown on the right). Abbreviations: NCCs: neural crest cells; SCPs: Schwann cell precursors; MSC: mesenchymal stem cells.

From Furlan, A., Adameyko, I., 2018. Schwann cell precursor: a neural crest cell in disguise? Dev. Biol. 444 (Suppl. 1), S25–S35.

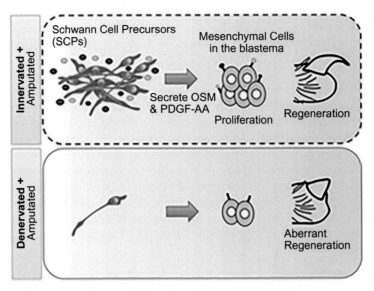

FIG. 2.22

Simplified illustration of the role of the SCPs in regeneration of the amputated mouse.

From Johnston, A.P., Yuzwa, S.A., Carr, M.J., Mahmud, N., Storer, M.A., Krause, M.P. et al., 2016.
Dedifferentiated Schwann cell precursors secreting paracrine factors are required for regeneration of the
mammalian digit tip. Cell Stem Cell 19, 433–448.

(platelet-derived growth factor AA), and other growth factors. Denervation before amputation of fingertips in mice leads to the diminished presence of SCPs and defective regeneration (Johnston et al., 2016) (Fig. 2.22).

In another example, regeneration of fractured mice mandibula is a function of mSSCs (mouse skeletal stem cells) and Schwann cells in a process where mSSCs adopt a primitive neural crest cell transcriptional program (Ransom et al., 2018). However, mandibular denervation by the inferior alveolar nerve (IAN) impairs mSSC proliferation and function, leading to impaired bone repair due to the lack of paracrine signaling with Schwan cells (Jones et al., 2019).

References

Abzhanov, A., Tzahor, E., Lassar, A. B., & Tabin, C. J. (2003). Dissimilar regulation of cell differentiation in mesencephalic (cranial) and sacral (trunk) neural crest cells in vitro. *Development, 130,* 4567–4579.

Adameyko, I., & Fried, K. (2016). The nervous system orchestrates and integrates craniofacial development: a review. *Front. Physiol., 7,* 49.

Adameyko, I., Lallemend, F., Aquino, J. B., Pereira, J. A., Topilko, P., Müller, T., et al. (2009). Schwann cell precursors from nerve innervation are a cellular origin of melanocytes in skin. *Cell, 139,* 366–379.

Adams, C. S., Korytko, A. I., & Blank, J. L. (2001). A novel mechanism of body mass regulation. *J. Exp. Biol.*, *204*, 1729–1734.

Anderson, M. J., Schimmang, T., & Lewandoski, M. (2016). An FGF3-BMP signaling axis regulates caudal neural tube closure, neural crest specification and anterior-posterior axis extension. *PLoS Genet.*, *12*(5), e1006018. 2010 May.

Andreae, L. C., & Burrone, J. (2015). Spontaneous neurotransmitter release shapes dendritic arbors via long-range activation of NMDA receptors. *Cell Rep.*, *10*, 873–882.

Aoto, J., Martinelli, D. C., Malenka, R. C., Tabuchi, K., & Südhof, T. C. (2013). Presynaptic neurexin-3 alternative splicing trans-synaptically controls postsynaptic AMPA-receptor trafficking. *Cell*, *154*, 75–88.

Ayala, R., Shu, T., & Tsai, L.-H. (2007). Trekking across the brain: the journey of neuronal migration. *Cell*, *128*, 29–43.

Baeckberg, M., Collin, M., Ovesjoe, M. L., & Meister, B. (2003). Chemical coding of GABAB receptor-immunoreactive neurones in hypothalamic regions regulating body weight. *J. Neuroendocrinol.*, *15*, 1–14.

Baggiolini, A., Varum, S., Mateos, J. M., Bettosini, D., John, N., Bonalli, M., et al. (2015). Premigratory and migratory neural crest cells are multipotent in vivo. *Cell Stem Cell*, *16*(3), 314–322. https://doi.org/10.1016/j.stem.2015.02.017.

Ballard, W. W. (1981). Morphogenetic movements and fate maps of vertebrates. *Am. Zool.*, *21*, 391–399.

Bellairs, R., & Osmond, M. (2014). Establishment of the embryonic body. In *Atlas of Chick Development* (3rd, pp. 29–43). New York: Academic Press.

Bernstein, B. W., & Bamburg, J. R. (2010). ADF/Cofilin: a functional node in cell biology. *Trends Cell Biol.*, *20*, 187–195.

Bhatt, S., Diaz, R., & Trainor, P. A. (2013). Signals and switches in mammalian neural crest cell differentiation. *Cold Spring Harb. Perspect. Biol.* https://doi.org/10.1101/cshperspect.a008326.

Blankenship, A. G., & Feller, M. B. (2009). Mechanisms underlying spontaneous patterned activity in developing neural circuits. *Nat. Rev. Neurosci.*, *11*, 18–29.

Bleckert, A., & Wong, R. O. (2011). Identifying roles for neurotransmission in circuit assembly: insights gained from multiple model systems and experimental approaches. *BioEssays*, *33*, 61–72.

Borodinsky, L. N., Root, C. M., Cronin, J. A., Sann, S. B., Gu, X., & Spitzer, N. C. (2004). Activity-dependent homeostatic specification of transmitter expression in embryonic neurons. *Nature*, *429*, 523–530.

Boulant, J. A. (2000). Role of the preoptic-anterior hypothalamus in thermoregulation and fever. *Clin. Infect. Dis.*, *31*(Suppl 5), S157–S161.

Bronner, M. E., & LeDouarin, N. M. (2012). Development and evolution of the neural crest: an overview. *Dev. Biol.*, *366*, 2–9.

Cabej, N. R. (2019b). *Epigenetic Principles of Evolution* (p. 7). London/San Diego, CA: Academic Press.

Chang, C., & Harland, R. M. (2007). Neural induction requires continued suppression of both Smad1 and Smad2 signals during gastrulation. *Development*, *134*, 3861–3872.

Chong, H. J., Young, N. M., Hu, D., Jeong, J., McMahon, A. P., Hallgrimsson, B., et al. (2012). Signaling by SHH rescues facial defects following blockade in the brain. *Dev. Dyn.*, *241*, 247–256.

Crair, M. C. (1999). Neuronal activity during development: permissive or instructive? *Curr. Opin. Neurobiol.*, *9*, 88–93.

Crawshaw, L., Grahn, D., Wollmuth, L., & Simpson, L. (1985). Central nervous regulation of body temperature in vertebrates: comparative aspects. *Pharmacol. Ther.*, *30*, 19–30.

Dagar, S., & Gottmann, K. (2019). Differential properties of the synaptogenic activities of the neurexin ligands neuroligin1 and LRRTM2. *Front. Mol. Neurosci.*, *12*, 269.

Daly, C., & Ziff, E. B. (1997). Post-transcriptional regulation of synaptic vesicle protein expression and the developmental control of synaptic vesicle formation. *J. Neurosci.*, *17*, 2365–2375.

de Rouvroit, C. L., & Goffinet, A. M. (2001). Neuronal migration. *Mech. Dev.*, *105*, 47–56.

Ding, X., Liu, S., Tian, M., Zhang, W., Zhu, T., Li, D., et al. (2017). Activity-induced histone modifications govern Neurexin-1 mRNA splicing and memory preservation. *Nat. Neurosci.*, *20*, 690–699.

Drescher, U., Bonhoeffer, F., & Müller, B. K. (1997). The Eph family in retinal axon guidance. *Curr. Opin. Neurobiol.*, *7*, 75–80.

Eggermont, J. J. (2015). Animal models of spontaneous activity in the healthy and impaired auditory system. *Front. Neural Circuits*, *9*, 19. https:/doi.org/10.3389/fncir.2015.00019.|.

Erickson, C. A., & Goins, T. L. (1995). Avian neural crest cells can migrate in the dorsolateral path only after they are specified as melanocytes. *Development*, *121*, 915–924.

Ericson, J., Morton, S., Kawakami, A., Roelink, H., & Jessell, T. M. (1996). Two critical periods of Sonic Hedgehog signaling required for the specification of motor neuron identity. *Cell*, *87*, 661–673.

Feller, M. B. (1999). Spontaneous correlated activity in developing neural circuits. *Neuron*, *22*, 653–656.

Feller, M. B. (2009). Retinal waves are likely to instruct the formation of eye-specific retinogeniculate projections. *Neural Dev.*, *4*, 24.

Flynn, K. C. (2013). The cytoskeleton and neurite initiation. *BioArchitecture*, *3*, 86–109.

Flynn, K. C., Hellal, F., Neukirchen, D., Jacob, S., Tahirovic, S., Dupraz, S., et al. (2012). ADF/Cofilin-mediated actin retrograde flow directs neurite formation in the developing brain. *Neuron*, *76*, 1091–1107.

Galis, F., & Metz, J. A. J. (2001). Testing the vulnerability of the phylotypic stage: on modularity and evolutionary conservation. *J. Exp. Zool.*, *291*, 195–204.

Giachello, C. N. G., & Baines, R. A. (2015). Inappropriate neural activity during a sensitive period in embryogenesis results in persistent seizure-like behavior. *Curr. Biol.*, *25*, 1–5.

Giachello, C. N., & Baines, R. A. (2017). Regulation of motoneuron excitability and the setting of homeostatic limits. *Curr. Opin. Neurobiol.*, *43*, 1–6.

Gorbman, A., & Davey, K. (1991). Endocrines. In C. L. Prosser (Ed.), *Neural and Integrative Animal Physiology* (4th, pp. 693–754). New York: Wiley-Liss.

Götz, M., & Huttner, W. B. (2005). The cell biology of neurogenesis. *Nat. Rev. Mol. Cell Biol.*, *6*, 777–788.

Halaas, J. L., Gajiwala, K. S., Maffei, M., Cohen, S. L., Chait, B. T., Rabinowitz, D., et al. (1995). Weight-reducing effects of the plasma protein encoded by the obese gene. *Science*, *269*, 543–546.

Hallböök, F. (1999). Evolution of the vertebrate neurotrophin and Trk receptor gene families. *Curr. Opin. Neurobiol.*, *9*, 616–621.

Hammel, H. T., Jackson, D. C., Stolwijk, J. A. J., Hardy, J. D., & Stroeme, S. B. (1963). Temperature regulation by hypothalamic proportional control with an adjustable set point. *J. Appl. Physiol.*, *18*, 1146–1154. https:/doi.org/10.1152/jappl.1963.18.6.1146.

Hanson, M. G., & Landmesser, L. T. (2004). Normal patterns of spontaneous activity are required for correct motor axon guidance and the expression of specific guidance molecules. *Neuron*, *43*, 687–701.

Hatanaka, Y., Zhu, Y., Torigoe, M., Kita, Y., & Murakami, F. (2016). From migration to settlement: the pathways, migration modes and dynamics of neurons in the developing brain. *Proc. Jpn. Acad. Ser. B Phys. Biol. Sci.*, *92*, 1–19.

Hemmati-Brivanlou, A., Kelly, O. G., & Melton, D. A. (1994). Follistatin, an antagonist of activin, is expressed in the Spemann organizer and displays direct neuralizing activity. *Cell*, *77*, 283–295.

Henion, P. D., & Weston, J. A. (1997). Timing and pattern of cell fate restrictions in the neural crest lineage. *Development*, *124*, 4351–4359.

Hiesinger, P. R., Zhai, R. G., Zhou, Y., Koh, T. W., Mehta, S. Q., Schulze, K. L., et al. (2006). Activity-independent prespecification of synaptic partners in the visual map of Drosophila. *Curr. Biol.*, *16*, 1835–1843.

His, W. (1868). *Untersuchungen über die erste Anlage des Wirbelthierleibes: die erste Entwickelung des Hühnchens im Ei. vol. 1* (p. 78). Leipzig: F.C.W. Vogel.

Homem, C. C. F., & Knoblich, J. A. (2012). *Drosophila* neuroblasts: a model for stem cell biology. *Development*, *139*, 4297–4310.

Homem, C. C. F., Repic, M., & Knoblich, J. A. (2015). Proliferation control in neural stem and progenitor cells. *Nat. Rev. Neurosci.*, *16*, 647–659.

Hu, D., & Marcucio, R. S. (2009). A SHH-responsive signaling center in the forebrain regulates craniofacial morphogenesis via the facial ectoderm. *Development*, *136*, 107–116.

Hu, D., Young, N. M., Xu, Q., Jamniczky, H., Green, R. M., Mio, W., et al. (2015). Signals from the brain induce variation in avian facial shape. *Dev. Dyn.*, *244*, 1133–1143.

Huang, X., & Saint-Jeannet, J. P. (2004). Induction of the neural crest and the opportunities of life on the edge. *Dev. Biol.*, *275*, 1–11.

Huberman, A. D., Feller, M. B., & Chapman, B. (2008). Mechanisms underlying development of visual maps and receptive fields. *Annu. Rev. Neurosci.*, *31*, 479–509.

Idelson, M. S., Ben-Jacob, E., & Hanein, Y. (2010). Innate synchronous oscillations in freely-organized small neuronal circuits. *PLoS One*, *5*(12), e14443.

Imai, H., Osumi-Yamashita, N., Ninomiya, Y., & Eto, K. (1996). Contribution of early emigrating midbrain crest cells to the dental mesenchyme of mandibular molar teeth in rat embryos. *Dev. Biol.*, *176*, 151–165.

Itoh, M., Kim, C.-H., Palardy, G., Oda, T., Jiang, Y.-J., Maust, D., et al. (2003). Mind bomb is a ubiquitin ligase that is essential for efficient activation of Notch signaling by Delta. *Dev. Cell*, *4*, 67–82.

Jessen, K. R., & Mirsky, R. (2005). The origin and development of glial cells in peripheral nerves. *Nat. Rev. Neurosci.*, *6*, 671–682.

Johnston, A. P., Yuzwa, S. A., Carr, M. J., Mahmud, N., Storer, M. A., Krause, M. P., et al. (2016). Dedifferentiated Schwann cell precursors secreting paracrine factors are required for regeneration of the mammalian digit tip. *Cell Stem Cell*, *19*, 433–448.

Jones, R. E., Salhotra, A., Robertson, K. S., Ransom, R. C., Foster, D. S., Shahour, H. N., et al. (2019). Skeletal stem cell-Schwann cell circuitry in mandibular repair. *Cell Rep.*, *28*, 2757–2766.e5.

Kalil, K., & Dent, E. W. (2014). Branch management: mechanisms of axon branching in the developing vertebrate CNS. *Nat. Rev. Neurosci.*, *15*, 7–18.

Kaneko, T., Macara, A. M., Li, R., Hu, Y., Iwasaki, K., Dunnings, Z., et al. (2017). Serotonergic modulation enables pathway-specific plasticity in a developing sensory circuit in Drosophila. *Neuron*, *95*, 623–638.e4.

Karzbrun, E., & Reiner, O. (2019). Brain organoids—a bottom-up approach for studying human neurodevelopment. *Bioengineering*, *2019*(6), 9.

Kaukua, N., Shahidi, M., Konstantinidou, C., Dyachuk, V., Kaucka, M., Furlan, A., et al. (2014). Glial origin of mesenchymal stem cells in a tooth model system. *Nature, 513*, 551–554.

Kimmel, C. B., Ballard, W. W., Kimmel, S. R., Ullmann, B., & Schilling, T. F. (1995). Stages of embryonic development of the zebrafish. *Dev. Dyn., 203*, 255–310.

Kirkby, L. A., Sack, G. S., Firl, A., & Feller, M. B. (2013). A role for correlated spontaneous activity in the assembly of neural circuits. *Neuron, 80*, 1129–1144.

Komiyama, A., & Suzuki, K. (1992). Age-related differences in proliferative responses of Schwann cells during Wallerian degeneration. *Brain Res., 573*, 267–275.

Kosse, C., Gonzalez, A., & Burdakov, D. (2015). Predictive models of glucose control: roles for glucose-sensing neurones. *Acta Physiol., 213*, 7–18.

Krispin, S., Nitzan, E., Kassem, Y., & Kalcheim, C. (2010). Evidence for a dynamic spatio-temporal fate map and early fate restrictions of premigratory avian neural crest. *Development, 137*, 585–595.

Kutsarova, E., Munz, M., & Ruthazer, E. S. (2017). Rules for shaping neural connections in the developing brain. *Front. Neural Circuits, 10*, 111.

Langenberg, T., Kahana, A., Wszalek, J. A., & Halloran, M. C. (2008). The eye organizes neural crest cell migration. *Dev. Dyn., 237*, 1645–1652.

Le Douarin, N. M., & Dupin, E. (2003). Multipotentiality of the neural crest. *Curr. Opin. Genet. Dev., 13*(5), 529–536. https:/doi.org/10.1016/j.gde.2003.08.002.

Lee, R., Clandinin, T., Lee, C., Chen, P.-L., Meinertzhagen, I. A., & Zipursky, S. L. (2003). The protocadherin flamingo is required for axon target selection in the Drosophila visual system. *Nat. Neurosci., 6*, 557–563.

Li, R., Liang, Y., Zheng, S., He, Q., & Yang, L. (2019). The atypical cadherin flamingo determines the competence of neurons for activity-dependent fine-scale topography. *Mol. Brain, 12*(1). https:/doi.org/10.1186/s13041-019-0531-7.

Lumsden, A. G. (1988). Spatial organization of the epithelium and the role of neural crest cells in the initiation of the mammalian tooth germ. *Development, 103*, 155–169.

Luo, R., Gao, J., Wehrle-Haller, B., & Henion, P. D. (2003). Molecular identification of distinct neurogenic and melanogenic neural crest sublineages. *Development, 130*, 321–330.

Malmersjö, S., Rebellato, P., Smedler, E., Planert, H., Kanatani, S., & Liste, I. (2013). Neural progenitors organize in small-world networks to promote cell proliferation. *Proc. Natl. Acad. Sci. U. S. A., 110*, 1524–1532.

Marcucio, R. S., Cordero, D. R., Hu, D., & Helms, J. A. (2005). Molecular interactions coordinating the development of the forebrain and face. *Dev. Biol., 284*, 48–61.

Marcucio, R. S., Young, N. M., Hu, D., & Hallgrimsson, B. (2011). Mechanisms that underlie co-variation of the brain and face. *Genesis, 49*(4), 177–189.

Martí, E., Bumcrot, D. A., Takada, R., & McMahon, A. P. (1995). Requirement of 19K form of Sonic Hedgehog for induction of distinct ventral cell types in CNS explants. *Nature, 375*, 322–325.

Meulemans, D., & Bronner-Fraser, M. (2004). Gene-regulatory interactions in neural crest evolution and development. *Dev. Cell, 7*, 291–299.

Mizuno, H., Hirano, T., & Tagawa, Y. (2007). Evidence for activity-dependent cortical wiring: formation of interhemispheric connections in neonatal mouse visual cortex requires projection neuron activity. *J. Neurosci., 27*, 6760–6770.

Morishita, H., & Hensch, T. K. (2008). Critical period revisited: impact on vision. *Curr. Opin. Neurobiol., 18*, 101–107.

Munyiri, F. N., & Ishikawa, Y. (2005). Endocrine changes associated with metamorphosis and diapause induction in the yellow-spotted longicorn beetle, *Psacothea hilaris. J. Insect Physiol., 50*, 1075–1081.

Munyiri, F. N., Shintani, Y., & Ishikawa, Y. (2004). Evidence for the presence of a threshold weight for entering diapause in the yellow-spotted longicorn beetle, *Psacothea hilaris*. *J. Insect Physiol.*, *50*, 295–301.

Nagamine, K., Ishikawa, Y., & Hoshizaki, S. (2016). Insights into how longicorn beetle larvae determine the timing of metamorphosis: starvation-induced mechanism revisited. *PLoS One*, *11*(8), e0162213. 2016 august 25;.

Nakashima, A., Ihara, N., Shigeta, M., Kiyonari, H., Ikegaya, Y., & Takeuchi, H. (2019). Structured spike series specify gene expression patterns for olfactory circuit formation. *Science*, *365*(6448), eaaw5030.

Nijhout, H. F., & Williams, C. M. (1974). Control of moulting and metamorphosis in the tobacco hornworm Manduca sexta (L.): Cessation of juvenile hormone secretion as trigger for pupation. *J. Exp. Biol.*, *61*, 493–501.

Oh, Y., Lai, J. S., Mills, H. J., Erdjument-Bromage, H., Giammarinaro, B., Saadipour, K., et al. (2019). A glucose-sensing neuron pair regulates insulin and glucagon in Drosophila. *Nature*, *574*, 559–564.

Okujeni, S., & Egert, U. (2019). Self-organization of modular network architecture by activity-dependent neuronal migration and outgrowth. *eLife*, *8*, e47996.

Pacak, K., & Palkovits, M. (2001). Stressor specificity of central neuroendocrine responses: implications for stress-related disorders. *Endocr. Rev.*, *22*, 502–548.

Pelleymounter, M. A., Cullen, M. J., Baker, M. B., Hecht, R., Winters, D., Boone, T., et al. (1995). Effects of the obese gene product on body weight regulation in ob/ob mice. *Science*, *269*, 540543.

Placzek, M. (1995). The role of the notochord and floor plate in inductive interactions. *Curr. Opin. Genet. Dev.*, *5*, 499–506.

Prasad, M. S., Charney, R. M., & García-Castro, M. I. (2019). Specification and formation of the neural crest: perspectives on lineage segregation. *Genesis*, *57*(1), e23276. 2019 Jan.

Raible, D., & Eisen, J. S. (1994). Restriction of neural crest cell fate in the trunk of the embryonic zebrafish. *Development*, *120*, 495–503.

Ransom, R. C., Carter, A. C., Salhotra, A., Leavitt, T., Marecic, O., Murphy, M. P., et al. (2018). Mechanoresponsive stem cells acquire neural crest fate in jaw regeneration. *Nature*, *563*, 514–521.

Rao, Y., & Wu, J. Y. (2001). Neuronal migration and the evolution of the human brain. *Nat. Neurosci.*, *4*, 860–862.

Reiner, O. (2013). LIS1 and DCX: implications for brain development and human disease in relation to microtubules. *Scientifica*, *2013*, 393975.

Rosenberg, S. S., & Spitzer, N. C. (2011). Calcium signaling in neuronal development. *Cold Spring Harb. Perspect. Biol.*, *3*(10). https:/doi.org/10.1101/cshperspect.a004259, a004259.

Rui, L. (2013). Brain regulation of energy balance and body weights. *Rev. Endocr. Metab. Disord.*, *14*(4). https:/doi.org/10.1007/s11154-013-9261-9. 2013 Dec.

Sander, K., Goodwin, B. C., Holder, N., & Wylie, C. C. (1983). The evolution of patterning mechanisms: gleanings from insect embryogenesis and spermatogenesis. In *Development and Evolution* (pp. 137–159). Cambridge: Cambridge University Press.

Sasai, Y., Lu, B., Steinbeisser, H., Geissert, D., Gont, L. K., & De Robertis, E. M. (1994). Xenopus chordin: A novel dorsalizing factor activated by organizer-specific homeobox genes. *Cell*, *79*, 779–790.

Schlosser, G. (2008). Do vertebrate neural crest and cranial placodes have a common evolutionary origin? *BioEssays*, *30*, 659–672.

Schneider, R. A., & Helms, J. A. (2003). The cellular and molecular origins of beak morphology. *Science*, *299*, 565–568.

Shaw, G. L., Harth, E., & Scheibel, A. B. (1982). Cooperativity in brain function: assemblies of approximately 30 neurons. *Exp. Neurol.*, *77*, 324–358.

Sheng, C., Javed, U., Gibbs, M., Long, C., Yin, J., Qin, B., et al. (2019). Experience-dependent structural plasticity targets dynamic filopodia in regulating dendrite maturation and synaptogenesis. *Nat. Commun.*, *9*, 3362.

Shimokita, E., & Takahashi, Y. (2011). Secondary neurulation: fate-mapping and gene manipulation of the neural tube in tail bud. *Dev. Growth Differ.*, *53*, 401–410.

Simoes-Costa, M., & Bronner, M. E. (2015). Establishing neural crest identity: a gene regulatory recipe. *Development*, *142*, 242–257.

Simoes-Costa, M., & Bronner, M. E. (2016). Reprogramming of avian neural crest axial identity and cell fate. *Science*, *352*, 1570–1573.

Slack, J. M., Holland, P. W., & Graham, C. F. (1993). The zootype and the phylotypic stage. *Nature*, *361*, 490–492.

Smith, W. C., & Harland, R. M. (1992). Expression cloning of noggin, a new dorsalizing factor localized to the Spemann organizer in Xenopus embryos. *Cell*, *70*, 829–840.

Soldatov, R., Kaucka, M., Kastriti, M. E., Petersen, J., Chontorotzea, T., Englmaier, L., et al. (2019). Spatiotemporal structure of cell fate decisions in murine neural crest. *Science*, *364*, 971.

Song, M., Martinowich, K., & Lee, F. S. (2017). BDNF at the synapse: why location matters. *Mol. Psychiatry*, *22*, 1370–1375.

Sperry, R. W. (1963). Chemoaffinity in the orderly growth of nerve fiber patterns and connections. *Proc. Natl. Acad. Sci. U. S. A.*, *50*, 703–710.

Spitzer, N. C. (2006). Electrical activity in early neuronal development. *Nature*, *444*, 707–712.

Spitzer, N. (2012). Activity-dependent neurotransmitter respecification. *Nat. Rev. Neurosci.*, *13*, 94–106.

Südhof, T. C. (2017). Synaptic neurexin complexes: a molecular code for the logic of neural circuits. *Cell*, *171*, 745–769.

Sun, Y., Huang, Z., Yang, K., Liu, W., Xie, Y., Yua, B., et al. (2011). Self-organizing circuit assembly through spatiotemporally coordinated neuronal migration within geometric constraints. *PLoS One*, *6*(11), e28156. https://doi.org/10.1371/journal.pone.0028156.

Suzuki, T., & Sato, M. (2014). Neurogenesis and neuronal circuit formation in the Drosophila visual center. *Dev. Growth Differ.*, *56*, 491–498.

Takeda, H., Nishimura, K., & Agata, K. (2009). Planarians maintain a constant ratio of different cell types during changes in body size by using the stem cell system. *Zool. Sci.*, *26*, 805e813.

Tarussio, D., Metref, S., Seyer, P., Mounien, L., Vallois, D., Magnan, C., et al. (2014). Nervous glucose sensing regulates postnatal β cell proliferation and glucose homeostasis. *J. Clin. Invest.*, *124*, 413–424.

Theveneau, E., & Mayor, R. (2012). Neural crest delamination and migration: from epithelium-to-mesenchyme transition to collective cell migration. *Dev. Biol.*, *366*, 34–54.

Thomas, A. J., & Erickson, C. A. (2009). FOXD3 regulates the lineage switch between neural crest-derived glial cells and pigment cells by repressing MITF through a non-canonical mechanism. *Development*, *136*, 1849–1858.

Tien, N.-W., & Kerschensteiner, D. (2018). Homeostatic plasticity in neural development neural development. *Neural Dev.*, *13*, 9.

Tierney, A. J. (1996). Evolutionary implications of neural circuit structure and function. *Behav. Process.*, *35*, 173–182.

Torborg, C. L., & Feller, M. B. (2005). Spontaneous patterned retinal activity and the refinement of retinal projections. *Prog. Neurobiol.*, *76*, 213–235.

Torres-Paz, J., Leclercq, J., & Rétaux, S. (2019). Maternally-regulated gastrulation as a source of variation contributing to cavefish forebrain evolution. *eLife*. https:/doi.org/10.7554/eLife.50160.

Trainor, P. A., Sobieszczuk, D., Wilkinson, D., & Krumlauf, R. (2002). Signalling between the hindbrain and paraxial tissues dictates neural crest migration pathways. *Development*, *129*, 433–442.

Truszkowski, T. L., & Aizenman, C. D. (2015). Neurobiology: setting the set point for neural homeostasis. *Curr. Biol.*, *25*, R1132–R1133.

Tucker, A. S., & Lumsden, A. (2004). Neural crest cells provide species-specific patterning information in the developing branchial skeleton. *Evol. Dev.*, *6*, 32–40.

Turrigiano, G. (2011). Homeostatic synaptic plasticity: local and global mechanisms for stabilizing neuronal function. *Cold Spring Harb. Perspect. Biol.* https:/doi.org/10.1101/cshperspect.a005736.

Tzschenke, B., & Basta, D. (2002). Early development of neuronal hypothalamic thermosensitivity in birds: influence of epigenetic temperature adaptation. *Comp. Biochem. Physiol. A Mol. Integr. Physiol.*, *131*, 825832.

Tzschenke, B., & Nichelmann, M. (1997). Influence of prenatal and postnatal acclimation on nervous and peripheral thermoregulation. *Ann. N. Y. Acad. Sci.*, *813*, 87–94 (abstract).

Valdes-Aleman, J., Fetter, R. D., Sales, E. C., Doe, C. Q., Landgraf, M., & Cardona, A. (2019). Synaptic specificity is collectively determined by partner identity, location and activity. *Neuron*, *109*(1). 105-122.e7.

Varoqueaux, F., Aramuni, G., Rawson, R. L., Mohrmann, R., Missler, M., Gottmann, K., et al. (2006). Neuroligins determine synapse maturation and function. *Neuron*, *51*, 741–754.

Vonhoff, F., & Keshishian, H. (2017). Activity-dependent synaptic refinement: new insights from Drosophila. *Front. Syst. Neurosci.*, *11*, 23.

Wagner, E., & Levine, M. (2012). FGF signaling establishes the anterior border of the Ciona neural tube. *Development*, *139*, 2351–2359.

Wan, Y., Wei, Z., Looger, L. L., Koyama, M., Druckmann, S., & Keller, P. J. (2019). Single-cell reconstruction of emerging population activity in an entire developing circuit. *Cell*, *179*, 355–372.e23.

Weng, R., & Cohen, S. M. (2015). Control of *Drosophila* Type I and Type II central brain neuroblast proliferation by bantam microRNA. *Development*, *142*(21), 3713–3720. https:/doi.org/10.1242/dev.127209.

West-Eberhard, M. J. (2003). *Developmental Plasticity and Evolution* (p. 462). Oxford and New York: Oxford University Press.

Wiedmer, P., Boschmann, M., & Klaus, S. (2004). Gender dimorphism of body mass perception and regulation in mice. *J. Exp. Biol.*, *207*, 2859–2866.

Wilkinson, M., Kane, T., Wang, R., & Takahashi, E. (2017). Migration pathways of thalamic neurons and development of thalamocortical connections in humans revealed by diffusion MR tractography. *Cereb. Cortex*, *27*, 5683–5695.

Williams, A. L., & Bohnsack, B. L. (2015). Neural crest derivatives in ocular development: discerning the eye of the storm. *Birth Defects Res. C Embryo Today*, *105*, 87–95.

Wilson, P. A., & Hemmati-Brivanlou, A. (1997). Vertebrate neural induction: inducers, inhibitors, and a new synthesis. *Neuron*, *18*, 699–710.

Wodarz, A., & Huttner, W. B. (2003). Asymmetric cell division during neurogenesis in Drosophila and vertebrates. *Mech. Dev.*, *120*, 1297–1309.

Wolfram, V., & Baines, R. A. (2013). Blurring the boundaries: developmental and activity-dependent determinants of neural circuits. *Trends Neurosci.*, *36*, 610–619.

Wolpert, L. (1991). *The Triumph of the Embryo* (pp. 183–187). Oxford: Oxford University Press.

Xu, H.-P., Furman, M., Mineur, Y. S., Chen, H., King, S. L., & Zenisek, D. (2011). An instructive role for patterned spontaneous retinal activity in mouse visual map development. *Neuron*, *70*, 1115–1127.

Xu, Y., O'Malley, B. W., & Elmquist, J. K. (2017). Brain nuclear receptors and body weight regulation. *J. Clin. Invest.*, *127*, 1172–1180.

Yamamoto, N., & López-Bendito, G. (2012). Shaping brain connections through spontaneous neural activity. *Eur. J. Neurosci.*, *35*, 1595–1604.

Yang, L., Li, R., Kaneko, T., Takle, K., Morikawa, R. K., Essex, L., et al. (2014). Trim9 regulates activity-dependent fine-scale topography in Drosophila. *Curr. Biol.*, *24*, 1024–1030.

Yuan, Q., Xiang, Y., Yan, Z., Han, C., Jan, L. Y., & Nung, Y. (2011). Light-induced structural and functional plasticity in Drosophila larval visual system. *Science*, *333*, 1458–1462.

Zhao, Z. D., Yang, W. Z., Gao, C., Fu, X., Zhangm, W., Zhou, Q., et al. (2017). A hypothalamic circuit that controls body temperature. *Proc. Natl. Acad. Sci. U. S. A.*, *114*, 2042–2047.

Zitnan, D., Sehnal, F., & Bryant, P. J. (1993). Neurons producing specific neuropeptides in the central nervous system of normal and pupariation-delayed Drosophila. *Dev. Biol.*, *156*, 117–135.

The inductive brain in animal development

The inductive brain in animal development

Even though the role of the nervous system in animal development has never been a special object of biological research, surprisingly extensive relevant evidence emanated from studies designed for other purposes. This evidence reveals the unrivaled and unique role of the nervous system in animal development. To the best of my knowledge, the inductive role of the nervous system in organogenesis and histogenesis was first emphasized by the Canadian developmental biologist (Hall, 1998a, b), when he pointed out that the incipient CNS immediately engenders a network of inductions that give rise to different cells, tissues, and organs of embryos and adults (Hall, 1998a, b).

1 Somitogenesis

Formation of somites takes place only after, and in relation to, the formation of the CNS (Hall, 1998a, b). No somites develop in the absence of axial organs, the neural tube-notochord. Differentiation of somites into separate compartments is induced by Shh and Noggin signals from the floorplate of the neural tube and by Wnts and Bmps from the dorsal neural tube and ectoderm (Cossu and Borello, 1999; Chang and Kioussi, 2018) (Fig. 3.1). Earlier, Munsterberg et al. (1995) demonstrated that the inducer effect of the dorsal neural tube and the floorplate-notochord could be mimicked, respectively, by administration of Shh and Wnts (Munsterberg and Lassar, 1995; Munsterberg et al., 1995).

Unilateral excision of the neural tube-notochord complex in chicken embryos, surprisingly, does not prevent the development of the limb and body wall muscles, but they begin to degenerate soon after E10 due to the lack of innervation. When the axial organs are extirpated at the 8–16-somite stage, no muscle cells differentiate in the operated side, while in the intact side, the development proceeds normally. When the separation of somites from the axial organs was made later (at the 17–19-somite stage), muscle cells were differentiated in the six first rostral somites and the specification of the respective dermatome, myotome, and sclerotome took place normally

The Inductive Brain in Development and Evolution. https://doi.org/10.1016/B978-0-323-85154-1.00004-7

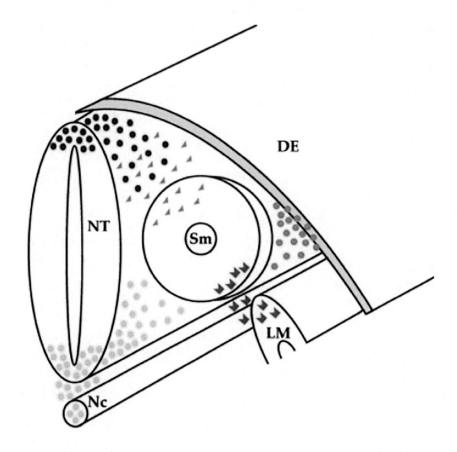

DE

NT

(Sm)

LM

Nc

- Wnt1
- Shh
- Wnt7a
- Bmp4
- Noggin

FIG. 3.1

A simplified scheme of signaling molecules in newly formed epithelial somite. Shh (*gray dots*), produced by notochord (Nc) and floor plate, acts on the ventral domain of newly formed epithelial somites, inducing sclerotome, and also on the dorsomedial domain, inducing medial dermomyotome. Wnt1 (*black dots*), produced by dorsal neural tube (NT), acts (with Shh) on the dorsomedial domain of newly formed somites (Sm), where Myf5 expression is observed soon after epaxial progenitors are specified. Wnt7a (*semilunar dots*), produced by dorsal ectoderm (DE), acts on the dorsolateral domain, where hypaxial progenitors are specified. BMP4 (*black polygons*), produced by lateral mesoderm (LM), prevents MyoD activation and early differentiation in the lateral domain of somites. Its action is counteracted by direct binding of Noggin (*gray triangles*) produced by dorsal neural tube.

From Cossu, G., Borello, U., 1999. Wnt signaling and the activation of myo- genesis in mammals. EMBO J. 18, 6867–6872.

(Rong et al., 1992). After 2 days in culture, somites excised from the 8- to 11-somite stage did not differentiate any muscle cells, whereas the somites separated from stage 14–15 embryos, muscle cells differentiated in somite pairs 3–10 (Rong et al., 1992).

2 Myogenesis

2.1 Myogenesis in *Drosophila*

Drosophila muscles originate from mesodermal derivatives, which include three myoblast types: muscle founder cells (FCs) and fusion-competent myoblasts (FCMs), which migrate to form larval body wall muscle, and adult muscle precursors (AMPs), which are set aside until metamorphosis (Dobi et al., 2015).

More than three decades ago, it was found that the patterning of a male segmental muscle (muscle of Lawrence) in *Drosophila* was determined by the information provided by motor or neurosecretory innervation rather than myoblasts themselves: "sex and segmental identity of the motor or neurosecretory neurons determine the development of muscle pattern" (Lawrence and Johnston, 1986). Severing of the local innervation prevented the formation of this male-specific muscle (Currie and Bate, 1995).

Local innervation is necessary for the proliferation and distribution of myoblasts as well as myogenesis (Currie and Bate, 1991) and development of the dorsoventral muscles (DVMs) in *Drosophila*. Denervation results in muscle patterning defects in *Drosophila*, with a failure to form dorsoventral muscles, indicating that denervation disturbed the differentiation of specific muscle fibers (Fernandes and Keshishian, 1998).

Drosophila has about 300 muscles. The basic myogenic unit of the fly is the segment (Hughes and Salinas, 1999), with each segment comprising 24–30 muscles and the second thoracic segment comprising two sets of indirect flight muscles (IFMs), which are the dorsal longitudinal muscles (DLMs) and dorsoventral muscles (DVMs). An already classic example of the instructive role of the nervous system in the development of particular muscles in insects is the case of the development during metamorphosis of the DEO1 (dorsal external oblique1) muscle fibers from the larval muscles in the moth *Manduca sexta* (Hegstrom et al., 1998). As larva, *M. sexta* has five muscle fibers (L1–L5) innervated by the motoneuron MN-12, originating in the ventral nerve cord of the CNS (Schulman et al., 2015). When the larva sheds the exoskeleton, the motoneuron prunes back of all but one of the larval muscle fibers, L1, which is the only one to survive and regrow into the adult DEO1 muscle. The rest of the larval muscles degenerate and are eliminated via apoptosis (Fig. 3.2). Investigators noticed that the pruning of the motoneuron from the larval muscle fibers was associated with the expression of the EcR-A (ecdysone receptor-A) isoform. Surprisingly, only the surviving larval muscle expressed the alternative isoform EcR-B1. To resolve the question of whether there is a correlation between the expression of the EcR-B1 isoform and the fate of the larval muscle fibers

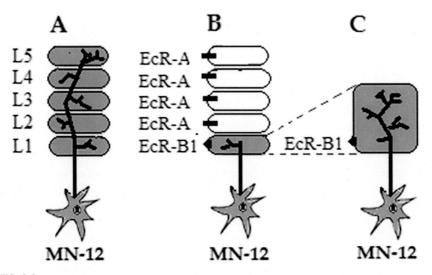

FIG. 3.2

Schematic representation of the development of the adult DEO1 from larval muscle fibers. (A) The motoneuron MN-12 innervates five larval muscle fibers (L1–L5). (B) At the beginning of the ecdysis, the axon terminal arbor prunes back from all muscle fibers, except for the proximal muscle fiber L1, which expresses the EcR-B1 isoform and survives, while the rest of the muscle fibers (L2–L5) secrete EcR-A, which cannot mediate the ecdysone myogenic action, are eliminated. (C) The surviving larval muscle fiber, L1, develops into the adult DEO1. Abbreviations: MN-12, motoneuron 12; EcRA, ecdysone receptor A; EcR-B1, ecdysone receptor B1.

From Cabej, N.R., 2019. Epigenetic Principles of Evolution. Academic Press, London, San Diego, Cambridge MA, Oxford UK, p. 235.

(elimination or regrowth into adult muscle), investigators performed unilateral denervation of the muscle and observed that on the intact side the administration of the hormone ecdysone (Ec) induced the expected upregulation of EcR-B1 whereas on the denervated side did not (Hegstrom et al., 1998), indicating that the motoneuron provided the information necessary for inducing expression of the EcR-B1 isoform. It is generally admitted that motor neuron arborization and innervation are crucial for muscle patterning in *Drosophila* (Schulman et al., 2015).

Myogenesis is triggered by the expression of myogenic regulatory factors: MyoD, Myf5, Myog (Myogenin), and MRF4 (muscle regulatory factor 4). In 1988, it was found that the mouse transcription factor MyoD1 (myoblast determination protein 1) converted fibroblasts and adipoblasts of certain lines into myoblasts (Tapscott et al., 1988), and soon afterward, it was shown that a human transcription factor, Myof-5, structurally very similar to the mouse MyoD1, also had myogenic activity and converted certain fibroblast lines into multinucleate myoblasts (Braun et al., 1989). Binding of MyoD transcription factors to thousands of binding sites genome-wide is followed by chromatin remodeling as a result of the recruitment of chromatin remodeling complexes and histone acetyltransferases. While activating the expression of numerous myogenesis-related genes, MyoD suppresses the expression of other nonmuscle genes through a negative feed-forward circuit, termed an incoherent feed-forward loop (Fong and Tapscott, 2013). Another basic regulatory myogenesis regulator is Myf5 (myogenic factor 5). Its expression is first observed in the neural tube, and its expression several days later in somites is induced by Shh signals from the neural tube. Shh secreted by the floor plate of the neural tube and notochord acts as an upstream signal for MyoD and Myf5 expression in the process of the differentiation of muscle progenitors in somites and the dermomyotome: "Shh has an inductive function in epaxial myogenic determination" (Borycki et al., 1999).

Downstream the Myf5 and MyoD acts Myog (Hasty et al., 1993), which contributes to the differentiation of myoblasts into myotubes and MRF4, which is expressed primarily in postnatal development and adult myogenesis and downregulates myogenin (Zhang et al., 1995). After maturation and innervation of myotubes, myogenin expression declines drastically (Kostrominova et al., 2000).

In a systems biology-inspired/based model of myogenesis, MyoD induces expression of the transcriptional repressor RP58, which represses expression of Id (inhibitors of DNA-binding protein), thus derepressing MyoD and Myog and promoting conversion of myoblasts into myotubes (Ito et al., 2012) (Fig. 3.3).

The process of transformation of myogenic progenitors, myoblasts, into myotubes and syncytial myofibers of skeletal muscles begins with the fusion of the mononuclear myoblasts after they stop proliferating. At the end of the proliferation stage, myoblasts begin the process of fusion, i.e., the union of the myoblasts known as FCs (founder cells) with FCMs (fusion-competent myoblast). The fusion process is evolutionarily conserved in animals of widely different groups like *Drosophila* and mammals. Despite the remarkable similarities of the process between both groups, differences are also observed in the molecular components of the fusion (Önel and Renkawitz-Pohl, 2009).

Adhesion of a FC with multiple FCMs transmits cell membrane signals to the actin cytoskeleton followed by the formation of a multiprotein complex, fusion-restricted myogenic-adhesive structure (FuRMAS), at the fusion site (Kesper et al., 2007) that activates the actin cytoskeleton and via elimination of the cell membrane remnants forms the fusion pore (Önel and Renkawitz-Pohl, 2009) through which cytoplasmic elements and vesicles move in both directions. At the end of the fusion process, the resulting cell has a second nucleus, which adopts the transcriptional profile of the FC (Deng et al., 2017) (Fig. 3.4).

3 Development of muscle tissue in vertebrates

In chick wings, motor neurons from the brachial plexus penetrate the limb bud and start secreting retinoic acid (RA) as early as stages 17–18 (Berggren et al., 2001; Hughes and Salinas, 1999). Even the RA secretion by limb bud mesenchyme is under the control of motor neurons (Berggren et al., 2001). Within the embryonic CNS, the

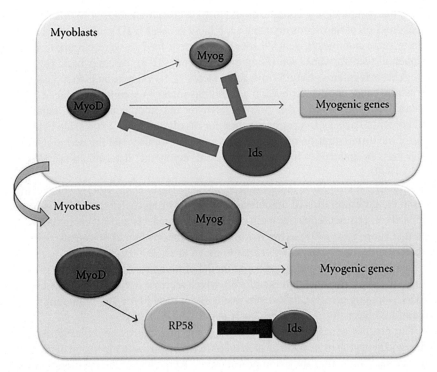

FIG. 3.3

Proposed myogenesis regulatory network by our systems approach in myoblasts; Id proteins are expressed and inhibit the myogenic bHLH factors. During myogenic differentiation, RP58 is promoted by MyoD and represses the Id transcription. Muscle-specific genes are then activated by myogenic bHLH factors.

From Ito, Y., Kayama, T., Asahara, H., 2012. A systems approach and skeletal myogenesis. Comp. Funct. Genomics 2012, 759407.

limb motor neurons in the spinal cord represent major sites of RALDH2 expression (Ross et al., 2000). An extremely strong expression of RA occurs in the cervical and lumbar regions of the spinal cord that correspond to the sites of the limb primordia and region-specific expression of genes in the limb bud (Colbert et al., 1993) (Fig. 3.5), and it is suggested that the hot spots of RA production are "a likely factor" in the formation of the limb zones (McCaffery and Dräger, 1994).

The penetration of nerves in the chick embryo wing buds coincides with the differentiation of the muscle fiber types: "It seems likely that innervation of the limbs is necessary for both muscle and bone development ... RA is a factor that could mediate this neuronal influence." Release of RA-synthesizing enzyme (RALDH-2) in chick embryo wing buds induces cell differentiation and muscle formation (Berggren et al., 2001).

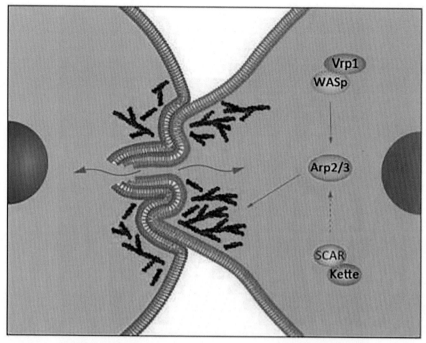

FIG. 3.4

Schematic representation of myoblast fusion. When they come in contact with the FC and FCM, their fusion machineries, actin monomers are assembled into filaments and form a dense F-actin focus on the FCM side and a thin layer on the FC side. The FCM actin focus invades the FC with multiple finger-like protrusions. N-cadherin is removed from the membrane to allow for pore formation. The formation of a fusion pore in the membrane occurs, allowing for the exchange of the cytoplasm between both cells.

From Deng, S., Azevedo, M., Baylies, M., 2017. Acting on identity: myoblast fusion and the formation of the syncytial muscle fiber. Semin. Cell Dev. Biol. 72, 45–55.

Ablation of different segments of the neural tube at various times may prevent the formation of muscles (Stern et al., 1995). Similarly, blockage of cholinergic activity (e.g., by succinylcholine) at 5–10 hpf may prevent muscle formation as it does in chick embryos (Hall and Herring, 1990). The role of RA in limb development is also indicated by the administration of citral, an inhibitor of RA synthesis, that causes defects in limb development (Tanaka et al., 1996).

Neural tube signals, such as Wnt-1 (Reshef et al., 1998), induce expression of Noggin, a BMP antagonist, which may induce somite myogenesis by allowing expression of MyoD and Myf5 in somite cells. Wnt-1 and WNT3a from the dorsal neural tube are involved in the migration of the cells of the dorsal part of the myotome, whereas Wnt and Shh signals induce expression of Myf5 and MyoD1 in

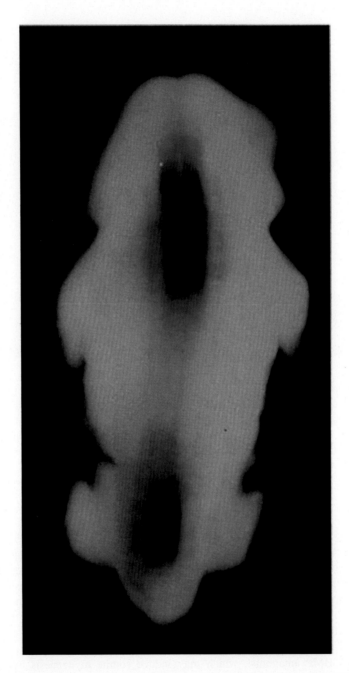

FIG. 3.5

Localization of RA in the cervical and lumbar regions of the spinal cord. Localization of RA in the cervical and lumbar regions of the spinal cord, corresponding to fore- and hindlimbs, in E 12.5 mouse embryo.

From Colbert, M.C., Linney, E., LaMantia, A., 1993. Local sources of retinoic acid coincide with retinoid-mediated transgene activity during embryonic development. Proc. Natl. Acad. Sci. U. S. A. 90, 6572–6576.

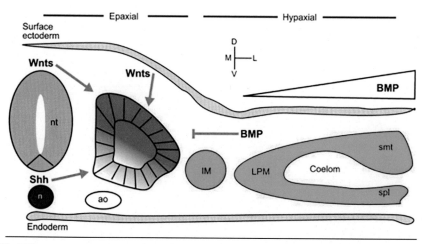

FIG. 3.6

Somite patterning and myotome formation. Spatial relationship between the epithelial somite and the surrounding structures. The mesodermal subtypes are shown, as well as the future epaxial and hypaxial domains. Each epithelial somite is patterned into dorsoventral, mediolateral, and anteroposterior compartments by signaling factors secreted by the surrounding tissues. Dorsally, Wnt signaling is required for dermomyotome specification, while BMP signaling produced by the lateral plate mesoderm (LPM) inhibits the differentiation of somitic lineages. Ventrally, Shh secreted from the midline plays a major role in sclerotome induction. Abbreviations: ao, dorsal aorta; IM, intermediate mesoderm; smt, somatopleura; spl, splanchnopleura; n, notochord; nt, neural tube; D, dorsal; L, lateral; M, medial; V, ventral.

From Chal, J., Pourquié, O., 2017. Making muscle: skeletal myogenesis in vivo and in vitro Development 144, 2104–2122; https://doi.org/10.1242/dev.151035.

migrating muscle precursors, thus determining their myogenic fate. Secretion by the neural tube of neurotrophin and Wnt1 converts cells of the central dermomyotome into cells of the myotome (Chang and Kioussi, 2018) (Figs. 3.6 and 3.7).

Skeletal muscles are composed of fast- and slow-twitch myofibers in varying proportions in different muscles. Myofibers differ widely in morphology and contractile physiology because of the different sets of proteins and enzymes they express, which in turn are determined by the firing patterns of the innervating motor neurons. Wnt signaling in adult slow myofibers is most likely initiated by Wnt molecules from motor neurons (Kuroda et al., 2013). The firing pattern of motor neurons innervating slow myofibers maintains a Ca^{2+} intracellular level that activates calcineurin, which facilitates the transport of NFAT (nuclear factor of activated T cell) transcription factors to the cell nucleus, thus stimulating the calcineurin-NFAT pathway: NFAT and MEF2 (myocyte enhancer factor) transcription factors bind DNA, inducing the slow-

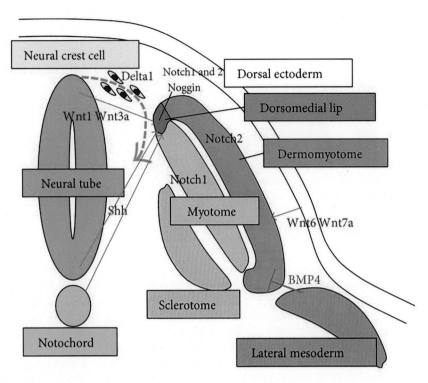

FIG. 3.7

Positive and negative signals from surrounding tissues for embryonic myogenesis.
Dermomyotome receives positive (Shh, Wnt1, Wnt3a, Wnt6, Wnt7a, Delta1, and Noggin) and
negative (BMP4) signals from surrounding tissues (dorsal neural tube, floor plate, notochord,
dorsal ectoderm, and lateral mesoderm) to form myotomes. This occurs at the Notch1/2-
positive dorsomedial lip of the dermomyotome.

From Kodaka, Y., Rabu, G., Asakura, A., 2017. Skeletal muscle cell induction from pluripotent stem cells. Stem
Cells Int. 2017, 1376151.

fiber protein synthesis program. The insufficiently frequent firing patterns of neurons
innervating fast-twitch fibers prevent activation of calcineurin and the calcineurin-
NFAT pathway. Experimental administration of cyclosporin, an inhibitor of calcineurin, leads to transformation of the slow-twitch into fast-twitch myofibers
(Chin et al., 1998) (Fig. 3.8).

The instructive role of the nervous system in the development of muscles is demonstrated in numerous experiments of denervation in vertebrates. Denervation in
skeletal muscles leads to an increase of 2–5-fold expression of MyoD and ~200-fold
of myogenin (Neville et al., 1992).

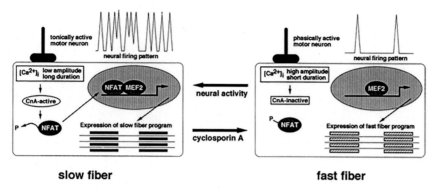

slow fiber   **fast fiber**

FIG. 3.8

Model for a calcineurin-dependent pathway linking specific patterns of motor nerve activity to distinct programs of gene expression that establish phenotypic differences between slow and fast myofibers. MEF2 is shown to represent the requirement for collaboration between activated NFAT proteins and muscle-restricted transcription factors in slow-fiber-specific gene transcription, but other proteins (not shown) are also likely to participate. Abbreviations: MEF2, myocyte enhancer factor-2; NFAT, nuclear factor of activated T cell transcription factors.

From Chin, E.R., Olson, E.N., Richardson, J.A., Yang, Q., Humphries, C., Shelton, J.M., et al., 1998. A calcineurin-dependent transcriptional pathway controls skeletal muscle fiber type. Genes Dev. 12, 2499–2509.

Degeneration of motor neurons leads not only to paralysis but also to atrophy or loss of muscle mass (Ferraro et al., 2012). The only known pathway of the transfer of information from the nervous system to muscles is via the neuromuscular junction that forms between the presynaptic terminal of the axon and the myotube. The presynaptic terminal releases the neurotransmitter acetylcholine (Ach) in the synaptic cleft, where it binds its preexisting receptor AChR in the central region of the myotome. AChR is expressed throughout the muscle, but it is concentrated primarily in the neuromuscular junction, in the prospective synaptic region. AChR is expressed in the muscle not only before the arrival but even in the absence of the axon. However, the axon is "required for the subsequent remodeling and selective stabilization of synaptic clusters that precisely appose post- to presynaptic elements" (Flanagan-Steet et al., 2005). Besides Ach, the presynaptic end of the axon secretes agrin in the synaptic basal lamina. Ach may postsynaptically both activate and decluster its receptor AChR, while agrin has the opposite effect of clustering AChRs in the myotube by concentrating them in the central region of the postsynaptic terminal of the myofiber (Misgeld et al., 2005) (Fig. 3.9). Agrin may even ectopically induce expression of AChR in the denervated myofibers (Ferraro et al., 2012).

In a series of experiments of denervation and chemical interruption of neuromuscular transmission in rats, it was demonstrated that the accumulation of AChR

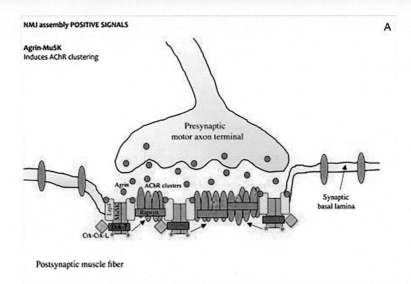

NMJ assembly POSITIVE SIGNALS

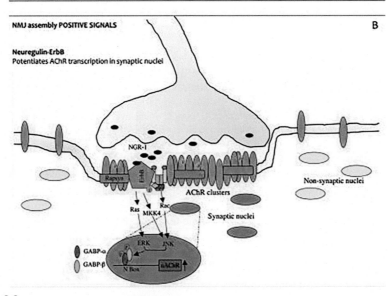

FIG. 3.9

The agrin-MuSK-Lrp4 and neuregulin-ErbB pathways induce NMJ (neuro-muscuar junction) assembly positive signals. (A) Agrin is released by the motor axon terminal and induces AChR clustering, phosphorylation, and stabilization at the postsynaptic membrane. Lrp4 associates with MuSK in the absence of agrin. Agrin binds to the preformed MuSK-Lrp4 complex by interacting with Lrp4 and promotes MuSK transphosphorylation and activation. Once phosphorylated, MuSK recruits the adapter protein Dok-7, which binds Crk and CrkL and stimulates further MuSK phosphorylation and kinase activity. This induces phosphorylation and stabilization of nascent AChR clusters. Rapsyn is a coeffector in the AChR assembly, which anchorates AChRs at the muscle membrane. (B) Neuregulin (NGR-1) is released by the nerve and induces AChR transcription in synaptic nuclei. NRG-1 acts by binding tyrosine kinase receptors ErbBs. ErbB phosphorylation induced by NRG stimulates ERK and JNK kinase activity, which phosphorylates GABP-α and GABP-β transcription factors. GABP-α heterodimerizes with GABP-β and binds DNA at the N-box, thereby enhancing transcription of AChR genes. Abbreviations: MuSK, muscle-specific kinase; NRG, neuregulin.

From Ferraro, E., Molinari, F., Berghella, L., 2012. Molecular control of neuromuscular junction development. J. Cachexia. Sarcopenia Muscle 3 (1), 13–23. https://doi.org/10.1007/s13539-011-0041-7.

subunits in the skeletal muscle is regulated by neural mechanisms. The concentration of the α, β, and δ subunits of the acetylcholine receptor in the innervated muscle soleus, assessed by the respective amounts of α, β, and δ subunit RNAs, is higher in the junctional region, whereas the concentration of the same subunits in the denervated soleus muscle is higher in the extrajunctional region, indicating neural induction of the acetylcholine receptor in the muscle (Goldman and Staple, 1989). In the course of embryonic development, AChR clusters coalesce to form oval-shaped plaques at the center of the muscle fiber (Guarino et al., 2020).

Ablation of the neural tube and notochord during early differentiation of chick embryo somites causes programed cell death in somites and prevents the formation of epaxial muscles, vertebrae, and ribs; the defect, however, is rescued by grafting Shh-producing cells as substitutes of the Shh secreted by the neural tube and notochord (Teillet et al., 1998). Either the neural tube or the notochord can prevent somitic apoptosis and induce muscle cell differentiation in myotomes (Rong et al., 1992).

In the absence of innervation, the Peking duck embryos develop no muscles (Creazzo and Sohal, 1983; Sohal and Holt, 1980). Chick embryo paralysis, implying repression of neural spontaneous activity, leads to a drastic reduction of muscle mass (Hall and Herring, 1990). The activity of motor neurons has a crucial role in determining the size of the length, number, and arrangements of sarcomeres in the developing muscles (Brennan et al., 2005). Based on their experimental work, Wenner and O'Donovan (2001) have shown that "many developing networks exhibit a transient period of spontaneous activity that is believed to be important developmentally…. In the spinal cord, spontaneous activity has been implicated in the development of limb muscles, bones and joints" (Wenner and O'Donovan, 2001). The causal relationship between spontaneous neural activity and muscle development indicates the role of spontaneous activity as a source of information for muscle development in vertebrates.

4 Neural determination of arterial vessel patterning

About six centuries ago, Dutch anatomist Andreas Vesalius (1514–1564) noticed and wrote that nerves and arteries generally run in parallel in the animal body (Gitler et al., 2004), a phenomenon that now is known as "neurovascular congruency" (Andreone et al., 2015).

During the last 3 decades, adequate solid evidence accumulated led to the conclusion that the nervous system is involved in vascularization patterns in animals (Carmeliet et al., 1996; Gale and Yancopoulos, 1999; Kawasaki et al., 1999; Liang et al., 2001). Empirical evidence indicates that the neural tube induces patterning of a peripheral vascular plexus, leading to the concept of the neural tube as the midline signaling center for vascular patterning in vertebrates (Hogan et al., 2004) (Fig. 3.10).

In 2002, Mukoyama et al. observed that in mice mutants lacking peripheral sensory axons or with disorganized peripheral nerve fibers, arterial differentiation and

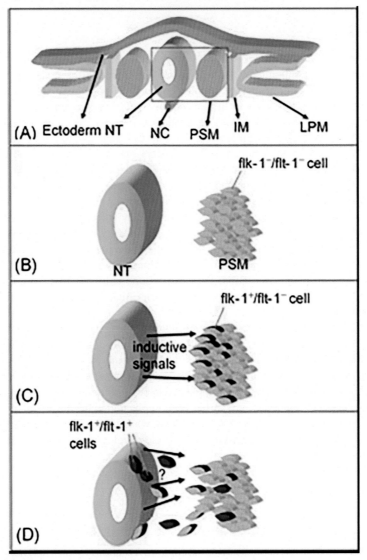

FIG. 3.10

Neural tube patterning of blood vessels—a model. (A) Diagram showing a cross-sectional view of a mouse embryo at 8.5 dpc. The boxed area is enlarged in B—D. (B) Initially, the presomitic mesoderm does not contain committed vascular precursor cells. (C) A VEGF-independent signal from the neural tube induces FLK1 expression in a subset of presomitic mesoderm cells. (D) FLK1-expressing angioblasts are now competent to respond to the neural tube-derived VEGFA-containing signal. The angioblasts now express additional vascular markers, such as FLT1, and they migrate and assemble the PNVP. Other unidentified signals emanating from the neural tube may also contribute to PNVP patterning in vivo. Abbreviations: dpc, day postcoitum; IM, intermediate mesoderm; LPM, lateral plate mesoderm; NC, notochord; NT, neural tube; PNVP, perineural vascular plexus; PSM, presomitic mesoderm.

From Hogan, K.A., Ambler, C.A., Chapman, D.L., Bautch, V.L., 2004. The neural tube patterns vessels developmentally using the VEGF signaling pathway. Development 131, 1503–1513.

remodeling were abnormal, while peripheral sensory neurons induced expression of arterial markers in cultured endothelial cells. They also found that the instructive role of innervation in arterial patterning and remodeling resulted from the secretion by nerve axons of the protein VEGF (vascular endothelial growth factor) and its receptor VEGFR2, which led them to the conclusion that "[p]eripheral nerves provide a template that determines the organotypic pattern of blood vessel branching and arterial differentiation in the skin, via local secretion of VEGF" (Mukoyama et al., 2002). Later, investigators observed that one of the VEGF isoforms, NRP1-binding isoform of VEGF, is a more potent inducer of arterial markers in vitro, whereas the almost normal arteriogenesis in mutants for VEGF and NRP1 binding of VEGF led them to the hypothesis that a second factor was involved in the neural control of arteriogenesis (Mukoyama et al., 2005).

The discovery came from studies on the primary capillary plexus (small blood vessels that form during early development preceding the onset of vasculogenesis) that forms in the developing limb skin that is soon invaded by sensory nerves. Under conditions of hypoxia, nerves are stimulated to secrete diffusible chemokine Cxcl1, which via its receptor Cxcr4, expressed in vessel cells, stimulates local cell migration and aligns capillary vessels with nerves. Subsequently, hypoxia-inducible nerve-derived VEGF-A stimulates arterial differentiation and arterial branching by inducing expression of specific arterial genes in the nerve-associated vessels (Li et al., 2013) (Fig. 3.11). Inactivation of Cxcl12-Cxcr4 perturbs nerve-vessel alignment and prevents arteriogenesis.

And finally, it is demonstrated that the instructive role of the murine nervous system in determining arterial branching patterns of the barrel region of the murine somatosensory cortex results from the neuronal electrical activity during early life (Lacoste et al., 2014). Enhancement of the neural activity resulting from sensory input promotes the development of the cerebrovascular network and, to the contrary, reduction of the sensory input through deafferentiation and reduction of neurotransmitter release decreases cerebral vascular density and branching. Based on a possible direct effect of neurons on vascular patterning by releasing the neurotransmitter glutamate by thalamocortical axons (TCAs) and an indirect mechanism through pathways that are activated by synaptic transmission involving cortical interneurons and glial cells, investigators believe that other mechanisms of the sensory-related neural activity may alter the cerebral vascular structure. Their conclusion is that "neural activity is necessary and sufficient to trigger alterations of vascular networks" (Lacoste et al., 2014). Thus, the neural activity, a computational neural process, rather than neural structure, is responsible for vascular patterning.

5 Specification of dermal lineages

During embryogenesis, the dorsal neural tube emanates Wnt ligands, Wnt-1 and Wnt-3, which bind protein receptors of the Frizzled (Fz) family. Mediators of the Wnts expressed in the neural tube are Frizzled (Fz) protein receptors. At the time

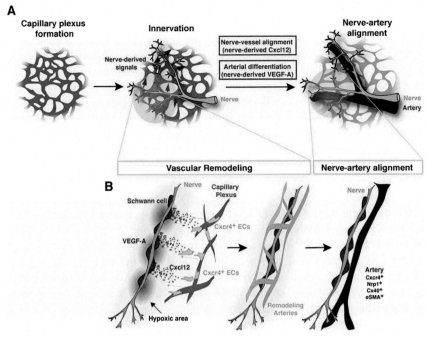

FIG. 3.11

Coordinate action of nerve-derived Cxcl12 and VEGF-A results in nerve-artery alignment.
(A) Schematic model for nerve-mediated vascular branching and arterial differentiation in
developing limb skin. Sensory nerves invade at approximately E11.5–13.5 after a primary
capillary plexus is established. Subsequently, the pattern of sensory nerves provides signals
(Cxcl12 and VEGF-A) that govern patterns of vascular branching and arterial differentiation
during vascular remodeling. As a result, the congruence of blood vessels and nerve patterns is
established in the skin. (B) This scheme shows how nerve-derived Cxcl12 and VEGF-A
control patterns of vascular branching and arterial differentiation. In this view, oxygen-starved
nerves may induce Cxcl12 and VEGF-A expression through activation of HIF-1 prior to
vascular remodeling. Cxcl12-Cxcr4 signaling functions as a long-range chemotactic guidance
cue to recruit vessels to align with nerves. Nerve-derived VEGF-A instructs arterial
differentiation in the nerve-associated vessels. Arterial differentiation presumably requires a
local action of VEGF-A to induce arterial marker expression.

*From Li, W., Kohara, H., Uchida, Y., James, J.M., Soneji, K., Cronshaw, D.G., et al., 2013. Peripheral nerve-
derived CXCL12 and VEGF-A regulate the patterning of arterial vessel branching in developing limb skin. Dev. Cell
24, 359–371.*

of the formation of the dermomyotome, the neural tube expresses both Wnt-1 and
Wnt-3a, but only Wnt-1 stimulates the formation of the dermomyotome compartment and the feather-inducing dermis. Removal of the neural tube and notochord
leads to the death of the chick medial somatic cells and the dorsal feather field.

Wnt-1 is also capable of restoring the dorsal feather field (Olivera-Martinez et al., 2001). The ventral neural tube releases primarily Noggin and Sonic hedgehog, inhibitors of Bmp4, which determine the ventral skin feather macropattern (Fliniaux et al., 2004).

Neurally induced dermatome is innervated by spinal nerve roots, and expression of Wnt-11 promotes the expansion of the medial dermomyotome and formation of the dermis. Expression of DKK (members of the Dickkopf family) inhibits Wnt signaling, leading to the formation of a thin incomplete dermis (Widelitz, 2008) (Fig. 3.12).

5.1 Role of the sympathetic innervation in the development and maintenance of hair follicle

Hair is among the best-studied mammalian appendages. Hair coat in animals contributes to thermal insulation and temperature regulation. Mammalian hair grows and emerges from the hair follicle, a dermal dynamic organ. The middle part of the hair follicle is surrounded by hair follicle stem cells (HFSCs) that form the hair bulge to which attaches the follicle muscle *arrector pili*, encircled by an elaborate sympathetic innervation with nerve endings loaded with neurotransmitter vesicles establishing synapse-like structures with the HFSCs (Shwartz et al., 2020) (Fig. 3.13).

Earlier observations showed that changes in hair growth are often correlated with changes in the sympathetic nervous system (Fig. 3.14). During the anagen phase, an increase in the number of the sympathetic nerve fibers secreting norepinephrine and β-adrenoreceptors in the bulge occurs, and administration of β-adrenoreceptor agonists, isoproterenol (Botchkarev et al., 1999) and procaterol (Shwartz et al., 2020), accelerates anagen entry. Pharmacological sympathectomy delays the activation of HFSCs and, to the same result, leads the loss of β2-adrenoreceptors (Shwartz et al., 2020).

Fan et al. (2018) demonstrated that in response to light stimuli, the mouse brain activates a neural pathway, the ipRGC-SCN (intrinsically photosensitive retinal ganglion cells-suprachiasmatic nucleus) autonomic nervous system circuit. Beginning from the third day after initial light stimulation, the sympathetic innervation by releasing neurotransmitter norepinephrine enhances Hh (Hedgehog) expression in HFSCs, thus promoting the proliferation of HFSCs, leading to increased hair growth and anagen entry (Fan et al., 2018). Stem cells in the upper bulge receive Shh from the sympathetic innervation (Brownell et al., 2011).

Schwartz et al. have shown that another environmental stimulus, the cold, that is responsible for goosebumps that develop in the skin in response to cold, when the cold persisting for relatively long periods of time, induces mice to enter anagen faster and produces new hair as a result of increased activity of the sympathetic innervation and increased proliferation of HFSCs (Shwartz et al., 2020).

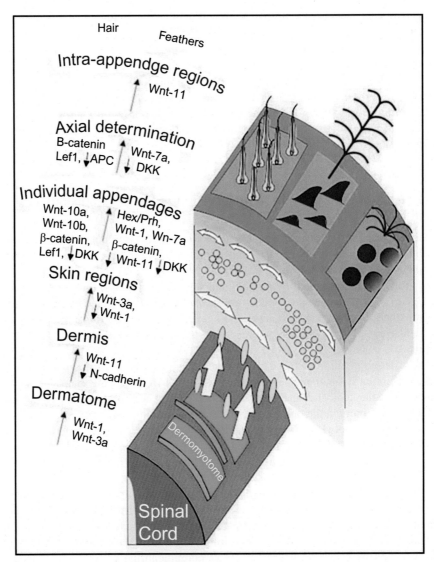

FIG. 3.12

Involvement of Wnt signaling in the specification of dermis. Schematic showing the involvement of Wnt signaling in dermis, tract, and skin appendage development as determined in mouse hairs (*left of arrows*) and chicken feathers (*right of red arrows* indicating developmental progression). Morphogenetic events that take place in the dermomyotome, dermis, feather tracts, and individual feather primordia are shown. *Spindles and circles* represent dermal cells. Two different tracts with different densities and shapes of skin appendages are shown. Abbreviations: DKK, any member of the Dickkopf family (inhibitors of Wnt signaling).

The figure is modified from ref. [22] (Chang, C.-H., Jiang, T.-X., Lin, C.-M., Burrus, L.W., Chuong, C.-M., Widelitz, R., 2004. Distinct Wnt members regulate the hierarchical morphogenesis of skin regions (spinal tract) and individual feathers. Mech. Dev. 121, 157–171) from Widelitz, R.B., 2008. Wnt signaling in skin organogenesis. Organogenesis 4, 123–133.

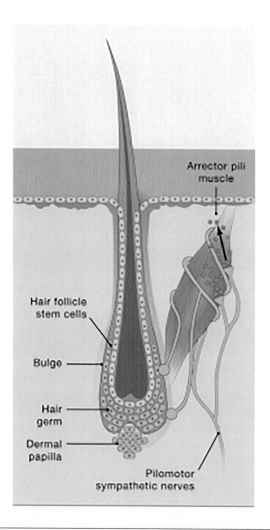

FIG. 3.13

Hair follicles, *arrector pili* muscles, and sympathetic nerves form a trilineage unit, which promotes piloerection or goosebumps as an acute response to cold. Arrector pili muscles attach to the bulge region of hair follicles where hair follicle stem cells (HFSCs) reside. Arrector pili muscles are enwrapped by sympathetic pilomotor nerves.

From Pascalau, R., Kuruvilla, R., 2020. A hairy end to a chilling event. Cell 182, 539–541.

6 Neural control and regulation of stem cell proliferation

Almost all animals during their development and adult life possess a stock of undifferentiated pluripotent cells capable of self-renewal via cell division and differentiation into other cell types to replace the normally lost cell types or produce cell types necessary for the regeneration of tissues and organs.

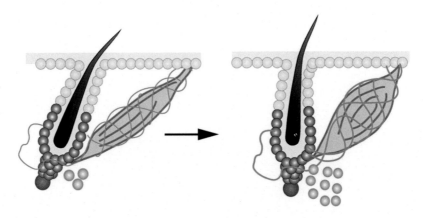

FIG. 3.14

Schematic showing sympathetic axons. Sympathetic axons extend to HFSCs while cell bodies are at the sympathetic ganglia.

From Shwartz, Y., Gonzalez-Celeiro, M., Chen, C., Pasolli, H.A., Sheu, S., Fan, S.M., et al., 2020. Cell types promoting goosebumps form a niche to regulate hair follicle stem cells. Cell 182 (3), 578. Aug 6; https://doi.org/10.1016/j.cell.2020.06.031.

In invertebrates such as flatworms of the genus *Planaria* (Planaridae family) and annelids, stem cells are known as neoblasts. They represent ~25% of the total number of cells in adult planarians (Friedländer et al., 2009). Neoblasts are pluripotent stem cells that sustain tissue homeostasis and regeneration in planaria that interestingly resemble germline cells; at a genetic level, they differ from the fact that only the latter express *nanos* (Rink, 2012).

For a long time, it was questioned whether neoblasts represented a population of pluripotent stem cells that in the process of regeneration produced all the planarian cell types or neoblast population contained lineage-restricted cells ready to use during the planarian regeneration. The issue was resolved empirically when it was demonstrated that transplantation of single clonogenic neoblasts (cNeoblasts) into a lethally irradiated planarian regenerated the otherwise dead worm (Wagner et al., 2011).

After planarian amputation, at stumps on two sides of amputation forms blastema, the regeneration bud. Formation of the blastema is associated with neoblast proliferation induced by signals from brain primordia (Fraguas et al., 2014), such as neuropeptide P and neuropeptide K (neurokinin K) (Baguñà et al., 1989). Experimental reduction of brain ganglia impairs blastema formation (Fraguas et al., 2014). Also, disruption of the ventral nerve cord in postpharyngeal planarian fragments leads to body plan abnormalities, suggesting the role of the ventral nerve cord as a source of regenerational information (Cebrià and Newmark, 2007; Oviedo et al., 2010).

Sublethally irradiated planaria can completely regenerate in the presence of a functioning EGF pathway. Key in the pathway is one of the EGF receptors, EGFR-3. Loss of egfr-3 expression or its putative ligand, neuregulin-7 (*nrg-7*), causes deficient neoblast repopulation and impairs regeneration. Neuregulin is a nerve-derived substance secreted by planarian PC2[+] neurons, but in experiments it was found that it is also produced by prog-1[+] epidermal follicle stem cells. Nevertheless, experiments showed that neuregulin-7-producing neurons may be involved in activation of the EGF pathway that is critical for neoblast repopulation of sublethally irradiated planarians (Lei et al., 2016).

The brain also regulates "both the absolute and relative numbers of different cell types in complex organs such as the brain during cell turnover, starvation, and regeneration" (Takeda et al., 2009) and reestablishes them in the process of regeneration. Neural signals also regulate the proportion of neoblasts that will differentiate into tissue- and organ-specific cell types (Takeda et al., 2009).

6.1 Regulation of stem cell niches

Stem cells for particular cell types are localized in particular sites in the vicinity of the cell types they will differentiate into, which are known as stem cell niches. The maintenance and proliferation of stem cells in *Drosophila*, among other things, depend on the nutrition state. The nutrition state is assessed in specific brain neurons of the fly. Neurons IR60c-Gal4 via their receptor IR76b, in response to a high concentration of amino acids in the hemolymph, suppress feeding (Croset et al., 2016). Glucose levels are assessed in a pair of glucose-sensing neurons, which regulate the level of insulin and glucagon (Oh et al., 2019). Physiological/subphysiological levels of amino acids, sugars, and triglycerides stimulate a group of brain neurons, IPCs (insulin-producing cells), to secrete insulin-like peptides (Dilp2, Dilp3, and Dilp5) that via their receptor InR regulate growth or maintain body size, including stem cells.

Mediators of the nutritional action on germline stem cells (GSCs) are brain-derived Dilps. Ablation of Dilps-producing neurons (IPCs), disruption of Dilps, or loss of the Dilp receptors lead to reduced GSC proliferation and reduced egg production (LaFever and Drummond-Barbosa, 2005). Experimental loss of IPC neurons also led to reduced cell division in GSC of *Drosophila* testes, and "it is likely that male GSC maintenance may be under the direct control of ILPs from brain cells" (Ueishi et al., 2009). Activation of insulin signaling suffices to suppress the stem cell loss and repopulate the GSC niche in starved *Drosophila* testes (McLeod et al., 2010).

A clear example of the role of the nervous system in the regulation of stem cells is offered by the experimental work of Katayama et al. (2006) on mice, demonstrating that the egress of hematopoietic stem and progenitor cells (HSPCs) residing in specific niches in the bone marrow (BM) and their migration via blood are induced by signals emanating from the nervous system. In response to G-CSF (granulocyte colony-stimulating factor), the sympathetic nervous system releases norepinephrine,

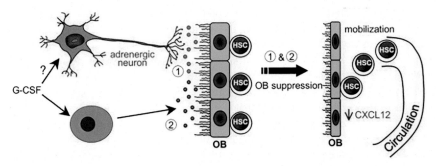

FIG. 3.15

Model for G-CSF-induced HSPC mobilization. G-CSF may activate the outflow of the sympathetic nervous system by influencing directly or indirectly autonomic neurons in sympathetic ganglions in the periphery. Released NE (1) and a yet unidentified signal (2) mediate osteoblast (OB) suppression, thereby reducing the synthesis of CXCL12. Posttranslational mechanisms (degradation/inactivation) may also contribute to lowering CXCL12 levels to those permissive for HSPC egress from their niche. We propose that OB suppression and CXCL12 reduction lead to HSPC mobilization. In addition, it is possible that adrenergic neurotransmission also regulates HSPC mobilization through other mechanisms, given the newly identified non-OB stem cell niches (Kiel et al., 2005).

From Katayama, Y., Battista, M., Kao, W-M., Hidalgo, A., Peired, A.J., Thomas, S.A., et al., 2006. Signals from the sympathetic nervous system regulate hematopoietic stem cell egress from bone marrow. Cell 124 (2), P407–P421. https://doi.org/10.1016/j.cell.2005.10.041.

thus inducing migration of HSPCs out of their bone marrow niche by suppressing osteoblast function and chemokine CXCL12. Administration of catecholamine antagonists (propranolol, 6-hydroxydopamine, etc.) results in dramatic reduction of HSPC mobilization, whereas catecholamine agonists (e.g., clenbuterol) promote HSPC egress (Katayama et al., 2006) (Fig. 3.15).

In the above context, a recent discovery of a brain mechanism of the translation of the light stimulus into a signal that activates the sympathetic nervous system inducing mouse hair follicle stem cell (HFSC) proliferation is noteworthy. The pathway starts with the reception of light stimulus in the neurons of the retinal ganglion, its transmission to the SCN (hypothalamic suprachiasmatic nucleus), and processing in further brain areas to activate the sympathetic innervation, which by enhancing Hh signaling and proliferation of HFSCs promotes hair growth (Fan et al., 2018) (Fig. 3.16).

7 Metamorphosis

Metamorphosis offers probably the most visible example of the role of the nervous system as the regulator and the source of information for the animal development. It is a common phenomenon in the life histories of many animals, from cnidarians to vertebrates (fish and amphibians).

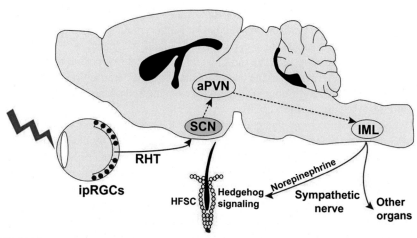

FIG. 3.16

Depiction of the neural pathway for light-stimulated sympathetic activity and HFSC activation. Abbreviations: aPVN, autonomic neurons of the paraventricular nucleus; HFSC, hair follicle stem cell; IML, intermediolateral cell column; ipRGCs, intrinsically photosensitive retinal ganglion cells; RHT, retinohypothalamic tract. Dashed lines indicate proposed neural relays for this pathway.

From Fan, S.M., Chang, Y.T., Chen, C.L., Wang, W.-H., Pan, M.-K., Chen, W.-P., et al., 2018. External light activates hair follicle stem cells through eyes via an ipRGC-SCN-sympathetic neural pathway. Proc. Natl. Acad. Sci. U. S. A. 115 (29), E6880–E6889. https://doi.org/10.1073/pnas.1719548115. [published correction appears in Proc. Natl. Acad. Sci. U S A. 2018 Dec 18;115 (51), E12121].

7.1 Metamorphosis in cnidarians

The larval form of the cnidarian *Hydractinia echinata*, under natural conditions, after growing to a 10,000-cell planula, stops cell proliferation. Detecting via sensory neurons in the anterior part of the body a chemical released by a bacterium, the planula stops swimming and settles while its cells once again start proliferating. The sensory neurons secrete a neuropeptide that "synchronizes the events of metamorphosis" (Schwoerer-Bohning et al., 1990).

Among neuronally derived neuropeptides directly involved in the regulation of metamorphosis of *H. echinata* are metamorphosin A (MMA) (Leitz et al., 1994), the "head activator" (Schaller, 1976) and two antagonistically acting neuropeptide types, GLWamides that stimulate metamorphosis, and RFamides that inhibit it (Katsukura et al., 2003).

All the inducers of cnidarian metamorphosis known so far are secreted by their nervous system.

7.2 Metamorphosis in insects

In insects, the key role in determining the timing and regulation of the process of metamorphosis is played by the neuropeptide prothoracicotropic hormone (PTTH), which regulates the pulsed production and secretion of ecdysone, growth cessation, and beginning of metamorphosis (Kawakami et al., 1990). PTTH is produced by two neurons in each of two brain hemispheres (McBrayer et al., 2007) (Figs. 3.17 and 3.18). These neurons innervate the prothoracic gland (Christensen et al., 2020), inducing it to secrete ecdysone, the molting and metamorphosis hormone. Ecdysone is released in the hemolymph at the end of the third instar, determining thus the critical weight threshold, beginning of metamorphosis, the end of the growth, and body size in insects (Shimell et al., 2018).

Hypomorphic mutant *gt* (gap gene giant) gene that affects the specification of the four PG (prothoracic gland) neurons prevents PTTH secretion by the PG due to the failure of PG neurons to innervate the PG. However, ablation of PG neurons causes more drastic effects, delay of pupariation by 4–6 additional days, and production of larger flies (Ghosh et al., 2010). Inactivation of PG neurons with tetanus toxin and inactivation of electrical activity of the PG neurons cause approximately a 24-h developmental delay (Shimell et al., 2018). Besides, three pairs of specific serotonergic neurons innervate the PG, providing relevant food-related information for ecdysone synthesis (Shimada-Niwa and Niwa, 2014).

7.3 Metamorphosis in vertebrates

Among vertebrates, fish and amphibians undergo this relatively abrupt life-history transition, characterized by morphological, physiological, and behavioral changes from the juvenile to the adult form.

Metamorphosis is regulated by brain signals via the hypothalamus–pituitary–thyroid (HPT) axis and hypothalamus–pituitary–adrenal (HPA) axis in fish and amphibians. The information for metamorphosis is generated in the neural circuits by processing the external (photoperiodic cues, temperature, etc.) and internal stimuli (growth beyond a certain body mass, etc.) from three brain nuclei in the hypothalamus paraventricular nucleus (PVN) (Fig. 3.19). The PVN outputs, neuropeptides thyrotropin-releasing hormone (TRH) and CRH (corticotropin-releasing hormone) (Denver, 1997a), by activating the HPT and HPA axes, stimulate the thyroid to secrete thyroid hormones (T3 and T4) and adrenal gland to produce cortisol, regulation of the metamorphosis, and remodeling of organs and tissues. In amphibians, the morphological changes taking place during metamorphosis are as profound as changes between different vertebrate classes (fish and amphibians). Distinctly from amphibians, teleost fish display less deep or sometimes only subtle changes but, depending on species, they can be sudden or gradual (McMenamin and Parichy, 2013). Regardless of the degree of morphological changes, all cases of metamorphosis in fish involve postembryonic remodeling of the body; the brain signals trigger activation of the HPT axis, whereas thyroid hormone and its receptors (THRs), TRα and TRβ, act as proximate inducers of metamorphosis. The larval stage may last from a few days to years. The transition to the juvenile/adult forms involves resorption of larval specific structures and remodeling of the juvenile/adult structures.

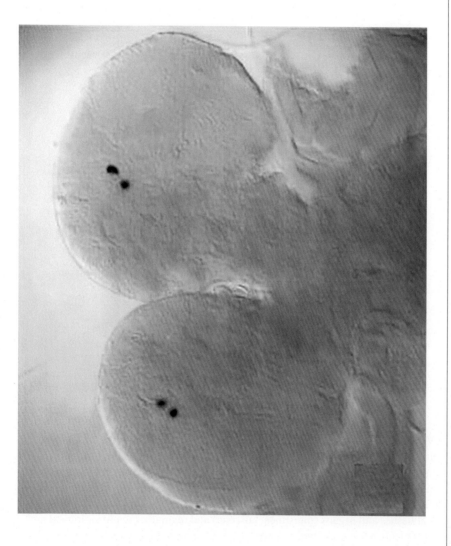

FIG. 3.17

Two PTTH-producing neuron pairs in the brain of a wandering third instar *Drosophila*.

From McBrayer, Z., Ono, H., Shimell, M., Parvy, J.-P., Beckstead, R.B., Warren, J.T. et al., 2007. Prothoracicotropic hormone regulates developmental timing and body size in *Drosophila*. Dev. Cell 13, 857–871. https://doi.org/10.1016/j.devcel.2007.11.003.

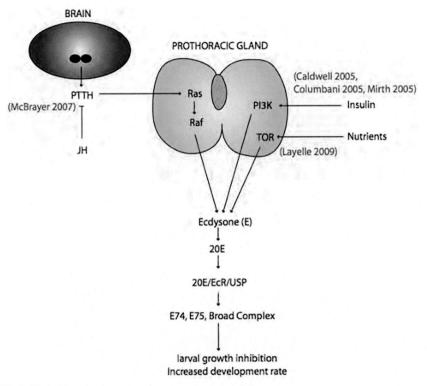

FIG. 3.18

The ecdysone pathway. Abbreviations: JH, Juvenile hormone; PTTH, prothoracicotropic hormone; TOR, target of rapamycin; 20E, 20-hydroxyecdysone; EcR, ecdysone receptor, USP, ultraspiracle; PG, prothoracic gland. References are in blue.

From Quinn, L., Lin, J., Cranna, N., Lee, J.E.A., Mitchell, N., Hannan, R., 2012. Steroid hormones in Drosophila: how ecdysone coordinates developmental signalling with cell growth and division. In: Steroids, Basic Science, InTech, Rijeka, Croatia.

PVN secretory neurons release neurohormone TRH in the portal circulation of the pituitary, stimulating it to secrete TSH (thyroid-stimulating hormone), which in turn induces the thyroid gland to secrete thyroid hormones, proximate regulators of metamorphosis in fish and amphibians (Fig. 3.20).

In the African clawed frog, *Xenopus laevis*, T3 via its receptor induces apoptosis of larval tissues and organs (intestines, tail, etc.) and regulates the physiological and morphological changes and body remodeling taking place in the course of metamorphosis. The hormone induces the expression of downstream genes (metalloproteinases, integrins, etc.) down to the metamorphic target genes. Thus, the *primary gene expression response* is followed by a *secondary gene expression response*. It is estimated that only in the tail resorption program, thyroid hormones upregulate ~35 genes and downregulate ~10 genes (Denver, 1997b).

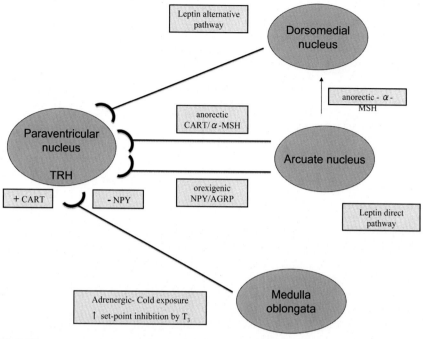

FIG. 3.19

Three brain areas provide their outputs to the hypothalamic TRH (thyroid-stimulating hormone-releasing hormone)-producing neurons of the paraventricular nucleus.

From Chiamolera, M.I., Wondisford, F.E., 2009. Thyrotropin-releasing hormone and the thyroid hormone feedback mechanism. Endocrinology 150(3), 1091–1096. https://doi.org/10.1210/en.2008-1795.

8 Development of endocrine glands

8.1 The anterior pituitary (adenohypophysis)

Development of the anterior pituitary initiates from the pituitary placode, which is located adjacent to the neural ectoderm of the ventral diencephalon, whose evagination forms the infundibular region (infundibulum). Wnt signaling from the ventral diencephalon, through β-catenin, promotes the establishment of the diencephalic pituitary organizing center (Osmundsen et al., 2017; Youngblood et al., 2018; Chukwurah et al., 2019). The pituitary organizer of the ventral diencephalon secretes FGF, BMP, SHH, and Notch (Chaa et al., 2004; Potok et al., 2008; Davis et al., 2013; Cheung et al., 2017; Camper et al., 2017). BMP signals inhibit the secretion of Shh in the infundibulum (Cheung et al., 2017) (Fig. 3.21). The balance of signaling pathways required for determining development, identity, growth, and shape is established by β-catenin signals from the ventral diencephalon (Camper et al., 2017).

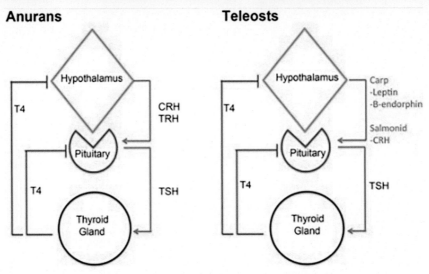

FIG. 3.20

Diagram depicting HPT-axis regulation in anurans and teleosts. In anurans, hypothalamic-derived CRH and/or TRH are involved in regulating pituitary TSHb expression and secretion that in turn regulates T4 production in the thyroid gland and consequently serum levels. Inversely, serum T4 then negatively regulates CRH/TRH and TSHb expression and secretion. In teleosts, HPT-axis regulation is more diverse and likely reflects species specificity.

From Campinho, M.A., 2019. Teleost metamorphosis: the role of thyroid hormone. Front Endocrinol (Lausanne).
10, 383. Published online 2019 Jun 14. https://doi.org/10.3389/fendo.2019.00383 (simplified).

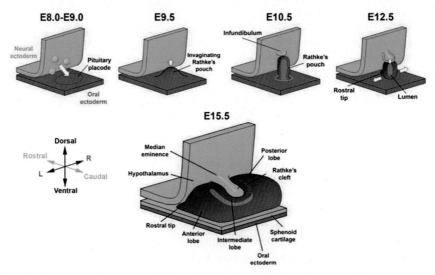

FIG. 3.21

The pituitary gland develops from distinct embryonic origins. Pituitary development in the mouse begins from E8.0 when the neural ectoderm adjacent to the pituitary placode produces signaling factors (yellow) to initiate the invagination of Rathke's pouch. The pouch will begin to pinch off from the oral ectoderm at E10.5 and completely separate by E12.5. Rathke's pouch gives rise to the anterior lobe, which continues to expand ventrally and laterally. The infundibulum evaginates downward from the neural ectoderm and forms the posterior lobe.

From Cheung, L.Y.M., Davis, S.W., Brinkmeier, M.L., Camper, S.A., Pérez-Millán, M.I., 2017. Regulation of
pituitary stem cells by epithelial to mesenchymal transition events and signaling pathways. Mol. Cell. Endocrinol.
445, 14–26.

Signals from the juxtaposed neural tissue induce the development of Rathke's pouch, the primordium of the anterior and intermediate pituitary lobe, from the ectodermal placode (Takuma et al., 1998; Davis et al., 2013). The role of the ventral diencephalon in developing Rathke's pouch and infundibulum is corroborated by the experimental observation that disruption of the transcription factor Nkx2 leads to the loss of the ventral forebrain and the complete absence of Rathke's pouch (Kelberman et al., 2009). Generally, FGF signaling from the diencephalic pituitary organizer promotes the survival of the pituitary progenitor cells in Rathke's pouch, while progenitor cells in the infundibulum give rise to pituicytes (Cheung et al., 2017).

8.2 Thymus

The CNS controls the thymus development both humorally via the neurohormone gonadotropin-releasing hormone (GnRH) and via the local sympathetic innervation. The fact that GnRH is required for thymus development led investigators to the idea that destruction of the anterior portion of the hypothalamus (AHTL) would negatively influence the thymic structure and function. Surprisingly, however, it led to the opposite result: it caused thymus hypertrophy and elevated levels of growth hormone. This unexpected effect was explained with the possibility that destruction of AHTL eliminated growth hormone-releasing inhibitory hormone (GHRIH) but not growth hormone-releasing hormone (GHRIH), inducing the increased production of growth hormone (GH) by the pituitary, thus leading to thymus hypertrophy (Utsuyama et al., 1997; Hirokawa et al., 1998; Hirokawa et al., 2001).

The sympathetic innervation in rats is present from the embryonic day 18 (ED18) (Leposavić et al., 1992) and in the mouse around ED17 to continually grow denser later until adulthood (Singh, 1984). Three decades ago, it was shown that chemical sympathectomy resulted in a loss of thymus mass and drastic changes in cell components of the gland (Kendall and al-Shawaf, 1991), suggesting a role of the sympathetic innervation in the development of the gland in mammalians. In the human fetus, the thymus is penetrated by innervation from the 18th gestational week (Anagnostou et al., 2007).

Besides the peripheral control, the CNS uses a neurohormonal mechanism of the regulation of thymus development involving gonadotropin-releasing hormone (GnRH) of two forms: GnRH1 synthesized in the brain and GnRH2 synthesized in the brain, but primarily in other organs (ovary, testis, mammary gland, etc.). These hormones influence the morphogenesis and functional maturation of the gland, including the immune functions (Melnikova et al., 2019) (Fig. 3.22).

Beginning from ED17, rat fetuses express the neurohormone GnRH receptor (GnRHR). In rat thymus, GnRHR was expressed in thymocytes but not in thymic stromal cells. Blockade of GnRHR by administration of GnRH antagonists in rat fetuses (in utero) suppressed the Con A-induced proliferative response of thymocytes (T cells) in adults while the following administration of GnRH restored the normal state in the gland (Melnikova et al., 2019).

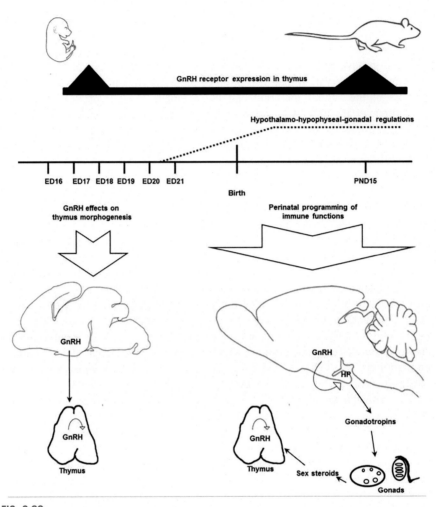

FIG. 3.22

Schematic representation of GnRH effects on the thymus development in fetal and early postnatal rats. During prenatal development up to ED20–21, before the establishment of the HPG endocrine regulations, hypothalamic GnRH can be released to general circulation and provides a direct effect on the morphogenesis of thymus. Since the development of blood–brain barrier and the establishment of the HPG axis in early postnatal life, GnRH is involved in the bidirectional programing of both neuroendocrine and immune functions via gonadotropins and sex steroids. In addition to the neuroendocrine regulation, the effects of GnRH synthesized in thymus can be realized by autocrine/paracrine mechanisms. The developmental pattern of the GnRH receptor expression in the thymus is in favor of this assumption. Abbreviations: ED, embryonic day; PND, postnatal day; HP, hypophysis.

From Melnikova, V.I., Lifantseva, N.V., Voronova, S.N., Zakharova, L.A., 2019. Gonadotropin-releasing hormone in regulation of thymic development in rats: profile of thymic cytokines. Int. J. Mol. Sci. 20 (16). 4033. https:// doi.org/10.3390/ijms20164033.

8.3 Pancreas

Pancreatic sympathetic denervation and the developmental blockade of the β-adrenergic signaling alter the shape and cytoarchitecture of Langerhans islets in the mice pancreas whereas stimulation of β-adrenergic signaling restores the islet structure (Borden et al., 2013). In mice, the differentiation of the β cells of the islets of Langerhans and expression of neuroendocrine markers coincides with the pancreas innervation during early development. Interestingly, while in mice β-cells are well innervated by the vagal cholinergic nerves and the sympathetic system, in humans, β-cells are barely innervated and only a few nerve fibers are seen in the adult islets (Proshchina et al., 2014). β-cell proliferation is stimulated by parasympathetic signals and inhibited by sympathetic signals (Moullé et al., 2019). The sympathetic activity in the pancreas is mediated by norepinephrine (NE) via the adrenergic receptor α2a AR, and the parasympathetic activity by acetylcholine (Ach) via its receptor M3R.

Neuronal instructional cues are indispensable for the development of β cells in the islets of Langerhans (Burris and Hebrok, 2007; Hsuan et al., 2018) (Fig. 3.23).

Initially the sympathetic fibers innervate α-cells in the periphery, but upon the maturation of islets they relocate to innervate β-cells (Cabrera-Vásquez et al., 2009). The maturation of β-cells is closely related to the islet maturation, and very early in the pancreas development, close contacts are established between the β-cells and the neural cell bodies (Burris and Hebrok, 2007).

Early during development, the pancreas anlage is invaded by neural crest cells, which differentiate into neurons and glial cells. These neurons, by inhibiting expression of Nkx2.2 in the immature insulin-producing cells, transform them into mature β-cells (Plank, 2011; Plank et al., 2011).

Even in adult life, the CNS maintains the homeostasis of the pancreatic islets by regulating their size (Nekrep et al., 2008; Thorens, 2014) (Fig. 3.24).

8.4 Adrenal gland

The dorsal aorta serves as a crucial morphogenetic station in the migration of the sympathoblast- and chromaffin cell-fated neural crest cells, but the mechanism that enables their ventral migration around the dorsal aorta and the aorto-gonadal-mesonephros remains unknown (Lumb and Schwarz, 2015). A separate sympathoadrenal (SA) cell lineage migrates from the neural crest between 18 and 24 somites to the adrenal medulla to give rise to sympathetic neurons and neuroendocrine chromaffin cells (Lumb and Schwarz, 2015) as well as another cell type that is intermediate between the neurons and chromaffin cells. The most recent evidence, however, suggests that there are two distinct lineages (sympathoblasts and chromatin cell precursors) rather than a single lineage of neural crest cells that migrate to the adrenal medulla. Evidence indicates that their specification takes place during the migration or even before the lineage leaves the neural tube (Shtukmaster et al., 2013).

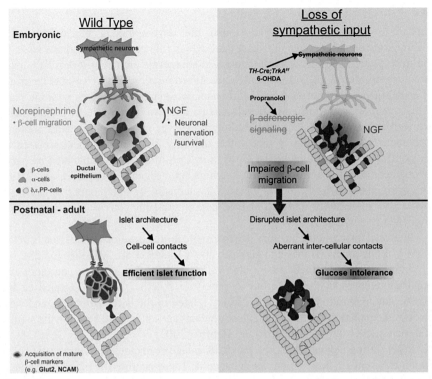

FIG. 3.23

Summary model of the developmental interactions between sympathetic neurons and pancreatic endocrine cells. Previous work has shown that sympathetic innervation of the embryonic pancreas occurs in response to target-derived NGF (nerve growth factor), which then supports the survival of innervating neurons. My work shows that sympathetic nerves release norepinephrine, which acts through β-adrenergic receptors on β-cells to contribute to endocrine cell migration and the formation of discrete islet clusters. I have further demonstrated that nerve-dependent establishment of islet architecture is essential for intercellular contacts within islets, β-cell maturation (i.e., Glut2 levels), and optimal glucose-stimulated insulin secretion.

From Borden, P., Houtz, J., Leach, S.D., Kuruvilla, R., 2013. Sympathetic innervation during development is necessary for pancreatic islet architecture and functional maturation. Cell Rep. 4, 287–301.

Recently, SCPs (Schwann cell precursors) are identified as another derivative of the neural crest cells that later, through an immature stage, differentiate into myelin and nonmyelin Schwann cells. SCPs use peripheral nerves as a migration route to the adrenal medulla. In mice, they appear first at ~E12 and are closely associated with neurons and nerves innervating the adrenal medulla and follow axon signals in their travel to the adrenal medulla. SCPs are necessary for the differentiation of neurons,

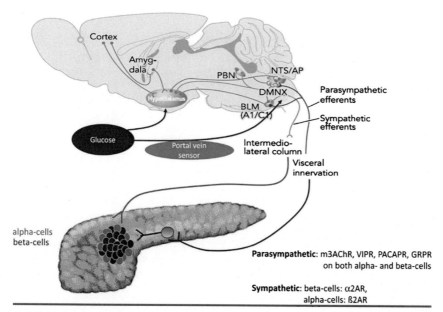

FIG. 3.24

Glucose control of the autonomic innervation of the endocrine pancreas. Glucose is detected by glucose-sensing cells located in peripheral locations such as the hepatoportal vein area and by specialized glucose-excited (GE) or glucose-inhibited (GI) neurons located in the hypothalamus or brainstem region. The brainstem regions containing glucose-responsive neurons are the nucleus tractus solitarius (NTS), the area postrema (AP), and the dorsal motor nucleus of the vagus (DMNX); these three nuclei form the dorsal vagal complex. Glucose-sensing neurons are also present in the basolateral medulla (BLM) regions containing A1 and C1 neurons. The peripheral, hypothalamic, and brainstem glucose-sensing neurons form a network of glucose-sensing cells connected to each other, and they monitor blood glucose concentrations at different locations. The integration of these signals is finally converted into the activation or inhibition of the sympathetic and parasympathetic nervous systems. These innervate peripheral organs involved in glucose homeostasis, liver, fat, muscles, and pancreatic islets. Parasympathetic activity stimulates glucagon secretion during hypoglycemia and insulin secretion in hyperglycemic conditions. Sympathetic activity stimulates glucagon secretion and inhibits insulin secretion in response to hypoglycemia.

Modified with permission from Ref. 52 (Marty, N., Dallaporta, M., Thorens, B., 2007. Brain glucose sensing, counterregulation, and energy homeostasis. Physiology 22, 241–251) from Thorens, B., 2014. Neural regulation of pancreatic islet cell mass and function. Diabetes Obes. Metab. 16, 87–95.

chromaffin cells, and the intermediate small intensely fluorescent (SIF) cells (Shtukmaster et al., 2013). Other evidence suggested that sympathetic, sensory, and melanocyte progenitors delaminate from the thoracic neural tube (NT) in successive, largely nonoverlapping waves and that at least certain NC progenitors are already fate-restricted within the neural tube before delamination and migration.

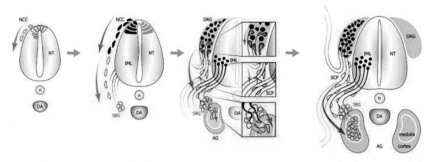

FIG. 3.25

Schematic showing the origin of chromaffin cells from nerve-associated SCPs. Abbreviations: IML, intermediolateral cell column; CC, central canal; AM, adrenal medulla; AG, adrenal gland; SRG, suprarenal ganglion; NCC, neural crest cells; NC, neural crest; NT, neural tube; n, notochord; DA, dorsal aorta; DRG, dorsal root ganglion; SCPs, Schwann cell precursors.

From Furlan, A., Dyachuk, V., Kastriti, M.E., Calvo-Enrique, L., Abdo, H., Hadjab, S., et al., 2017. Multipotent peripheral glial cells generate neuroendocrine cells of the adrenal medulla. Science 357 (6346), eaal3753. Jul 7.

Peripheral nerves serve as niches and transportation routes for chromatin cell progenitors. Most adrenalin-producing chromaffin cells differentiate from an intermediate progenitor cell type derived from nerve-associated Schwann cell precursors. These precursors are directed to the adrenal primordium by the preganglionic axons of the neurons of the IML (intermediolateral cell column) of the spinal cord, while SEMA proteins act as guidance cues for the migrating chromaffin cell precursors (Furlan et al., 2017; Lumb et al., 2018) (Fig. 3.25).

9 Sensory organs of the vertebrate head

9.1 The head placodes

The preplacodal region (PPR) is a transitional thickened ectodermal structure that develops exclusively in the head region as a horseshoe-shaped structure adjacent to the neural plate and the neural crest (Singh and Groves, 2016; Alsina and Whitfield, 2017) and is absent from the trunk region. It develops in the interface between the neural and surface ectoderm and because of their interaction (Cheung et al., 2017). The anlagen of the vertebrate head, sensory organs, eyes, ears, olfactory sensory neurons and olfactory epithelium, anterior pituitary, etc. differentiate from head placodes (Fig. 3.26).

PPR evolved separately from the neural crest and before the onset of the neural crest in vertebrates (Šestak et al., 2013). Formation of the PPR may be induced from the same maternal factors that determine the formation of the neural plate (Baker and Bronner-Fraser, 2001), but signals from the neural plate and ectoderm may also be involved.

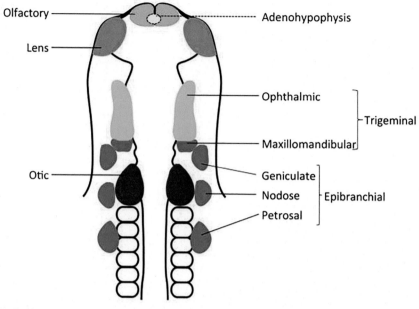

FIG. 3.26

Location of cranial placodes in the embryonic vertebrate head. Schematic representation of various types of cranial placodes in a 10-somite stage chick embryo. Individual placodes develop in morphologically distinct domains along the neural tube in the head region. The adenohypophyseal placode develops ventral to the forebrain and is indicated here with a *dotted line.*

Modified from Streit, A., 2004. Early development of the cranial sensory nervous system: from a common field to individual placodes. Dev Biol. 276, 1–15; From Singh, S., Groves, A.K., 2016. The molecular basis of craniofacial placode development. Wiley Interdiscip. Rev. Dev. Biol. 5 (3), 363–376.

9.2 Eye

Eye primordia develop from neural precursors of the eye field extending along the anterior neural plate (ANP) midline or the forebrain (the latter diencephalon and the dorsally developing telencephalon) and form by the signaling pathways that regionalize the development of the CNS (Cavodeassi and Houart, 2012). While the eye field core consists of mesenchymal cells, its margin differentiates into a layer of neuroepithelial cells under the action of a relatively small number of EFTFs (eye field transcription factors) (Bazin-Lopez et al., 2015) (Fig. 3.27), which ensure that the eye field remains.

Neuroepithelial organization of the basal cells of the eye field begins prior to the formation of optic vesicles and the neighboring neural tissues. In zebrafish, after the onset of the optic vesicles, secretion by the dorsal neural and nonneural tissues determines the regionalization of the nasal and temporal domains (Ivanovitch et al., 2013). The anterior neural plate (ANP) then folds to form two lateral bulges, the optic

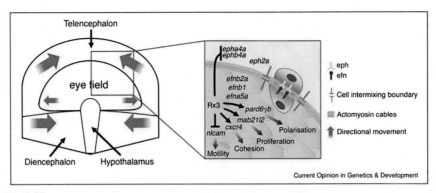

FIG. 3.27

Eye field cells have different behaviors to, and do not intermix with, cells in adjacent neural plate domains. (*Left*) Schematic representation of prospective forebrain territories at the neural plate stage highlighting the eye field-telencephalon boundary. Eye field cells start to evaginate laterally (*small green arrows*) at the same time that most anterior neural plate cells are still converging toward the midline (*large green arrows*). The *inset* highlights the eye field-telencephalon boundary: Rx3 regulates genes that influence cell behaviors in the eye field.

From Bazin-Lopez, N., Valdivia, L.E., Wilson, S.W., Gestri, G., 2015. Watching eyes take shape. Curr. Opin. Genet. Dev. 32, 73–79.

vesicles, whereas the other regions of the neural plate converge to form solid structures like the telencephalon and diencephalon. This takes place at the same time with the formation of the neural tube (Ivanovitch et al., 2013). Removal of the telencephalon (anterior part of the neural plate) before 16 hpf (hours postfertilization) in zebrafish (Moreno-Marmol et al., 2018) may prevent the formation of retina and eye (Rétaux and Casane, 2013).

The eye field shows apicobasal polarity earlier than neighboring neural structures, and the essential molecular components involved in the development of the polarity are proteins laminins and pard6γb, which are downstream elements of a signal cascade starting with transcription factor Rx3, which, along with the activation of the Eph/Ephrin pathway, are primarily responsible for the discreetness of the eye field from the surrounding tissues (Bazin-Lopez et al., 2015). Additionally, Rx3 maintains low levels of the Ig-domain protein Nlcam, thus facilitating the lateral migration of eye field cells and formation of optic vesicles (Brown et al., 2010), while high expression of neighboring neural structures, telencephalon and diencephalon, promotes medially convergent cell migration, leading to the formation of compact brain structures (Bazin-Lopez et al., 2015). Disruption of pard6γb and laminin-γ 1 prevents the formation of the optic vesicle (Ivanovitch et al., 2013).

The optic vesicle is surrounded by extracellular matrix components laminin, fibronectin, and collagen IV (Kwan, 2014). It soon comes in contact with, and apparently adheres to, the head ectoderm, leading to the formation of the crystalline lens and the double-layered optic cup via invagination (Hosseini et al., 2014). At the site

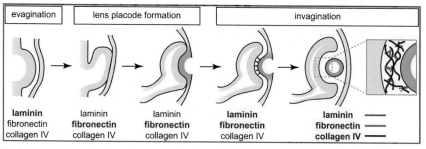

FIG. 3.28

Overview of optic cup morphogenesis and extracellular matrix components. In this schematic of a generic vertebrate, the optic vesicle undergoes a series of dramatic shape changes to form the optic cup: it first evaginates from the brain neuroepithelium, makes close contact with prelens ectoderm as the lens placode forms, and then invaginates with the lens to form the optic cup. Laminin, fibronectin, and collagen IV are all present throughout optic cup morphogenesis; when listed in *red*, it has been demonstrated to be functionally required during that stage of optic cup formation. (*Inset*) Closer view of the retina-lens interface at the end of optic cup morphogenesis. Note that ECM is present not just at the retina-lens interface but also surrounding the eye and lens structures throughout optic cup morphogenesis. *Blue*, laminin; *orange*, fibronectin; *purple*, collagen.

From Kwan, K.M., 2014. Coming into focus: the role of extracellular matrix in vertebrate optic cup morphogenesis. Dev. Dyn. 243, 1242–1248.

of the contact, the ectoderm thickens as a result of the deposition of the extracellular matrix rather than cell proliferation or cell size increase (Huang et al., 2011) (Fig. 3.28) and forms the crystalline lens. In the interaction of the prospective retina with the lens, essential is the FGF, which is presented by heparan sulfate (Kwan, 2014).

Development of the crystalline lens is associated with its invagination and the retina in a process that is controlled by the retina, which generates the necessary force through filopodia. This is proved by the fact that the retina invaginates even in the absence of the crystalline lens (Kwan, 2014). The complete development of the lens crystalline is followed by its separation of the overlaying ectoderm.

9.3 Inner ear

Development of the inner ear begins with the formation of the otic placode in the head region from the common preplacodal region (PPR) under the influence of FGF-8 and FGF-3 signals from the hindbrain rhombomere 4 (Leger and Brand, 2002). In chick, FGF-3 is expressed only in the ventral hindbrain adjacent to the otic anlage (Olaya-Sánchez et al., 2017). The otic placode then undergoes a process of closure, reminiscent of the process of the folding and closure of the neural plate

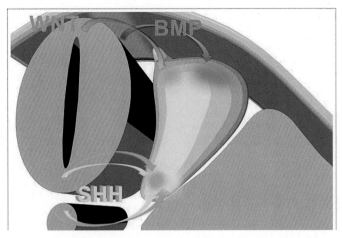

FIG. 3.29

Diagram of the secreted growth factors that pattern the otocyst. Two factors are known to act primarily on the dorsal otocyst during its early patterning. WNTs, secreted by the roof plate of the neural tube, act on the dorsomedial region of the otocyst, and BMPs, similarly secreted by the roof plate, but perhaps also by the cristae of the developing otocyst (not shown), act on the dorsolateral region of the otocyst. By contrast, SHH, secreted by the floor plate of the neural tube and the notochord, acts primarily on the ventral otocyst. Crosstalk occurs among these three signaling molecules to orchestrate the dorsoventral patterning of the otocyst.
From Ohta, S., Schoenwolf, G.C., 2018. Hearing crosstalk: the molecular conversation orchestrating inner ear dorsoventral patterning. Wiley Interdiscip. Rev. Dev. Biol. 7 (1), 10.1002/wdev.302. 2018 Jan.

in the development of the neural tube during the primary neurulation. This process of the folding of the otic placode results in the formation of the otocyst (Fig. 3.29) and the otic cup (Basch et al., 2016). Sonic hedgehog (Shh) signals from the neural plate and notochord are required for the ventral ear structure induction and differentiation of the auditory cells (Bok et al., 2005; Riccomagno et al., 2002), whereas hindbrain signals restrict Shh expression to the ventral and medial regions of the ear epithelium (Riccomagno et al., 2002).

Development of the otocyst and the otic cup requires signals from adjacent structures, the neural tube and notochord. The dorsal neural tube secretes Wnt, which regulates in a dose-dependent manner expression of transcription factors Dlx5 and Gbx2 in the dorsal otocyst, which are necessary for the development of the semicircular canals (Ohta and Schoenwolf, 2018). The ventral neural tube and notochord secrete Shh, which acts on the ventral otocyst, thus determining the dorsoventral polarity of the otocyst (Ohta and Schoenwolf, 2018). Wnt and Fgf3 gradients from the developing hindbrain (Fritzsch et al., 2002) help establish the medial and lateral identity of the otocyst and otic cup (Basch et al., 2016) (Fig. 3.30).

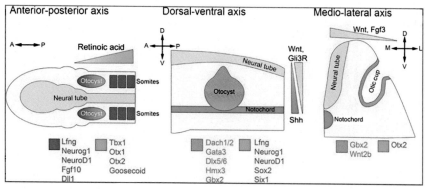

FIG. 3.30

Cardinal axis determination of the amniote inner ear. As the otic placode invaginates to form the otic cup and eventually closes to form the otocyst, the earliest axes established are the M–L axis and the A–P axis, with the D–V axis determined shortly afterward. A posterior source of RA provides a gradient that allows for the expression of posterior and anterior otic genes. The neural tube provides a source of Wnt to create a dorsalizing gradient, and these dorsalizing signals are augmented by a gradient of inhibitory Gli3R, while the notochord sets up a gradient of Shh, establishing a ventral character. During the otic placode and otic cup stages, Wnt and Fgf3 gradients from the neural tube help establish a medial and a lateral identity.

From Basch, M.L., Brown, II R.M., Jen, H-I., Groves, A.K., 2016. Where hearing starts: the development of the mammalian cochlea. J. Anat. 228 (2), 233–254.

10 Limb development

Inductive signals, Hh and FGF, from the midline structures, the neural tube and notochord, initiate a myogenic program in somites, but somitic cells remain undifferentiated. These cells are defined as migratory muscle progenitor cells when they begin delaminating from somites (Nord et al., 2019) and the ventral regions of dermomyotomes (Tani-Matsuhana et al., 2018) to settle in limb/fin bud where they form the dorsal and ventral premuscle masses. The nerve axons arrive at the limb bud concurrently with myogenic cells, and "the presence of nerve in the limb muscle mass at the time of onset of primary myogenesis, indicates that crosstalk between nerve and myogenic cells in mammalian embryos may occur much earlier than previously appreciated" (Hurren et al., 2015) (Fig. 3.31).

At the brachial and pelvic levels of the spinal cord, motoneurons that innervate limbs form lateral motor columns (LMCs) and the sites of the forelimb and hindlimb initiation are determined according to the expression patterns of, respectively, *Hox6* and *Hox10* genes in the spinal cord (Dasen et al., 2005).

Early experimental evidence indicated that the nervous system is substantially involved in limb development and regeneration. For instance, temporary denervation in embryos and adults of the frog *Rana pipiens* caused a decrease in limb size

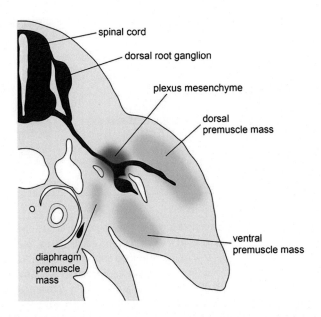

FIG. 3.31

Schematic drawing of the plexus mesenchyme in the E10.5 mouse. The plexus mesenchyme-expressing Gdnf is of lateral plate mesodermal origin and does not involve muscle precursor cells.

Based on Wright, D.E., Snider, W.D., 1996. Focal expression of glial cell line-derived neurotrophic factor in developing mouse limb bud. Cell Tissue Res. 286, 209–217. https://doi.org/10.1007/s004410050689; from Hirasawa, T., Kuratani, S., 2018. Evolution of the muscular system in tetrapod limbs. Zool. Lett. 4, 27. Published 2018 Sep 20. https://doi.org/10.1186/s40851-018-0110-2.

(Dietz, 1987, 1989), and the missing parts of fish fins regenerate in about 2 weeks, but denervation prevented blastema formation and regeneration (Nakatani et al., 2007).

The observation that denervation caused in limbs developmental defects that resembled those caused by the absence of retinoic acid (RA) led to the hypothesis of a causal relationship between the nervous system and RA in the tetrapod limb development. This appeared to be true when it was found that the nervous system was the main producer of RA during the early development, at the time of the development of the limb bud. It was also observed that the hot spots of the production of RA were cervical and lumbar regions of the CNS that correspond to the areas where the forelimbs and hindlimbs respectively develop. The hot spots, in all likelihood, are involved in determining the limb anlage zones: "The hot spots appear with the formation of the limbs and persist until limb innervation is about complete ... the hot spots are a likely factor in the formation of the limb zones" (McCaffery and Dräger, 1994). Not only spatially but temporally as well, the onset of the hotspots of the RA synthesis in the CNS coincides with the location of the limb buds and the spatial

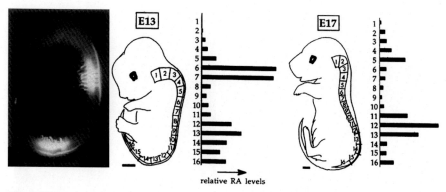

relative RA levels

FIG. 3.32

(*Left*) Fluorescent view of E12.8 spinal cord labeled with DiAsplO from the limbs (*Left*) to illustrate origins of limb innervations (E13 and E17, bars = 100 μm). (*Right*) Retinoic acid (RA) amounts released from 16 consecutive spinal cord samples of E13 and E17 mice cultured overnight. Tissue was dissected free from meninges and contained no dorsal root ganglia. Comparisons were done separately for each age. Only the developmental changes in spatial patterns should be noted (E13 and E17, bars = 100 μm).

From McCaffery, P., Dräger, U.C., 1994. Hot spots of retinoic acid synthesis in the developing spinal cord. Proc. Natl. Acad. Sci. U. S. A. 91, 7194–7197.

patterns and amounts of RA change over time, reflecting the temporal order of the development of the forelimbs and hindlimbs (Fig. 3.32). Thus, at the onset of the limb bud, the central nervous system is the primary producer of RA (Niederreither et al., 2002a,b), and its concentration gradually declines with the increased distance from the neural tube (somites to lateral plate) (Maden et al., 1998).

RA regulates the ordered expression of *Hox* genes along the AP axis during embryonic development in vertebrates. Initially, no RA is produced in the limb anlage; only when the motor axons and blood vessels invade the growing chick limb bud, the adjacent mesenchyme begins expressing RA-synthesizing enzyme, retinaldehyde dehydrogenase (RALDH2). Within the CNS, motor neurons are the primary source of RALDH2 (Ross et al., 2000), and RA production by the limb mesenchyme is at least partly under the control of the motor axons and vasculature (Berggren et al., 2001). This is verified with experiments on chick forelimb (wing) buds where motor denervation leads to a drastic decrease of RALDH2 in the limb mesenchyme (Fig. 3.33). Denervation of calf muscle in adult mice increases myogenin and MyoD expression to levels that are, respectively, 40- and 15-fold higher than those found in the innervated muscle (Eftimie et al., 1991), and denervation in young rats increased myogenin protein 14-fold above the control (Kostrominova et al., 2000).

Initially, the forelimb bud (primordial wing) in chick embryos does not produce RA, but the motor neurons in the adjacent brachial plexus do (Berggren et al., 1999). Later, motor neurons invading the chick developing bud begin synthesis of RA/RALDH2. Motor neurons release RALDH2 (RA-synthesizing enzyme), inducing

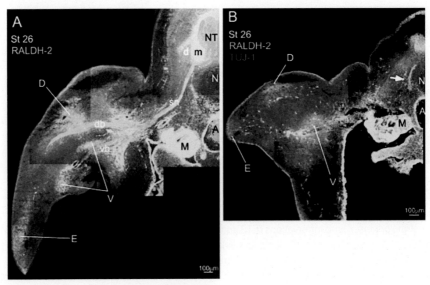

FIG. 3.33

Decrease in wing mesenchymal retinaldehyde dehydrogenase type 2 immunoreactivity (RALDH-2-IR) seen after motor denervation. (A) Stage (St) 26 control wing section. Dorsal (D), ventral (V), and distal (E) mesenchymal RALDH-2-IR is seen in relation to the dorsal (db) and ventral (vb) nerve branches. Abbreviations: NT, neural tube; m, motor neurons; d, dorsal root ganglia; sn, spinal nerve; A, aorta; M, mesonephros. (B) Decrease in the area of RALDH-2-IR after motor denervation in a stage 26 embryo that had been denervated at stage 16. Sensory nerves stained with TUJ-1 are present in axial regions (TUJ-1 stain, short arrow) but very sparse in the wing.

From Berggren, K., Ezerman, E.B., McCaffery, P., Forehand, C.J., 2001. Expression and regulation of the retinoic acid synthetic enzyme RALDH-2 in the embryonic chicken wing. Dev. Dyn. 222, 1–16.

muscle cell differentiation and muscle formation, and the development of muscle fiber types in chicks coincides with the penetration of nerves in the limb bud (Berggren et al., 2001; Hirasawa and Kuratani, 2018). Also, the coculture of motor neurons with RA-responsible cells showed that cells in contact with only the processes of motor neurons responded, demonstrating that neuronal axons were the source of RA in culture (Berggren et al., 2001). Hence, "RA is a factor that could mediate this neuronal influence. Between stages 23 and 30, localized regions of RALDH-2 expression develop in the mesenchyme along the vasculature and nerve branches; this RALDH-2 is partially under the control of the blood vessels and motor axons as they enter the wing" (Berggren et al., 2001).

The instructive role of RA and the RA-inducing innervation in the development of limb buds was demonstrated in experiments in which cultures of developing chick limb buds treated with citral, an inhibitor of RA formation, upon grafting to the stumps of host embryos, led to the development of defective wings (Tanaka et al., 1996).

The limb-inducing activity of RA is mediated by *Hox* genes, which in turn induce secretion of SHHs, FGFs, and formation of the zone of polarizing activity (ZPA) and apical ectodermal ridge (AER) (Zakany and Duboule, 2007; Gillis et al., 2009).

It is noteworthy that by the early 1960s, the use of thalidomide as a cure against cancer in breastfeeding and pregnant women led to serious defects in the embryos of the treated patients: defects in the development of the sensory nerves, especially axon degeneration, and abnormal cell proliferation in undifferentiated mesenchyme reflected morphologically in dysmelia, characterized among other things by shortening or even lack of limbs. Thalidomide is a sensory toxin, and the distribution of limb defects coincides exactly with the segmental sensory supply (Soper et al., 2019); hence, "dysmelic deformities of the limbs are secondary to toxic embryonic neuropathy" (McCredie et al., 1984).

The role of the nervous system in the development of limbs in tetrapods would be incomplete if one would neglect the neurohormonal activity of the CNS during the latter stages of limb development. Signal cascades starting in the CNS are involved in the formation and specification of limb tissues. So, for example, a cascade starting with the secretion of GHRH (growth hormone-releasing hormone), via the pituitary GH (growth hormone), stimulates the proliferation of prechondrocytes in the epiphyseal growth plates of developing bones (Wolpert et al., 1998).

Further corroborating evidence on the inductive activity of the nervous system in limb development comes from studies on the influence of the denervation in limb regeneration. The nerve-dependent nature of appendage regeneration is demonstrated in echinoderms such as starfish, amphibians, and fish (Li et al., 2019). In fish, denervation performed at the early stages of fin regeneration prevents regeneration as a result of inhibition of blastema formation (Nakatani et al., 2007). In zebrafish, denervation prevents the formation of a functional AEC (apical epithelial cap) and blastema, leading to an impaired regeneration. The investigators believe that the role of the innervation in fin regeneration is to release in the stump a "factor x," which induces expression of *Fgf*, *Wnt*, *Shh*, *lef1*, and *krt8* genes at levels appropriate for complete regeneration (Simões et al., 2014) (Fig. 3.34). Denervation greatly reduces the amount of RALDH2 compared with controls (Berggren et al., 2001). It also leads to a rapid and substantial increase in muscle regulatory factors. So, e.g., unilateral denervation of rat hindlimb muscles leads to rapid 150–200-fold higher levels of myogenin than the innervated contralateral hindlimb (Voytik et al., 1993).

Later studies showed that the instructive "factor x" was a well-known mitogenic substance secreted by neurons, neuregulin1 (NRG1), whose level declines drastically in the blastemas of the denervated limbs. The mediator of NRG1's effect in limb regeneration is its receptor Erb2. Administration of NRG1 in denervated axolotl limbs 6 days after amputation, when inflammation has ended, supported blastema growth and regeneration up to the point of digit formation (Farkas et al., 2016; Farkas and Monaghan, 2017).

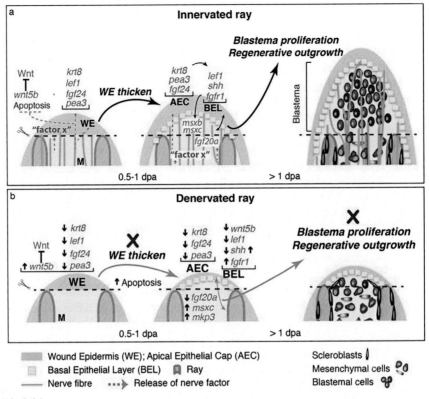

FIG. 3.34

Contribution of innervation to fin regeneration. (A) Illustration of the putative role of nerves in zebrafish fin regeneration. The WE (wound epidermis) is formed after fin amputation in a process that is independent of nerves. Innervation may be important to the subsequent thickening of the WE and the establishment of the AEC, which contributes to the formation and outgrowth of the blastema and to the progression of the regenerative process. The role of nerves may be to release factor(s) ("factor x") that regulate the expression of target genes in the WE, such as pea3, fgf24, and lef1, which are important to the thickening and maintenance of the WE and to the communication established with the underlying cells. At the same time, nerves may be involved in the inhibition of wnt5b in the WE, as well as in the pathways that lead to apoptosis. "Factor x" may also be directly released by nerves into the stump to induce cell proliferation and dedifferentiation. (B) Illustration of the cellular and molecular changes that occurred upon fin denervation. In the absence of innervation, fins do not establish a functional AEC, and several signaling pathways are affected. Wnt5b, an inhibitor of Wnt signaling and regeneration, is upregulated at 0.5 dpa but is downregulated at 1 dpa. During this period, krt8 and pea3, which are essential to WE maintenance, as well as fgf24 and lef1, important for the communication established with the underlying cells, are downregulated. At the same time, apoptosis activity is increased in the WE. The Fgf signaling molecules fgfr1, msxc, and mkp3 are upregulated, while fgf20a is downregulated. Shh is early downregulated at 0.5 dpa but is then upregulated at 1 dpa. These signaling defects result in a breakdown of communication with mesenchymal cells and impairment of the formation of the blastema and fin regeneration.

From Simões, M.G., Bensimon-Brito, A., Fonseca, M., Farinho, A., Valério, F., Sousa, S., et al., 2014.
Denervation impairs regeneration of amputated zebrafish fins. BMC Dev. Biol. 14, 49.

11 Heart

Heart muscle cells, cardiomyocytes, stop cell cycle and proliferation shortly after birth. This transition depends on the sympathetic innervation. During embryonic development, sympathetic innervation plays a continuous but changing role in determining the size of cardiomyocytes, and sympathetic denervation leads to increased *Meis1* expression, indicating its role in the regulation of cardiac myocyte proliferation and, consequently, the cardiac myocyte number and the heart muscle size and structure. Lesioning the sympathetic innervation in cardiac myocyte cultures delays their cell cycle withdrawal time. The action of the sympathetic innervation in the heart muscle is mediated by adrenergic receptors (Kreipke and Birren, 2015).

Solid evidence has shown that the neurotransmitter serotonin, 5-hydroxytryptamine (5-HT), is another regulator of cardiomyocyte differentiation and proliferation. It stimulates differentiation and proliferation of cardiac myocytes during embryonic development and regulates and maintains the cardiac structure in adults (Nebigil et al., 2000; Nebigil and Maroteaux, 2001). The mediator of the action of serotonin is its receptor 5-HT(2B)R, which uses the pathway of the tyrosine kinase receptor ErbB-2 in the ventricular myocardium, making the possible proliferation of differentiated myocytes (Nebigil and Maroteaux, 2001). Inactivation of the gene for serotonin receptor 5-HT(2B) causes embryonic and neonatal death because of heart defects. Among defects observed in the surviving mice are ventricular hypoplasia as a result of the reduced proliferative capacity of myocytes and dilation (because of the reduced contractility and structural deficits at the intercellular junctions between cardiac myocytes), lack of heart trabeculae, and several histopathological changes (Nebigil et al., 2000).

In the heart, serotonin stimulates DNA synthesis and division of valve interstitial cells (VICs) not only via serotonergic signaling but also by potentiating TGFβ signaling (Buskohl et al., 2012; Goldberg et al., 2017) (Fig. 3.35). Overexpression of serotonin receptor (5-HT2B) leads to cardiac hypertrophy and proliferation of mitochondria (Nebigil et al., 2003).

Neonatal mammalians, as opposed to adult mammalians, exhibit an innate regenerative response to heart injury, but neonatal mice sympathectomized with the neurotoxin 6-HDOA (6-hydroxydopamine) show only a defective repair response by developing instead a scar and adipocyte deposition in the cardiac muscle. This clearly demonstrates the critical role of sympathetic innervation in neonatal cardiac muscle regeneration (White et al., 2015). Similarly, inactivation of the parasympathetic branch leads to impairment of the regenerative potential of the neonatal heart muscle: the mechanical denervation of the left vagus nerve in neonatal mice cripples cardiomyocyte proliferation and prevents regeneration of the resected cardiac apex (tip of the left ventricle). A similar reduction of the proliferative potential of cardiomyocytes and neonatal heart regeneration is observed in neonatal mice and zebrafish after pharmacological inhibition of the cholinergic nerve function. It is demonstrated that nerves are required for cardiomyocyte proliferation and heart regeneration of both zebrafish and neonatal mice (Mahmoud et al., 2015).

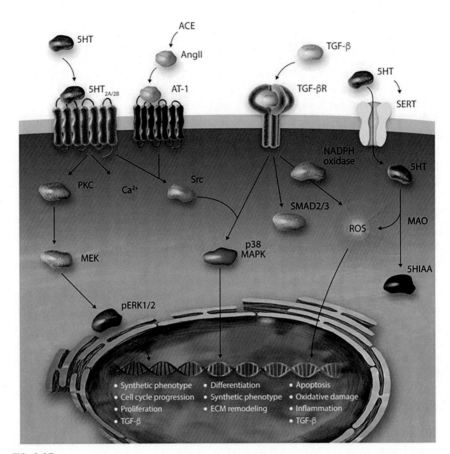

FIG. 3.35

Schematic representation of serotonergic signaling in valve interstitial cells (VICs). Abbreviations: 5HIAA, 5-hydroxyindoleacetic acid; 5HT, serotonin; ACE, angiotensin-converting enzyme; AngII, angiotensin II; AT-1, angiotensin receptor 1; MAPK, mitogen-activated protein kinase; MEK, mitogen-activated protein kinase; pERK1/2, phosphorylated p42/44 mitogen-activated protein kinases; PKC, protein kinase C; ROS, reactive oxygen species; SERT, serotonin transporter; MAO, monoamine oxidase.

From Goldberg, E., Grau, J.B., Fortier, J.H., Salvati, E., Levy, R.J., Ferrari, G., 2017. Serotonin and catecholamines in the development and progression of heart valve diseases. Cardiovasc. Res. 113 (8), 849–857.

Sympathetic denervation by administration of the neurotoxin (6-OHDA), while the parasympathetic branch was left intact, in rats caused pathological changes, inflammatory cell infiltration, and increased collagen deposition in the heart. However, coadministration of 6-OHDA and the protectant of the sympathetic myocardial nerves, mecobalamin, alleviated myocardial damage (Jiang et al., 2015). The role of the peripheral innervation in heart myogenesis is also corroborated by studies on mammalian heart regeneration that demonstrate the critical role of both its sympathetic and parasympathetic branches in cardiac regeneration and homeostasis.

12 Development of airways

The gas exchange in *Drosophila*, like most insects, is carried out by the tracheal system, a network of tubules delivering oxygen into cells throughout the insect body. In *Drosophila*, the tracheal development starts at the embryonic stage 10 with the formation of 10 pairs of tracheal placodes on both sides of the body, between the second thoracal and eighth abdominal segments (Hayashi and Kondo, 2018). Formation of the tracheal system in *Drosophila*, as a special case of tubulogenesis, requires Dalmatian signals from the fly's nervous system (Kerman and Andrew, 2006). Evolution of the airways system is conserved from invertebrates to vertebrates, and its development is under neural control (Bower et al., 2014).

The central nervous system controls the development and remodeling of the tracheal system by inducing differentiation of tracheal components both systemically and via local neurons in response to the nutrient and body oxygenation state (Linneweber et al., 2014). Selective ablation of nerves in *Drosophila* embryos by inducing expression of the RicinA toxin disturbs proper migration of tracheoblasts and elaboration of the tracheal airways in tubular segments as a result of the loss of the neuronal guidance (Bower et al., 2014).

Among the factors involved in tracheal branching in *Drosophila*, the best known and apparently the most important is hypoxia, low oxygen level in tissues, which leads to an interesting coupling of nutrition with the tracheal remodeling (Linneweber et al., 2014; Marxreiter and Thummel, 2014). Under both hypoxia and appropriate nutrient supply, eight brain insulin-producing neurons increase production of Ilps (insulin-like peptides) and the pigment-dispersing factor (Pdf), which is similar to the vertebrate vasoactive intestinal peptide (VIP), inducing fly's tracheal remodeling (Fig. 3.36). Specific deletion of neurons in *Drosophila* embryos also perturbs airway development (Bower et al., 2014).

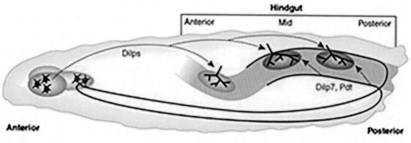

FIG. 3.36

Cells in the brain, represented by *blue circles*, secrete Dilps in response to nutrients. Two nerves from the ventral ganglion of the brain, shown in *red*, extend to the intestine and innervate the mid and posterior regions of the hindgut. These nerves release localized Dilp7 and Pdf (*red arrows*) that, in combination with secreted Dilps, modulate tracheal plasticity in the intestine in response to nutrients.

From Marxreiter, S., Thummel, C.S., 2014. Will branch for food–nutrient-dependent tracheal remodeling in Drosophila branching. EMBO J. 33 (3), 179–180.

The lung anlage in mammalians develops as an outgrowth of the endoderm into the adjacent mesenchyme and is composed of pluripotent cells. Pulmonary neuroendocrine cells (PNECs) are the first differentiated cells that appear in the anlage (Pan et al., 2006; Linnoila, 2006; Cutz et al., 2008) of the primitive airway epithelium around E16 in mice (Warburton et al., 2010). PNECs are hypoxia-sensitive cells found as individual cells and in clusters of about 30 cells, with the latter being innervated (Tata and Rajagopal, 2017; Nikolić et al., 2018; Whitsett et al., 2019), suggesting that they may be involved in airway branching.

Airway branching is neurally guided and requires intact local innervation, but neurotransmitter signaling is dispensable (Bower et al., 2014). One-sided selective denervation of the cultured left lung by laser ablation of neurons (94% of neurons destroyed) halted lung branching and resulted in 30% reduction of cell proliferation and by two-thirds of the number of endothelial cells, while the budding proceeded normally in the right intact cultured lung, demonstrating that the lung innervation regulates airway branching as well as epithelial and endothelial proliferation (Bower et al., 2014). Given the proximity of nerve terminals to nascent epithelial buds, "neural impacts on airway branching may be mediated via secreted growth factors, structural cues provided by the nerves themselves, or signaling directly between neurons and epithelium or augmented by an intermediary tissue, such as the vasculature" (Bower et al., 2014) (Fig. 3.37).

13 Development of the submandibular gland

Three paired salivary glands, the submandibular gland (SMG), sublingual gland (SLG), and parotid gland (PG), produce 90% of the saliva secreted in the oral cavity. Saliva production is controlled by the local parasympathetic innervation. During development, the parasympathetic ganglion innervation associates with the SMG duct and occurs in parallel with the epithelial branching (Coughlin, 1975; Ferreira and Hoffman, 2013). Recently, it was observed that acetylcholine (Ach) released by parasympathetic nerves, via muscarinic M1 receptor and by increasing secretion of epidermal growth factor (EGFR), induces expansion of keratin 5-positive (K5+) progenitor cells in SMG and is believed to regulate differentiation and maturation of SMG ductal progenitors. The parasympathetic innervation of the submandibular gland maintains K5+ progenitors (Knox et al., 2010; Szymaniak et al., 2017), whereas blockade of the parasympathetic mandibular ganglion inhibits the development of the SMG epithelium and decreases the number of end buds (Knox et al., 2010).

The instructive role of the parasympathetic innervation in the development of the SMG is also suggested by the fact that parasympathectomy prevents regeneration of injured salivary glands. The therapeutic irradiation injury diminishes parasympathetic function in the gland, induces apoptosis of epithelial cells, and reduces neurturin (NRTN) production, leading to increased neuronal apoptosis (neurturin secretion is induced by the neurohormone vasoactive intestinal peptide (VIP) from

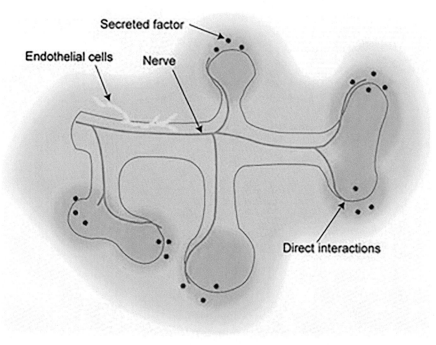

FIG. 3.37

Diagram of possible mechanisms by which nerves may regulate airway branching. Diagram of possible mechanisms by which nerves may regulate airway branching, including via secreted growth factors, direct cellular interactions, or intermediary interactions with other tissues.

From Bower, D.V., Lee, H.-K., Lansford, R., Zinn, K., Warburton, D., Fraser, S.E., et al., 2014. Airway branching has conserved needs for local parasympathetic innervation but not neurotransmission. BMC Biol. 12, 92.

the parasympathetic innervation) (Fig. 3.38). In cases of irradiation injury, treatment with neurturin promotes regeneration of the gland by stimulating neuron survival and function, reducing neuronal apoptosis, and stimulating epithelial regeneration (Knox et al., 2013). Irradiation of the salivary gland impairs the parasympathetic function, leading to epithelial apoptosis and neuronal apoptosis, but administration of glial cell line-derived neurotrophic (GDNF) factors restores the parasympathetic function, resulting in increased epithelial regeneration (Knox et al., 2013).

14 Liver development

In humans, innervation of the embryo begins by the eighth gestational week (Tiniakos et al., 2008; Terada, 2015). To the best of my knowledge, there is no evidence on neural control of the development of the liver, but this is not surprising given the fact that no experiments to investigate the possibility have been performed.

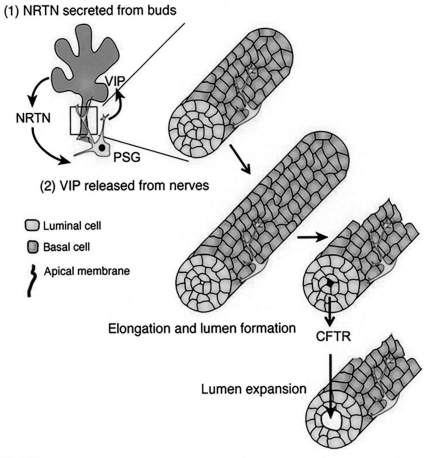

(1) NRTN secreted from buds

VIP

NRTN

PSG

(2) VIP released from nerves

☐ Luminal cell
◐ Basal cell
} Apical membrane

Elongation and lumen formation

CFTR

Lumen expansion

FIG. 3.38

Schematic of induction of the ductal tubulogenesis in the submandibular gland. Abbreviations: CFTR, cystic fibrosis transmembrane conductance regulator; NRTN, neurotrophic factor neurturin; PSG, parasympathetic submandibular ganglion; VIP, vasoactive intestinal peptide.

From Nedvetsky, P.I., Emmerson, E., Finley, J.K., Ettinger, A., Cruz-Pacheco, N., Prochazka, J., et al., 2014. Parasympathetic innervation regulates tubulogenesis in the developing salivary gland. Dev. Cell 30, 449–462.

Being the only organ capable of complete regrowth after 70% removal (Preziosi and Monga, 2017), liver regeneration is extensively studied; hence, the wealth of information accumulated on the mechanisms of regulation of liver regeneration may provide important clues about the developmental pathways of liver development.

Neural mechanisms involved in liver regeneration comprise antagonistic actions of the parasympathetic and sympathetic nervous system and the central neurohormonal mechanisms.

In rats the hepatic branch of the vagus nerve stimulates activation of hepatic progenitor cells and proliferation of hepatocytes after partial hepatectomy; it also stimulates the growth of bile duct epithelial cells. These parasympathetic actions are mediated by the binding of acetylcholine to the muscarinic acetylcholine receptor type 3 (M3 receptor) of these cells (Cassiman et al., 2002). The parasympathetic nervous system promotes the accumulation of hepatic progenitor cells and reduces liver damage caused by drug-induced hepatitis. Similar results produced inhibition of the sympathetic nervous system in cases of damages caused by chronic hepatotoxin exposure (Oben et al., 2003).

The sympathetic liver denervation reduces or reverses obesity-induced hepatic steatosis (excessive fat in the liver) (Hurr et al., 2019), whereas SNS (sympathetic nervous system) inhibition promotes the accumulation of hepatic oval cells (HOCs) and reduces liver damage (Oben and Diehl, 2004; Li et al., 2004) (Fig. 3.39). SNS stimulation promotes hepatic steatosis and hepatic fibrosis, whereas inhibition of SNS improves steatosis (Amir et al., 2020).

Administration of human growth hormone (hGH) after partial hepatectomy increases hepatocyte mitotic activity, leading to increased liver mass in rats (Abu Rmilah et al., 2020). Norepinephrine is a neurotransmitter and a hormone secreted

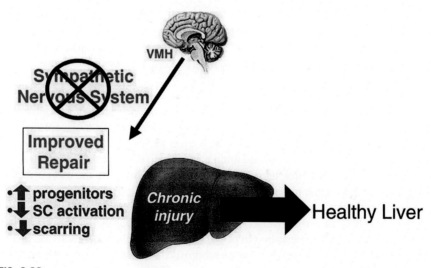

FIG. 3.39

Schematic representation of the effect of SNS inhibition on liver repair via SNS regulation of HSCs (hepatic stellate cells) and OCs (oval cells). SNS inhibition either at its origin in the hypothalamic ventral medial nucleus (VMH) or peripherally will lead to a reduction of HSC activation, an increase of progenitor cell numbers, a consequent reduction of scarring, and perhaps return of the injured liver to health.

From Oben, J.A., Diehl, A.M., 2004. Sympathetic nervous system regulation of liver repair. Anat. Rec. 280A, 874–883.

by adrenal glands, locus coeruleus nuclei in the pons, and from postganglionic sympathetic neurons (Abu Rmilah et al., 2020). Its binding to α1- and β-adrenergic receptors increases the mitogenic activity of hepatocytes and general activation of these receptors (Wen et al., 2016; Abu Rmilah et al., 2020).

15 Kidney

By the middle of the last century, Clifford Grobstein demonstrated that the mouse embryonic spinal cord separated by a cell-nonpenetrable filter membrane induced differentiation of the metanephric mesenchyme into epithelialized nephrons (Grobstein, 1953) and formation of renal tubules in the mesenchyme (Grobstein, 1956). In the early 1990s, it was found that at the level of gene expression, the mesenchymal-epithelial transition is related with the expression of several Wnt (Wnt-1, Wnt-3, Wnt-4, and Wnt-7b) glycoproteins in the spinal cord that stimulate expression of Wnt-4 in the metanephric mesenchyme (Fig. 3.40).

Later, it was reported that the ureter bud can induce the formation of kidney tubuli, but soon became clear that it was a relatively weak inducer, whereas a dorsal part of the embryonic spinal cord was a very potent one (Vainio et al., 1999). Even the weak inductive effect of the ureter bud in kidney differentiation was tentatively explained with the possible presence of mesenchymal neurons that were left in the ureter bud during microsurgical separation, as suggested by the terminal endings of neurites found around the condensates of the mesenchyme (Sariola et al., 1988).

Metanephrogenic mesenchyme from the mouse E11.5 in culture with the homologous spinal cord, or in contact with Wnt-4, forms kidney tubules and glomeruli (Kispert et al., 1998). The spinal cord also induces cultures of the metanephric ridge to differentiate into kidney epithelial cells (Karp et al., 1994). It has been concluded that "only embryonic spinal cord and brain were effective, whereas the ureter bud did not induce …. These studies suggest that embryonic neurons are the most effective inducers of nephrogenic mesenchyme in vitro" (Sariola et al., 1989).

An inbred mouse strain displays a high incidence of uni- and bilateral kidney agenesis resulting in early postnatal death. It was found that the condition resulted from the failure of the ureteric bud to penetrate the metanephric mesenchyme (hence the name FUBI of the strain), leading to apoptosis of the metanephric cells and kidney agenesis. Comparative studies of coculturing FUBI metanephros with homologous spinal cord stimulated their differentiation into epithelial kidney cells, while coculture of metanephros with homologous ureteric bud frequently failed to induce differentiation of the metanephros cells (Kamba et al., 2001).

When embryoid bodies composed of embryonic stem cells and spinal cords were transplanted beneath the kidney capsule, they underwent massive tubulogenesis and many blood vessels had integrated into the transplanted tissue and the new glomeruli (Taguchi et al., 2014; Taguchi and Nishinakamura, 2017).

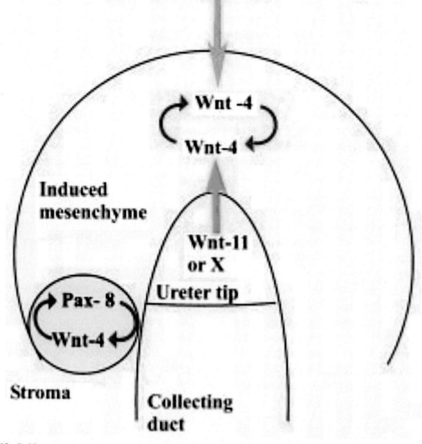

FIG. 3.40

A model showing how Wnt-4 operates as "a second" tubule inductive signal. A ureter-derived signal, in the form of Wnt-11 or X, leads to induction of expression of Wnt-4, which autoregulates itself and triggers tubule morphogenesis. Wnt-4 signaling involves Pax 8. The spinal cord acts as an inducer tissue as it expresses a panel of Wnts, including Wnt-4, which triggers the autoregulated Wnt-4 gene expression to induce tubules.

From Vainio, S.J., Itäranta, P.V., Peräsaari, J.P., Uusitalo, M.S., 1999. Wnts as kidney tubule inducing factors.
Int. J. Dev. Biol. 43, 419–423.

16 Gonad development

16.1 In planarians

As hermaphroditic organisms, planarians have ovary and testes as well as accessory reproductive organs, but they lack true genital organs. The onset of the sexual maturation and development of reproductive structures in these worms depends on signals coming from neurons of the brain and ventral nerve cord in the form of neuropeptide NPY-8. Elevated secretion of NPY-8 activates its receptor NPYR-1, and the fact that NPYR-1 knockdown planarians display a phenotype that is similar to *npy-8* gene knockdown proves that it is the mediator of the reproductive functions of the neuropeptide (Saberi et al., 2016; Collins III et al., 2010). In corroboration of the above conclusion comes the experimental fact that extirpation of the brain and nerve cords causes testes atrophy (Collins III et al., 2010). The neuropeptide NPY-8 also regulates the formation of accessory reproductive organs (seminal vesicles, spermiducts, oviducts, etc.) (Collins III et al., 2010).

The development of gonads in planarians is neurally induced, and germ cells differentiate from neoblasts (planarian stem cells) in response to neural signals. Asexually reproducing planarians detach part of their body (tail). In this case, secretion of the neurohormone NPY-8 is suppressed, gonads do not develop, and germ cells do not differentiate (Collins et al., 2015) (Fig. 3.41). Investigators have emphasized the similarity of neoblasts and germ cells, as far as the gene expression is concerned, and apparently, the most important difference between them is that the latter express *nanos*.

A negative feedback mechanism is discovered in planaria that regulates the brain–body proportions (Fig. 3.42). The mechanism consists of a set point in the brain size where wnt11–6 through β-catenin-1 activates *notum*. In turn, notum secretion inhibits wnt11–6 so that it cannot induce differentiation of neoblasts into neurons. Wnt11–6 prevents the differentiation of neoblasts into neurons in a β-catenin-1-independent mode (Hill and Petersen, 2015). This way, the relative levels of wnt11–6 and notum determine the number of neoblasts that must differentiate into brain neurons.

16.2 In Drosophila

A recent publication revealed the existence in *Drosophila melanogaster* of a brain-gonad axis, involving two types of brain neurons, tyramine/octopamine neurons (OPNs) and insulin-producing neurons (IPCs) (Dhiman et al., 2019). A subset of ~137 OPNs in the *Drosophila* brain produce a highly conserved endocytic protein, Monensin sensitive 1 (Mon1), which stimulates ovarian development. It was observed that the downregulation of mon1 in OPNs leads to delays in the development of the fly's egg chambers. *mon1* mutants are characterized by the very small ovary, ovarian defects, absence of egg chambers, and reduced secretion of

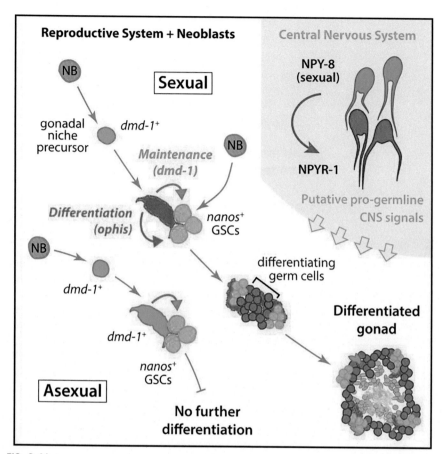

FIG. 3.41

Schematic of the developmental mechanisms involved in planarian testis formation. Cells in both sexual and asexual worms are required for specification of nanos GSCs. In sexual planarians, these cells express *ophis*, which is required for further differentiation of GSCs into mature gametes. NPY-8 signaling, which occurs in the CNS, systematically promotes latter stages of germ cell maturation. Abbreviations: NB, neuroblasts; GSCs, germline stem cells.

From Saberi, A., Jamal, A., Beets, I., Schoofs, L., Newmark, P.A., 2016. GPCRs direct germline development and somatic gonad function in planarians. PLoS Biol. 14 (5), e1002457. 2016 May.

neurohormonal insulin-like peptides (IPCs). On the contrary, induction of the *mon1* expression in OPNs rescues the mutant phenotype and increases ILP production. All this speaks about close cooperation between both neuron types, between OPNs and IPCs, in the control and regulation of the ovary development in *Drosophila* (Dhiman et al., 2019).

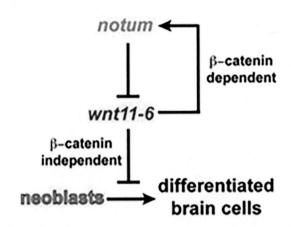

FIG. 3.42

Notum/wnt11–6 feedback regulation dampens brain cell differentiation to achieve proper brain–body scaling. Model for regulatory pathway influencing brain size. wnt11–6 inhibits neoblast production of differentiated brain cells (including cintillo+, gad+, trpA+, and tph+ neurons) by suppressing formation or division of neural progenitors (ap2+ smedwi-1+, coe+ smedwi-1+, and lhx1/5–1+ smedwi-1+ cells) from pluripotent neoblasts. wnt11–6 signals directly or indirectly through β-catenin-dependent canonical Wnt signaling activate expression of notum in neurons at the anterior brain commissure. wnt11–6 is likely to signal independently of beta-catenin-1 in the control of neoblasts to suppress brain size. Notum encodes a secreted protein that inhibits wnt11–6 function to promote the ongoing synthesis of brain cells from neoblasts. Levels of wnt11–6 and notum signaling control numbers of neoblasts committing to brain cell fates to influence the size of the brain.

From Hill, E.M., Petersen, C.P., 2015. Wnt/Notum spatial feedback inhibition controls neoblast differentiation to regulate reversible growth of the planarian brain. Development 142, 4217–4229.

17 Frontonasal morphogenesis

The cellular basis of the development of the vertebrate face is laid by the cranial neural crest (CNC) cells that populate the frontal part of the head (Kaucka et al., 2018), giving rise to facial mesenchymal cells subsequently differentiating into chondrocytes that shape the nasal capsule, the facial skeleton, and the overall facial shape (Kaucka et al., 2017).

A signaling center regulating the growth and patterning of the upper jaw, generally known as the frontonasal ectodermal zone (FEZ), was identified in chick embryos (Hu et al., 2003) and mammalian embryos (Hu and Marcucio, 2009). The FEZ regulates morphogenesis of the distal tip of the upper jaw in vertebrates (Hu et al., 2003) by releasing into the ectomesenchyme several secreted proteins, primarily SHH, FGF8, BMPs (bone morphogenetic proteins), etc. It was also observed that widely different morphologies of the facial shape in birds and mammalians were a result of differences in their FEZs. So, e.g., birds have only a single

FEZ spanning the FNP (frontonasal process), but in mice, each of the median nasal processes has a FEZ, which explains the widely different facial appearances of the birds and mammalians. Suppression of the *Shh* expression in the brain prevents FEZ formation (Marcucio et al., 2005, 2015).

In parallel with the recognition of FEZ in frontonasal morphogenesis, it was reported that the expression of the Shh in the frontonasal process was induced by Shh signals coming from the brain. Experimental inhibition of Shh expression in chick neuroectoderm disturbed gene expression patterns in the forebrain and the adjacent mesenchyme, leading to the development of truncated upper beaks (Marcucio et al., 2005), whereas activation of the Shh pathway in the chick embryo brain leads within 72 h to the development of a drastically shorter mammalian-like upper beak, demonstrating that brain signals induce Shh expression in FEZ and the resulting change in the frontonasal process expression. Thus, the brain signals represent the ultimate cause of the appearance of the mammalian face phenotype (Hu and Marcucio, 2009) (see Fig. 2.19).

A simplified signal cascade for the development of the upper beak anlage in chicks (Marcucio et al., 2011) would look as follows:

Shh signaling from the diencephalon → Shh expression in telencephalon
→ Shh expression in the facial ectoderm
(upon arrival of the cranial neural crest cells)
→ Development of the upper jaw anlage.

Recent evidence shows that Shh signals reach the facial ectoderm even earlier from the floorplate of the nervous system, as demonstrated by evidence that Shh selective ablation from the floorplate leads to the loss of the nasal septum cartilage (Kaucka et al., 2018).

17.1 Teeth

Tooth development is related to the activation of a gene regulatory network (GRN) comprising a set of more than 50 genes (including 12 transcription factors) expressed in the enamel knots of the odontogenic epithelium (Salazar-Ciudad and Jernvall, 2002). However, what triggers activation of the odontogenic GRN is not known.

Accumulating evidence during the last three decades indicates that local innervation is indispensable for tooth development, leading to the idea that innervation may have an instructive role in tooth development. In the early 1990s, Tuisku and Hildebrand (1994) observed that in cichlid fish, *Tilapia mariae*, nerve branches are seen just under the odontogenic epithelium and reach individual tooth primordia before the onset of tooth formation, whereas denervation arrests de novo formation of replacement tooth germs in the lower jaw of the fish. This led investigators to the hypothesis that mandibular innervation by the trigeminal nerve fibers "may have a primary initiating or instructive role in early odontogenesis" (Tuisku and Hildebrand, 1994). Similarly, Luukko (1997), who previously denied the essential

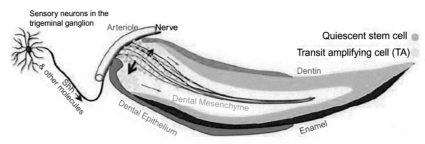

FIG. 3.43

Schematic diagram of our model of the NVB (neurovascular bundle) niche and the in vivo origin of incisor MSCs. The NVB provides a niche to support the continuous turnover of incisor mesenchyme. Sensory neurons in the trigeminal ganglion produce Shh, which is transported through axons into the incisor mesenchyme. Shh activates Gli1 expression in the quiescent stem cells surrounding the arterioles near the cervical loop region and regulates the odontogenic differentiation process. These quiescent stem cells continuously give rise to actively dividing TA (transit-amplifying) cells, which then differentiate into odontoblasts and all other dental mesenchymal derivatives to support the rapid cellular turnover of the incisor.

From Zhao, H., Feng, J., Seidel, K., Shi, S., Klein, O., Sharpe, P., et al., 2014a. Secretion of Shh by a neurovascular bundle niche supports mesenchymal stem cell homeostasis in the adult mouse incisor. Cell Stem Cell 14 (2), 160–173. https://doi.org/10.1016/j.stem.2013.12.013; Zhao, H., Feng, J., Seidel, K., Shi, S., Klein, O., Sharpe, P., et al. Secretion of shh by a neurovascular bundle niche supports mesenchymal stem cell homeostasis in the adult mouse incisor [published correction appears in Cell Stem Cell. 2018 Jul 05; 23 (1), 147]. Cell Stem Cell 2014b; 14 (2), 160–173. https://doi.org/10.1016/j.stem.2013.12.013.

role of the mandibular innervation in the formation of tooth primordia, based on his experiments on the rat teeth, admitted that "local neurons may participate in the regulation of mammalian tooth formation" (Luukko, 1997).

Studies in mouse incisors showed that the sensory nerve fibers secrete Shh (Sonic hedgehog) protein, which activates Gli1-expressing cells, a subpopulation of mesenchymal stromal/stem cells (MSCs) that differentiate into odontoblasts and pulp cells (Fig. 3.43).

Glia-expressing cells in the dental epithelium surround the neurovascular bundle and receive the Shh signal from the sensory nerve (Ishida et al., 2011). They proliferate and differentiate to populate the entire dental pulp with various cell types, including ameloblasts, cells involved in forming dental enamel (Hosoya et al., 2020) (Fig. 3.44).

Sensory nerve injury leads to apoptosis of dental pulp stromal/stem cells (Liu et al., 2020). It was found that MSCs differentiate from Schwan cells and Schwann cell precursors (SCPs) that are recruited from nerves (Kaukua et al., 2014).

Experiments on tooth regeneration in the salamander, *Ambystoma mexicanum*, provided further evidence on the indispensable role of the nervous system in tooth bud induction; denervation of the mandible prevented Shh expression and inhibited the invagination of the dental lamina and tooth regeneration. The local innervation performs its ontogenetic function by secreting Fgf and Bmp (Makanae et al., 2020).

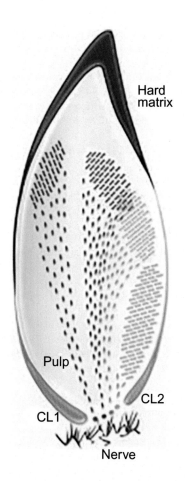

FIG. 3.44

Illustration of clonally organized pulp and odontoblasts (note correlation between colors of odontoblasts and adjacent pulp cells).

From Kaukua, N., Shahidi, M., Konstantinidou, C., Dyachuk, V., Kaucka, M., Furlan, A., et al., 2014. Glial origin of mesenchymal stem cells in a tooth model system. Nature 513, 551–554.

References

Abu Rmilah, A. A., Zhou, W., & Nyberg, S. L. (2020). Hormonal contribution to liver regeneration. *Mayo Clin. Proc.*, *4*, P315–P338.

Alsina, B., & Whitfield, T. T. (2017). Sculpting the labyrinth: morphogenesis of the developing inner ear. *Semin. Cell Dev. Biol.*, *65*, 47–59.

Amir, M., Yu, M., He, P., & Srinivasan, S. (2020). Hepatic autonomic nervous system and neurotrophic factors regulate the pathogenesis and progression of non-alcoholic fatty liver disease. *Front Med (Lausanne).*, *7*, 62.

Anagnostou, V. K., Doussis-Anagnostopoulou, I., Tiniakos, D. G., Karandrea, D., Agapitos, E., Karakitsos, P., et al. (2007). Ontogeny of intrinsic innervation in the human thymus and spleen. *J. Histochem. Cytochem.*, *55*, 813–820.

Andreone, B. J., Lacoste, B., & Gu, C. (2015). Neuronal and vascular interactions. *Annu. Rev. Neurosci.*, *38*, 25–46. https://doi.org/10.1146/annurev-neuro-071714-033835.

Baguñà, J., Saló, E., & Romero, R. (1989). Effects of activators and antagonists of the neuropeptides substance P and substance K on cell proliferation in planarians. *Int. J. Dev. Biol.*, *33*, 261–266.

Baker, C. V., & Bronner-Fraser, M. (2001). Vertebrate cranial placodes I. Embryonic induction. *Dev. Biol.*, *232*, 1–61.

Basch, M. L., Brown, R. M., II, Jen, H.-. I., & Groves, A. K. (2016). Where hearing starts: the development of the mammalian cochlea. *J. Anat.*, *228*(2), 233–254.

Bazin-Lopez, N., Valdivia, L. E., Wilson, S. W., & Gestri, G. (2015). Watching eyes take shape. *Curr. Opin. Genet. Dev.*, *32*, 73–79.

Berggren, K., McCaffery, P., Dräger, U., & Forehand, C. J. (1999). Differential distribution of retinoic acid synthesis in the chicken embryo as determined by immunolocalization of the retinoic acid synthetic enzyme, RALDH-2. *Dev. Biol.*, *210*, 288–304.

Berggren, K., Ezerman, E. B., McCaffery, P., & Forehand, C. J. (2001). Expression and regulation of the retinoic acid synthetic enzyme RALDH-2 in the embryonic chicken wing. *Dev. Dyn.*, *222*, 1–16.

Bok, J., Bronner-Fraser, M., & Wu, D. K. (2005). Role of the hindbrain in dorsoventral but not anterio-posterior axial specification of the inner ear. *Development*, *132*, 2115–2124.

Borden, P., Houtz, J., Leach, S. D., & Kuruvilla, R. (2013). Sympathetic innervation during development is necessary for pancreatic islet architecture and functional maturation. *Cell Rep.*, *4*, 287–301.

Borycki, A.-G., Brunk, B., Tajbakhsh, S., Buckingham, M., Chiang, C., & Emerson, C. P. (1999). Sonic hedgehog controls epaxial muscle determination through Myf5 activation. *Development*, *126*, 4053–4063.

Botchkarev, V. A., Peters, E. M., Botchkareva, N. V., Maurer, M., & Paus, R. (1999). Hair cycle-dependent changes in adrenergic skin innervation, and hair growth modulation by adrenergic drugs. *J. Invest. Dermatol.*, *113*, 878–887.

Bower, D. V., Lee, H.-K., Lansford, R., Zinn, K., Warburton, D., Fraser, S. E., et al. (2014). Airway branching has conserved needs for local parasympathetic innerva- tion but not neurotransmission. *BMC Biol.*, *2014*(12), 92.

Braun, T., Buschhausen-Denker, G., Bober, E., Tannich, E., & Arnold, H. H. (1989). A novel human muscle factor related to but distinct from MyoD1 induces myogenic conversion in 10T1/2 fibroblasts. *EMBO J.*, *8*(3), 701–709.

Brennan, C., Mangoli, M., Dyer, C. E. F., & Ashworth, R. (2005). Acetylcholine and calcium signaling regulates muscle fibre formation in the zebrafish embryo. *J. Cell Sci.*, *118*, 5181–5190.

Brown, K. E., Keller, P. J., Ramialison, M., Rembold, M., Stelzer, E. H. K., Loosli, F., et al. (2010). Nlcam modulates midline convergence during anterior neural plate morphogenesis. *Dev. Biol.*, *339*, 14–25.

Brownell, I., Guevara, E., Bai, C. B., Loomis, C. A., & Joyner, A. L. (2011). Nerve-derived sonic hedgehog defines a niche for hair follicle stem cells capable of becoming epidermal stem cells. *Cell Stem Cell*, *8*(5), 552–565. https://doi.org/10.1016/j.stem.2011.02.021.

Burris, R. E., & Hebrok, M. (2007). Pancreatic innervation in mouse development and β-cell regeneration. *Neuroscience*, *150*, 592–602.

Buskohl, P. R., Sun, M. J., Sun, M. L., Thompson, R. P., & Butcher, J. T. (2012). Serotonin potentiates transforming growth factor-beta3 induced biomechanical remodeling in avian embryonic atrioventricular valves. *PLoS One*, *7*, e42527.

Cabrera-Vásquez, S., Navarro-Tableros, V., Sánchez-Soto, C., Gutiérrez-Ospina, G., & Hiriart, M. (2009). Remodelling sympathetic innervation in rat pancreatic islets ontogeny. *BMC Dev. Biol.*, *9*, 34.

Camper, S. A., Daly, A. Z., Stallings, C. E., & Ellsworth, B. S. (2017). Hypothalamic β-catenin is essential for FGF8-mediated anterior pituitary growth: links to human disease. *Endocrinology*, *158*, 3322–3324.

Carmeliet, P., Ferreira, V., Breier, G., Pollefeyt, S., Kieckens, L., Gertsenstein, M., et al. (1996). Abnormal blood vessel development and lethality in embryos lacking a single VEGF allele. *Nature*, *380*, 435–439.

Cassiman, D., Libbrecht, L., Sinelli, N., Desmet, V., Denef, C., & Roskams, T. (2002). The vagal nerve stimulates activation of the hepatic progenitor cell compartment via muscarinic acetylcholine receptor type 3. *Am. J. Pathol.*, *161*(2), 521–530. https://doi.org/10.1016/S0002-9440(10)64208-3.

Cavodeassi, F., & Houart, C. (2012). Brain regionalization: of signaling centers and boundaries. *Dev. Neurobiol.*, *72*, 218–233.

Cebrià, F., & Newmark, P. A. (2007). Morphogenesis defects are associated with abnormal nervous system regeneration following roboA RNAi in planarians. *Development*, *134*, 833–837.

Chaa, K. B., Douglas, K. R., Potok, M. A., Liang, H., Jones, S. N., & Camper, S. A. (2004). WNT5A signaling affects pituitary gland shape. *Mech. Dev.*, *121*, 183–194.

Chang, C.-N., & Kioussi, C. (2018). Location, location, location: signals in muscle specification. *J. Dev. Biol.*, *6*(2), 11. 2018 Jun;.

Cheung, L. Y. M., Davis, S. W., Brinkmeier, M. L., Camper, S. A., & Pérez-Millán, M. I. (2017). Regulation of pituitary stem cells by epithelial to mesenchymal transition events and signaling pathways. *Mol. Cell. Endocrinol.*, *445*, 14–26.

Chin, E. R., Olson, E. N., Richardson, J. A., Yang, Q., Humphries, C., Shelton, J. M., et al. (1998). A calcineurin-dependent transcriptional pathway controls skeletal muscle fiber type. *Genes Dev.*, *12*, 2499–2509.

Christensen, C. F., Koyama, T., Nagy, S., Danielsen, E. T., Texada, M. J., Halberg, K. A., et al. (2020). Ecdysone-dependent feedback regulation of prothoracicotropic hormone controls the timing of developmental maturation. *Development*, *147*(14). https://doi.org/10.1242/dev.188110, dev188110.

Chukwurah, E., Osmundsen, A., Davis, S. W., & Lizarraga, S. B. (2019). All together now: modeling the interaction of neural with Non-neural systems using organoid models. *Front. Neurosci.*, *13*, 582.

Colbert, M. C., Linney, E., & LaMantia, A. (1993). Local sources of retinoic acid coincide with retinoid-mediated transgene activity during embryonic development. *Proc. Natl. Acad. Sci. U. S. A.*, *90*, 6572–6576.

Collins, J. J., III, Hou, X., Romanova, E. V., Lambrus, B. G., Miller, C. M., Saberi, A., et al. (2010). Genome-wide analyses reveal a role for peptide hormones in planarian germline development. *PLoS Biol.*, *8*(10), e1000509. 2010 Oct.

Cossu, G., & Borello, U. (1999). Wnt signaling and the activation of myo- genesis in mammals. *EMBO J.*, *18*, 6867–6872.

Coughlin, M. D. (1975). Early development of parasympathetic nerves in the mouse submandibular gland. *Dev. Biol.*, *43*, 123–139.

Creazzo, T. L., & Sohal, G. S. (1983). Neural control of embryonic acetylcholine receptor and skeletal muscle. *Cell Tissue Res.*, *228*, 1–12.

Croset, V., Schleyer, M., Arguello, J. R., Gerber, B., & Benton, R. (2016). A molecular and neuronal basis for amino acid sensing in the *Drosophila* larva. *Sci. Rep.*, *6*, 34871. Published 2016 Dec 16 https://doi.org/10.1038/srep34871.

Currie, D. A., & Bate, M. (1991). The development of adult abdominal muscles in *Drosophila*: myoblasts express *twist* and are associated with nerves. *Development*, *113*(1), 91–102.

Currie, D. A., & Bate, M. (1995). Innervation is essential for the development and differentiation of a sex-specific adult muscle in *Drosophila melanogaster*. *Development*, *121*, 2549–2557.

Cutz, E., Yeger, H., Pan, J., & Ito, T. (2008). Pulmonary neuroendocrine cell system in health and disease. *Curr. Resp. Med. Rev.*, *4*, 174–186.

Dasen, J. S., Tice, B. C., Brenner-Morton, S., & Jessell, T. M. (2005). A hox regulatory network establishes motor neuron pool identity and target-muscle connectivity. *Cell*, *23*(3), P477–P491. https://doi.org/10.1016/j.cell.2005.09.009.

Davis, S. W., Ellsworth, B. S., Peréz Millan, M. I., Gergics, P., Schade, V., Foyouzi, N., et al. (2013). Pituitary gland development and disease: from stem cell to hormone production. *Curr. Top. Dev. Biol.*, *106*, 1–47.

Deng, S., Azevedo, M., & Baylies, M. (2017). Acting on identity: myoblast fusion and the formation of the syncytial muscle fiber. *Semin. Cell Dev. Biol.*, *72*, 45–55.

Denver, R. J. (1997a). Thyroid hormone-dependent gene expression program for *Xenopus* neural development. *J. Biol. Chem.*, *272*, 8179–8188.

Denver, R. J. (1997b). Environmental stress as a developmental cue: corticotropin-releasing hormone is a proximate mediator of adaptive phenotypic plasticity in amphibian metamorphosis. *Horm. Behav.*, *31*, 169–179.

Dhiman, N., Shweta, K., Tendulkar, S., Deshpande, G., Ratnaparkhi, G. S., & Ratnaparkhi, A. (2019). *Drosophila* Mon1 constitutes a novel node in the brain-gonad axis that is essential for female germline maturation. *Development*, *146*. https://doi.org/10.1242/dev.166504, dev166504.

Dietz, F. R. (1987). Effect of peripheral nerve on limb development. *J. Orthop. Res.*, *5*, 576–585.

Dietz, F. R. (1989). Effect of denervation on limb growth. *J. Orthop. Res.*, *7*, 292–303.

Dobi, K. C., Schulman, V. K., & Baylies, M. K. (2015). Specification of the somatic musculature in *Drosophila*. *Wiley Interdiscip. Rev. Dev. Biol.*, *4*(4), 357–375. https://doi.org/10.1002/wdev.182.

Eftimie, R., Brenner, H. R., & Buonanno, A. (1991). Myogenin and MyoD join a family of skeletal muscle genes regulated by electrical activity. *Proc. Natl. Acad. Sci. U. S. A.*, *88*, 1349–1353.

Fan, S. M., Chang, Y. T., Chen, C. L., Wang, W.-H., Pan, M.-K., Chen, W.-P., et al. (2018). External light activates hair follicle stem cells through eyes via an ipRGC-SCN-sympathetic neural pathway. *Proc. Natl. Acad. Sci. U. S. A.*, *115*(29), E6880–E6889. https://doi.org/10.1073/pnas.1719548115 (published correction appears in Proc. Natl. Acad. Sci. U S A. 2018 Dec 18;115(51):E12121).

Farkas, J. E., & Monaghan, J. R. (2017). A brief history of the study of nerve dependent regeneration. *Neurogenesis (Austin).*, *4*(1), e1302216.

Farkas, J., Freitas, P., Bryant, D. M., Whited, J. L., & Monaghan, J. R. (2016). Neuregulin-1 signaling is essential for nerve-dependent axolotl limb regeneration. *Development*, *143*(15), 2724–2731. https://doi.org/10.1242/dev.133363.

Fernandes, J. J., & Keshishian, H. (1998). Nerve-muscle interactions during flight muscle development in *Drosophila*. *Development*, *125*, 1769–1779.

Ferraro, E., Molinari, F., & Berghella, L. (2012). Molecular control of neuromuscular junction development. *J. Cachexia. Sarcopenia Muscle*, *3*(1), 13–23. https://doi.org/10.1007/s13539-011-0041-7.

Ferreira, J. N., & Hoffman, M. P. (2013). Interactions between developing nerves and salivary glands. *Organogenesis*, *9*(3), 199–205. https://doi.org/10.4161/org.25224.

Flanagan-Steet, H., Fox, M. A., Meyer, D., & Sanes, J. R. (2005). Neuromuscular synapses can form *in vivo* by incorporation of initially aneural postsynaptic specializations. *Development*, *132*(20), 4471–4481. https://doi.org/10.1242/dev.02044.

Fliniaux, I., Viallet, J. P., & Dhouailly, D. (2004). Signaling dynamics of feather tract formation from the chick somatopleura. *Development*, *131*, 3955–3966.

Fong, A. P., & Tapscott, S. J. (2013). Skeletal muscle programming and re-programming. *Curr. Opin. Genet. Dev.*, *23*(5), 568–573. https://doi.org/10.1016/j.gde.2013.05.002.

Fraguas, S., Barberan, S., Iglesias, M., Rodriguez-Esteban, G., & Cebria, F. (2014). egr-4, a target of EGFR signaling, is required for the formation of the brain primordia and head regeneration in planarians. *Development*, *141*, 1835–1847.

Friedländer, M. R., Adamidi, C., Han, T., Lebedeva, S., Isenbarger, T. A., Hirst, M., et al. (2009). High-resolution profiling and discovery of planarian small RNAs. *Proc. Natl. Acad. Sci. U. S. A.*, *106*, 11546–11551. https://doi.org/10.1073/pnas.0905222106.

Fritzsch, B., Beisel, K. W., Jones, K., Farinas, I., Maklad, A., Lee, J., et al. (2002). Development and evolution of inner ear sensory epithelia and their innervation. *J. Neurobiol.*, *53*, 143–156.

Furlan, A., Dyachuk, V., Kastriti, M. E., Calvo-Enrique, L., Abdo, H., Hadjab, S., et al. (2017). Multipotent peripheral glial cells generate neuroendocrine cells of the adrenal medulla. *Science*, *357*(6346), eaal3753. Jul 7.

Gale, N. W., & Yancopoulos, G. D. (1999). Growth factors acting via endothelial cell-specific receptor tyrosine kinases: VEGFs, angiopoietins, and ephrins in vascular development. *Genes Dev.*, *13*, 1055–1066.

Ghosh, A., McBrayer, Z., & O'Connor, M. B. (2010). The *Drosophila* gap gene giant regulates ecdysone production through specification of the PTTH-producing neurons. *Dev. Biol.*, *347*(2), 271–278. https://doi.org/10.1016/j.ydbio.2010.08.011.

Gillis, J. A., Dahn, R. D., & Shubin, N. H. (2009). Shared developmental mechanisms pattern the vertebrate gill arch and paired fin skeletons. *Proc. Natl. Acad. Sci. U. S. A.*, *106*, 5720–5724.

Gitler, A. D., Lu, M. M., & Epstein, J. A. (2004). PlexinD1 and semaphorin signaling are required in endothelial cells for cardiovascular development. *Dev. Cell*, *7*, 107–116. https://doi.org/10.1016/j.devcel.2004.06.002.

Goldberg, E., Grau, J. B., Fortier, J. H., Salvati, E., Levy, R. J., & Ferrari, G. (2017). Serotonin and catecholamines in the development and progression of heart valve diseases. *Cardiovasc. Res.*, *113*(8), 849–857.

Goldman, D., & Staple, J. (1989). Spatial and temporal expression of acetylcholine receptor RNAs in innervated and denervated rat soleus muscle. *Neuron*, *3*, 219–228.

Grobstein, C. (1953). Inductive epitheliomesenchymal interaction in cultured organ rudiments of the mouse. *Science*, *118*, 52–55.

Grobstein, C. (1956). Trans-filter induction of tubules in mouse metanephrogenic mesenchyme. *Exp. Cell Res.*, *10*, 424–440.

Guarino, S. R., Canciani, A., & Forneris, F. (2020). Dissecting the extracellular complexity of neuromuscular junction organizers. *Front. Mol. Biosci.*, *6*, 156. Published 2020 Jan 10 https://doi.org/10.3389/fmolb.2019.00156.

Hall, B. K. (1998). *Evolutionary Developmental Biology* (2nd ed., p. 134). London: Chapman & Hall.

Hall, B. K. (1998b). *Evolutionary Developmental Biology* (2nd ed., p. 164). London: Chapman & Hall.

Hall, B. K., & Herring, S. W. (1990). Paralysis and growth of the musculoskeletal system in the embryonic chick. *J. Morphol.*, *206*, 45–56.

Hasty, P., Bradley, A., Morris, J. H., Edmondson, D. G., Venuti, J. M., Olson, E. N., et al. (1993). Muscle deficiency and neonatal death in mice with a targeted mutation in the myogenin gene. *Nature*, *364*, 501–506. https://doi.org/10.1038/364501a0.

Hayashi, S., & Kondo, T. (2018). Development and function of the *Drosophila* tracheal system. *Genetics*, *209*, 367–380.

Hegstrom, C. D., Riddiford, L. M., & Truman, J. W. (1998). Spatial restriction of expression during metamorphosis of muscle in the moth, *Manduca sexta*. *J. Neurosci.*, *18*, 1786–1794.

Hill, E. M., & Petersen, C. P. (2015). Wnt/Notum spatial feedback inhibition controls neoblast differentiation to regulate reversible growth of the planarian brain. *Development*, *142*, 4217–4229.

Hirasawa, T., & Kuratani, S. (2018). Evolution of the muscular system in tetrapod limbs. *Zoological Lett.*, *4*, 27. Published 2018 Sep 20 https://doi.org/10.1186/s40851-018-0110-2.

Hirokawa, K., Utsuyama, M., & Kobayashi, S. (1998). Hypothalamic control of development and aging of thymus. *Mech. Ageing Dev.*, *100*, 177–185.

Hirokawa, K., Utsuyama, M., & Kobayashi, S. (2001). Hypothalamic control of thymic function. *Cell Mol Biol.(Noisy-le-grand)*, *47*, 97–102.

Hogan, K. A., Ambler, C. A., Chapman, D. L., & Bautch, V. L. (2004). The neural tube patterns vessels developmentally using the VEGF signaling pathway. *Development*, *131*, 1503–1513.

Hosoya, A., Shalehin, N., Takebe, H., Shimo, T., & Irie, K. (2020). Sonic hedgehog signaling and tooth development. *Int. J. Mol. Sci.*, *21*, 1587.

Hosseini, H. S., Beebe, D. C., & Taber, L. A. (2014). Mechanical effects of the surface ectoderm on optic vesicle morphogenesis in the chick embryo. *J. Biomech.*, *47*, 3837–3846.

Hsuan, Y., Yang, C., Kawakami, K., & Stainier, D. Y. R. (2018). A new mode of pancreatic islet innervation revealed by live imaging in zebrafish. *Elife*, *7*, e34519.

Hu, D., & Marcucio, R. S. (2009). A SHH-responsive signaling center in the forebrain regulates craniofacial morphogenesis via the facial ectoderm. *Development*, *136*(1), 107–116.

Hu, D., Marcucio, R., & Helms, J. A. (2003). A zone of frontonasal ectoderm regulates patterning and growth in the face. *Development*, *130*(9), 1749–1758.

Huang, J., Rajagopal, R., Liu, Y., Dattilo, L. K., Shaham, O., Ashery-Padan, R., et al. (2011). The mechanism of lens placode formation: a case of matrix-mediated morphogenesis. *Dev. Biol.*, *355*(1), 32–42.

Hughes, S. M., & Salinas, P. C. (1999). Control of muscle fibre and motoneuron diversification. *Curr. Opin. Neurobiol.*, *9*, 54–64.

Hurr, C., Simonyan, H., Morgan, D. A., Rahmouni, K., & Young, C. N. (2019). Liver sympathetic denervation reverses obesity-induced hepatic steatosis. *J. Physiol.*, *597*, 4565–4580.

Hurren, B., Collins, J. J., Duxson, M. J., & Deries, M. (2015). First neuromuscular contact correlates with onset of primary myogenesis in rat and mouse limb muscles. *PLoS One*, *10*(7), e0133811. Published 2015 Jul 24 https://doi.org/10.1371/journal.pone.0133811.

Ishida, K., Murofushi, M., Nakao, K., Morita, R., Ogawa, M., & Tsuji, T. (2011). The regulation of tooth morphogenesis is associated with epithelial cell proliferation and the expression of Sonic hedgehog through epithelial-mesenchymal interactions. *Biochem. Biophys. Res. Commun.*, *405*, 455–461.

Ito, Y., Kayama, T., & Asahara, H. (2012). A systems approach and skeletal myogenesis. *Comp Funct Genomics.*, *2012*, 759407.

Ivanovitch, K., Cavodeassi, F., & Wilson, S. W. (2013). Precocious acquisition of neuroepithelial character in the eye field underlies the onset of eye morphogenesis. *Dev. Cell*, *27*(3), 293–305.

Jiang, Y.-H., Jiang, P., Yang, J.-L., Ma, D.-F., Lin, H.-Q., Su, W.-G., et al. (2015). Cardiac dysregulation and myocardial injury in a 6-hydroxydopamine-induced rat model of sympathetic denervation. *PLoS One*, *10*(7), e0133971.

Kamba, T., Higashi, S., Kamoto, T., Shisa, H., Yamada, Y., Ogawa, O., et al. (2001). Failure of ureteric bud invasion: a new model of renal agenesis in mice. *Am. J. Pathol.*, *159*(6), 2347–2353. https://doi.org/10.1016/S0002-9440(10)63084-2.

Karp, S. L., Ortiz-Arduan, A., Li, S., & Neilson, E. G. (1994). Epithelial differentiation of metanephric mesenchymal cells after stimulation with hepatocyte growth factor or embryonic spinal cord. *Proc. Natl. Acad. Sci. U. S. A.*, *91*, 5286–5290.

Katayama, Y., Battista, M., Kao, W.-M., Hidalgo, A., Peired, A. J., Thomas, S. A., et al. (2006). Signals from the sympathetic nervous system regulate hematopoietic stem cell egress from bone marrow. *Cell*, *124*(2), P407–P421. https://doi.org/10.1016/j.cell.2005.10.041.

Katsukura, Y., Ando, H., David, C. N., Grimmelikhuijzen, C. J. P., & Sugiyama, T. (2003). Control of planula migration by LWamide and RFamide neuropeptides in *Hydractinia echinata*. *J. Exp. Biol.*, *207*, 1803–1810.

Kaucka, M., Zikmund, T., Tesarova, M., Gyllborg, D., Hellander, A., Jaros, J., et al. (2017). Oriented clonal cell dynamics enables accurate growth and shaping of vertebrate cartilage. *Elife*, *6*. https://doi.org/10.7554/eLife.25902. (2017).

Kaucka, M., Petersen, J., Tesarova, M., Szarowska, B., Kastriti, M. E., Xie, M., et al. (2018). Signals from the brain and olfactory epithelium control shaping of the mammalian nasal capsule cartilage. *Elife*, *7*, e34465.

Kaukua, N., Shahidi, M., Konstantinidou, C., Dyachuk, V., Kaucka, M., Furlan, A., et al. (2014). Glial origin of mesenchymal stem cells in a tooth model system. *Nature*, *513*, 551–554.

Kawakami, A., Kataoka, H., Oka, T., Mizoguchi, A., Kimura-Kawakami, M., Adachi, T., et al. (1990). Molecular cloning of the *Bombyx mori* prothoracicotropic hormone. *Science*, *247*, 1333–1335. https://doi.org/10.1126/science.2315701.

Kawasaki, T., Kitsukawa, T., Bekku, Y., Matsuda, Y., Sanbo, M., Yagi, T., et al. (1999). A requirement for neuropilin-1 in embryonic vessel formation. *Development*, *126*, 4895–4902.

Kelberman, D., Rizzoti, K., Lovell-Badge, R., Robinson, I. C. A. F., & Dattani, M. T. (2009). Genetic regulation of pituitary gland development in human and mouse. *Endocr. Rev.*, *30*, 790–829.

Kendall, M. D., & al-Shawaf, A. A. (1991). Innervation of the rat thymus gland. *Brain Behav. Immun.*, *5*(1), 9–28. https://doi.org/10.1016/0889-1591(91)90004-t.

Kerman, B. K., & Andrew, D. J. (2006). Analysis of Dalmatian suggests a role for the nervous system in *Drosophila* embryonic trachea and salivary duct development. In *47th Drosophila Research Conference Abstracts* (p. 275).

Kesper, D. A., Stute, C., Buttgereit, D., Kreiskother, N., Vishnu, S., Fischbach, K. F., et al. (2007). Myoblast fusion in Drosophila melanogaster is mediated through a fusion-restricted myogenic-adhesive structure (FuRMAS) Dev. *Dynamis, 236*, 404–415.

Kiel, M. J., Yilmaz, O. H., Iwashita, T., Terhorst, C., & Morrison, S. J. (2005). SLAM family receptors distinguish hematopoietic stem and progenitor cells and reveal endothelial niches for stem cells. *Cell, 121*, 1109–1121.

Kispert, A., Vainio, S., & McMahon, A. P. (1998). Wnt-4 is a mesenchymal signal for epithelial trans- formation of metanephric mesenchyme in the developing kidney. *Development, 125*, 4225–4234.

Knox, S. M., Lombaert, M. A., Reed, X., Vitale-Cross, L., Gutkind, J. S., & Hoffman, M. P. (2010). Parasympathetic innervation maintains epithelial progenitor cells during salivary organogenesis. *Science, 329*, 1645–1647.

Knox, S. M., Lombaert, I. M., Haddox, C. L., Abrams, S. R., Cotrim, A., Wilson, A. J., et al. (2013). Parasympathetic stimulation improves epithelial organ regeneration. *Nat. Commun., 4*, 1494. https://doi.org/10.1038/ncomms2493.

Kostrominova, T. Y., Macpherson, P. C. D., Carlson, B. M., & Goldman, D. (2000). Regulation of myogenin protein expression in denervated muscles from young and old rats. *Am. J. Physiol. Regul. Integr. Comp. Physiol., 279*(1), R179–R188.

Kreipke, R. E., & Birren, S. J. (2015). Innervating sympathetic neurons regulate heart size and the timing of cardiomyocyte cell cycle withdrawal. *J. Physiol., 593*(23), 5057–5073.

Kuroda, K., Kuang, S., Taketo, M. M., & Rudnicki, M. A. (2013). Canonical Wnt signaling induces BMP-4 to specify slow myofibrogenesis of fetal myoblasts. *Skelet. Muscle., 3*, 5. https://doi.org/10.1186/2044-5040-3-5.

Kwan, K. M. (2014). Coming into focus: the role of extracellular matrix in vertebrate optic cup morphogenesis. *Dev. Dyn., 243*, 1242–1248.

Lacoste, B., Comin, C. H., Ben-Zvi, A., Kaeser, P. S., Xu, X., da Costa, L. F., et al. (2014). Sensory-related neural activity regulates the structure of vascular networks in the cerebral cortex. *Neuron, 83*, 1117–1130.

LaFever, L., & Drummond-Barbosa, D. (2005). Direct control of germline stem cell division and cyst growth by neural insulin in *Drosophila. Science, 309*(5737), 1071–1073.

Lawrence, P. A., & Johnston, P. (1986). The muscle pattern of a segment of Drosophila may be determined by neurons and not by contributing myoblasts. *Cell, 45*, 505–513.

Leger, S., & Brand, M. (2002). Fgf8 and Fgf3 are required for zebrafish ear placode induction, maintenance and inner ear patterning. *Mech. Dev., 119*, 91–108.

Lei, K., Vu, H. T.-K., Mohan, R. D., McKinney, S. A., Seidel, C. W., & Alexander, R. (2016). Egf signaling directs neoblast repopulation by regulating asymmetric cell division in planarians. *Dev. Cell, 38*(4), P413–P429. https://doi.org/10.1016/j.devcel.2016.07.012.

Leitz, T., Morand, K., & Mann, M. (1994). A novel peptide controlling development of the lower metazoan (Coelenterata, Hydrozoa). *Dev. Biol., 163*, 440–446.

Leposavić, G., Mićić, M., Ugresić, N., Bogojević, M., & Isaković, K. (1992). Components of sympathetic innervation of the rat thymus during late fetal and postnatal development: histofluorescence and biochemical study. Sympathetic innervation of the rat thymus. *Thymus, 19*, 77–87.

Li, Z., Torbenson, M., Yang, S., Lin, H., Smedh, U., Moran, T. H., et al. (2004). Hepatic fibrogenesis requires sympathetic neurotransmitters. *Gut, 53*, 438–445.

Li, W., Kohara, H., Uchida, Y., James, J. M., Soneji, K., Cronshaw, D. G., et al. (2013). Peripheral nerve-derived CXCL12 and VEGF-A regulate the patterning of arterial vessel branching in developing limb skin. *Dev. Cell*, *24*, 359–371.

Li, Z., Meyers, C. A., Chang, L., Lee, S., Li, Z., Tomlinson, R., et al. (2019). Fracture repair requires TrkA signaling by skeletal sensory nerves. *J. Clin. Invest.*, *129*, 5137–5150.

Liang, D., Chang, J. R., Chin, A. J., Smith, A., Kelly, C., Weinberg, E. S., et al. (2001). The role of vascular endothelial growth factor (V J., EGF) in vasculogenesis, angiogenesis, and hematopoiesis in zebrafish development. *Mech. Dev.*, *108*, 29–43.

Linneweber, G. A., Jacobson, J., Busch, K. E., Hudry, B., Christov, C. P., Dormann, D., et al. (2014). Neuronal control of metabolism through nutrient-dependent modulation of tracheal branching. *Cell*, *156*(1-2), 69–83. https://doi.org/10.1016/j.cell.2013.12.008.

Linnoila, R. I. (2006). Functional facets of the pulmonary neuroendocrine system. *Lab. Invest.*, *86*, 425–444.

Liu, A. Q., Zhang, L. S., Fei, D. D., Guo, H., Wu, M.-. L., Liu, J., et al. (2020). Sensory nerve-deficient microenvironment impairs tooth homeostasis by inducing apoptosis of dental pulp stem cells. *Cell Prolif.*, *53*(5). https://doi.org/10.1111/cpr.12803, e12803.

Lumb, R., & Schwarz, Q. (2015). Sympathoadrenal neural crest cells: the known, unknown and forgotten? *Dev. Growth Differ.*, *57*, 146–157.

Lumb, R., Tata, M., Xu, X., Joyce, A., Marchant, C., Harvey, N., et al. (2018). Neuropilins guide preganglionic sympathetic axons and chromaffin cell precursors to establish the adrenal medulla. *Development*, *145*, dev162552.

Luukko, K. (1997). Immunohistochemical localization of nerve fibres during development of embryonic rat molar using peripherin and protein gene product 9.5 antibodies. *Arch. Oral Biol.*, *42*, 189–195.

Maden, M., Sonneveld, E., van der Saag, P. T., & Gale, E. (1998). The distribution of endogenous reti- noic acid in the chick embryo: implications for developmental mechanisms. *Development*, *125*, 4133–4144.

Mahmoud, A. I., O'Meara, C. C., Gemberling, M., Zhao, L., Bryant, D. M., & Zheng, R. (2015). Nerves regulate cardiomyocyte proliferation and heart regeneration. *Dev. Cell*, *34*(4), 387–399.

Makanae, A., Tajika, Y., Nishimura, K., Saito, N., Tanaka, J.-I., & Satoh, A. (2020). Neural regulation in tooth regeneration of *Ambystoma mexicanum. Sci. Rep.*, *10*, 9323. https://doi.org/10.1038/s41598-020-66142-2.

Marcucio, R. S., Cordero, D. R., Hu, D., & Helms, J. A. (2005). Molecular interactions coordinating the development of the forebrain and face. *Dev. Biol.*, *284*, 48–61.

Marcucio, R. S., Young, N. M., Hu, D., & Hallgrimsson, B. (2011). Mechanisms that underlie co-variation of the brain and face. *Genesis*, *49*, 177–189.

Marcucio, R., Hallgrimsson, B., & Young, N. M. (2015). Facial morphogenesis: physical and molecular interactions between the brain and the face. *Curr. Top. Dev. Biol.*, *115*, 229–319.

Marxreiter, S., & Thummel, C. S. (2014). Will branch for food–nutrient-dependent tracheal remodeling in *Drosophila* branching. *EMBO J.*, *33*(3), 179–180.

McBrayer, Z., Ono, H., Shimell, M., Parvy, J.-P., Beckstead, R. B., Warren, J. T., et al. (2007). Prothoracicotropic hormone regulates developmental timing and body size in *Drosophila. Dev. Cell*, *13*, 857–871. https://doi.org/10.1016/j.devcel.2007.11.003.

McCaffery, P., & Dräger, U. C. (1994). Hot spots of retinoic acid synthesis in the developing spinal cord. *Proc. Natl. Aad. Sci. USA*, *91*, 7194–7197.

McCredie, J., North, K., & de Longh, R. (1984). Thalidomide deformities and their nerve supply. *J. Anat.*, *139*, 397–410.

McLeod, C. J., Wang, L., Wong, C., & Jones, D. L. (2010). Stem cell dynamics in response to nutrient availability. *Curr. Biol.*, *20*(23), 2100–2105.

McMenamin, S. K., & Parichy, D. M. (2013). Metamorphosis in teleosts. *Curr. Top. Dev. Biol.*, *103*, 127–165. https://doi.org/10.1016/B978-0-12-385979-2.00005-8.

Melnikova, V. I., Lifantseva, N. V., Voronova, S. N., & Zakharova, L. A. (2019). Gonadotropin-releasing hormone in regulation of thymic development in rats: profile of thymic cytokines. *Int. J. Mol. Sci.*, *20*(16), 4033. https://doi.org/10.3390/ijms20164033.

Misgeld, T., Kummer, T. T., Lichtman, J. W., & Sanes, J. R. (2005). Agrin promotes synaptic differentiation by counteracting an inhibitory effect of neurotransmitter. *Proc. Natl. Acad. Sci. U. S. A.*, *102*(31), 11088–11093.

Moreno-Marmol, T., Cavodeassi, F., & Bovolenta, P. (2018). Setting eyes on the retinal pigment epithelium. *Front. Cell Dev. Biol.*, *6*, 145.

Moullé, V. S., Tremblay, C., Castell, A., Vivot, K., Ethier, M., Fergusson, G., et al. (2019). The autonomic nervous system regulates pancreatic ß-cell proliferation in adult male rats. *Am. J. Physiol. Endocrinol. Metab.*, *317*, E234–E243.

Mukoyama, Y. S., Shin, D., Britsch, S., Taniguchi, M., & Anderson, D. J. (2002). Sensory nerves determine the pattern of arterial differentiation and blood vessel branching in the skin. *Cell*, *109*, 693–705.

Mukoyama, Y., Gerber, H.-P., Ferrara, N., Gu, C., & Anderson, D. J. (2005). Peripheral nerve-derived VEGF promotes arterial differentiation via neuropilin 1-mediated positive feedback. *Development*, *132*, 941–952.

Munsterberg, A. E., & Lassar, A. B. (1995). Combinatorial signals from the neural tube, floor plate and notochord induce myogenic bHLH gene expression in the somite. *Development*, *121*, 651–660.

Munsterberg, A. E., Kitajewski, J., Bumcrot, D. A., McMahon, A. P., & Lassar, A. B. (1995). Combinatorial signaling by Sonic hedgehog and Wnt family members induces myogenic bHLH gene expression in the somite. *Genes Dev.*, *9*, 2911–2922.

Nakatani, Y., Kawakami, A., & Kudo, A. (2007). Cellular and molecular processes of regeneration, with special emphasis on fish fins. *Dev. Growth Differ.*, *49*, 145–154.

Nebigil, C. G., & Maroteaux, L. (2001). A novel role for serotonin in heart. *Trends Cardiovasc. Med.*, *11*, 329–335.

Nebigil, C. G., Choi, D.-S., Dierich, A., Hickel, P., Le Meur, M., Messaddeq, N., et al. (2000). Serotonin 32B receptor is required for heart development. *Proc. Natl. Acad. Sci. U. S. A.*, *97*, 9508–9513. https://doi.org/10.1073/pnas.97.17.9508.

Nebigil, C. G., Jaffré, F., Messaddeq, N., Hickel, P., Monassier, L., Launay, J. M., et al. (2003). Overexpression of the serotonin 5-HT2B receptor in heart leads to abnormal mitochondrial function and cardiac hypertrophy. *Circulation*, *107*, 3223–3229.

Nekrep, N., Wang, J., Miyatsuka, T., & German, M. S. (2008). Signals from the neural crest regulate beta-cell mass in the pancreas. *Development*, *135*, 2151–2160.

Neville, C. M., Schmidt, M., & Schmidt, J. (1992). Response of myogenic determination factors to cessation and resumption of electrical activity in skeletal muscle: a possible role for myogenin in denervation supersensitivity. *Cell. Mol. Neurobiol.*, *12*, 511–527. https://doi.org/10.1007/BF00711232.

Niederreither, K., Vermot, J., Fraulob, V., Chambon, P., & Dollé, P. (2002a). Retinaldehyde dehydro- genase 2 (RALDH2)-independent patterns of retinoic acid synthesis in the mouse embryo. *Proc. Natl. Acad. Sci. U. S. A.*, *99*(25), 16111–16116.

Niederreither, K., Vermot, J., Schuhbaur, B., Chambon, P., & Dollé, P. (2002b). Embryonic synthesis of retinoic acid is required for forelimb growth and anteroposterior patterning in the mouse. *Development*, *129*, 3563–3574.

Nikolić, M. Z., Sun, D., & Rawlins, E. L. (2018). Human lung development: recent progress and new challenges. *Development*, *145*. https://doi.org/10.1242/dev, dev163485.

Nord, H., Dennhag, N., Tydinger, H., & von Hofsten, J. (2019). The zebrafish HGF receptor met controls migration of myogenic progenitor cells in appendicular development. *PLoS One*, *14*(7), e0219259. Published 2019 Jul 9 https://doi.org/10.1371/journal.pone.0219259.

Oben, J. A., & Diehl, A. M. (2004). Sympathetic nervous system regulation of liver repair. *Anat. Rec.*, *280A*, 874–883.

Oben, J. A., Roskams, T., Yang, S., Lin, H., Sinelli, N., Li, Z., et al. (2003). Sympathetic nervous system inhibition increases hepatic progenitors and reduces liver injury. *Hepatology*, *38*, 664–673.

Oh, Y., Lai, J. S.-Y., Mills, H. J., Erdjument-Bromage, H., Giammarinaro, B., Saadipour, K., et al. (2019). A glucose-sensing neuron pair regulates insulin and glucagon in *Drosophila*. *Nature*, *574*(7779), 559–564. https://doi.org/10.1038/s41586-019-1675-4.

Ohta, S., & Schoenwolf, G. C. (2018). Hearing crosstalk: the molecular conversation orchestrating inner ear dorsoventral patterning. *Wiley Interdiscip. Rev. Dev. Biol.*, *7*(1). https://doi.org/10.1002/wdev.302. 2018 Jan.

Olaya-Sánchez, D., Sánchez-Guardado, L.Ó., Ohta, S., Chapman, S. C., Schoenwolf, G. C., Puelles, L., et al. (2017). Fgf3 and Fgf16 expression patterns in the developing chick inner ear. *Brain Struct. Funct.*, *222*, 131–149.

Olivera-Martinez, I., Thélu, J., Teillet, M.-A., & Dhouailly, D. (2001). Dorsal dermis development depends on a signal from the dorsal neural tube, which can be substituted by Wnt-1. *Mech. Dev.*, *100*, 233–244.

Önel, S.-F., & Renkawitz-Pohl, R. (2009). FuRMAS: triggering myoblast fusion in *Drosophila*. *Dev. Dyn.*, *238*, 1513–1525.

Osmundsen, A. M., Keisler, J. L., Taketo, M. M., & Davis, S. W. (2017). Canonical WNT signaling regulates the pituitary organizer and pituitary gland formation. *Endocrinology*, *158*, 3339–3353.

Oviedo, N. J., Morokuma, J., Walentek, P., Kema, I. P., Gu, M. B., Ahn, J.-M., et al. (2010). Long-range neural and gap junction protein-mediated cues control polarity during planarian regeneration. *Dev. Biol.*, *339*, 188–199.

Pan, J., Copland, I., Post, M., Yeger, H., & Cutz, E. (2006). Mechanical stretch-induced serotonin release from pulmonary neuroendocrine cells: implications for lung development. *Am. J. Physiol. Lung Mol. Physiol.*, *290*, L185–L193.

Plank, J. L. (2011). *Cell Autonomous and Non-Cell Autonomous Regulation of Beta Cell Mass Expansion* (p. 64). Nashville, Tennessee: Dissert.Vanderbilt University.

Plank, J. L., Mundell, N. A., Frist, A. Y., LeGrone, A. W., Kim, T., Musser, M. A., et al. (2011). Influence and timing of arrival of murine neural crest on pancreatic beta cell development and maturation. *Dev. Biol.*, *349*, 321–330.

Potok, M. A., Cha, K. B., Hunt, A., Brinkmeier, M. L., Leitges, M., Kispert, A., et al. (2008). WNT signaling affects gene expression in the ventral diencephalon and pituitary gland growth. *Dev. Dyn.*, *237*, 1006–1020.

Preziosi, M. E., & Monga, S. P. (2017). Update on the mechanisms of liver regeneration. *Semin. Liver Dis.*, *37*(2), 141–151. https://doi.org/10.1055/s-0037-1601351.

Proshchina, A. E., Krivova, Y. S., Barabanov, V. M., & Saveliev, S. V. (2014). Ontogeny of neuro-insular complexes and islets innervation in the human pancreas. *Front. Endocrinol. (Lausanne)*, *5*, 57.

Reshef, R., Maroto, M., & Lassar, A. B. (1998). Regulation of dorsal cell fates: BMPs and noggin control the timing and pattern of myogenic regulator expression. *Genes Dev.*, *12*, 290–303.

Rétaux, S., & Casane, D. (2013). Evolution of eye development in the darkness of caves: adaptation, drift, or both? *EvoDevo*, *4*, 26.

Riccomagno, M. M., Martinu, L., Mulheisen, M., Wu, D. K., & Epstein, D. J. (2002). Specification of the mammalian cochlea is dependent on sonic hedgehog. *Genes Dev.*, *16*, 2365–2378.

Rink, J. C. (2012). Stem cell systems and regeneration in planaria. *Dev. Genes Evol.*, *223*(1-2), 67–84.

Rong, P. M., Teillet, M.-A., Ziler, C., & Le Duarin, N. M. (1992). The neural tube/notochord complex is necessary for vertebral but not limb and body wall striated muscle differentiation. *Development*, *115*, 657–672.

Ross, S. A., McCaffery, P. J., Dräger, U. C., & De Luca, L. M. (2000). Retinoids in embryonal development. *Physiol. Rev.*, *80*, 1021–1054.

Saberi, A., Jamal, A., Beets, I., Schoofs, L., & Newmark, P. A. (2016). GPCRs direct germline development and somatic gonad function in planarians. *PLoS Biol.*, *14*(5), e1002457. 2016 May.

Salazar-Ciudad, I., & Jernvall, J. (2002). A gene network model accounting for development and evolution of mammalian teeth. *Proc. Natl. Acad. Sci. U. S. A.*, *99*, 8116–8120.

Sariola, H., Holm, K., & Heinke-Fahle, S. (1988). Early innervation of the metanephric kidney. *Development*, *104*, 589–599.

Sariola, H., Ekblom, P., & Henke-Fahle, S. (1989). Embryonic neurons as in vitro inducers of differ entiation of nephrogenic mesenchyme. *Dev. Biol.*, *132*, 271–281.

Schaller, H. C. (1976). Action of the head activator as growth hormone in hydra. *Cell Differ.*, *5*, 1–11.

Schulman, V. K., Dobim, K. C., & Baylies, M. K. (2015). Morphogenesis of the somatic musculature in *Drosophila melanogaster*. *Wiley Interdiscip. Rev. Dev. Biol.*, *4*(4), 313–334. https://doi.org/10.1002/wdev.180.

Schwoerer-Bohning, B., Kroiher, M., & Müller, W. A. (1990). Signal transmission and covert prepattern in the metamorphosis of *Hydractinia echinata*. *Rouxs Arch. Dev. Biol.*, *198*, 245–251.

Šestak, M. S., Božičević, V., Bakarić, R., Dunjko, V., & Domazet-Lošo, T. (2013). Phylostratigraphic profiles reveal a deep evolutionary history of the vertebrate head sensory systems. *Front. Zool.*, *10*(1), 18. https://doi.org/10.1186/1742-9994-10-18. Published 2013 Apr 12.

Shimada-Niwa, Y., & Niwa, R. (2014). Serotonergic neurons respond to nutrients and regulate the timing of steroid hormone biosynthesis in *Drosophila*. *Nat. Commun.*, *5*, 5778. 2014 Dec 15.

Shimell, M., Pan, X., Martin, F. A., Ghosh, A. C., Leopold, P., O'Connor, M. B., et al. (2018). Prothoracicotropic hormone modulates environmental adaptive plasticity through the control of developmental timing. *Development*, *145*(6), dev159699. Published 2018 Mar 14. https://doi.org/10.1242/dev.159699.

Shtukmaster, S., Schier, M. C., Huber, K., Krispin, S., Kalcheim, C., & Unsicker, K. (2013). Sympathetic neurons and chromaffin cells share a common progenitor in the neural crest in vivo. *Neural Dev.*, *8*, 12.

Shwartz, Y., Gonzalez-Celeiro, M., Chen, C., Pasolli, H. A., Sheu, S., Fan, S. M., et al. (2020). Cell types promoting goosebumps form a niche to regulate hair follicle stem cells. *Cell*, *182*(3), 578. Aug 6; https://doi.org/10.1016/j.cell.2020.06.031.

Simões, M. G., Bensimon-Brito, A., Fonseca, M., Farinho, A., Valério, F., Sousa, S., et al. (2014). Denervation impairs regeneration of amputated zebrafish fins. *BMC Dev. Biol.*, *14*, 49.

Singh, U. (1984). Sympathetic innervation of fetal mouse thymus. *Eur. J. Immunol.*, *14*, 757–759.

Singh, S., & Groves, A. K. (2016). The molecular basis of craniofacial placode development. *Wiley Interdiscip. Rev. Dev. Biol.*, *5*(3), 363–376.

Sohal, G. H., & Holt, R. K. (1980). Role of innervation on the embryonic development. *Cell Tissue Res.*, *210*, 383–393.

Soper, J. R., Fiona Bonar, S., O'Sullivan, D. J., McCredie, J., & Willert, H.-G. (2019). Thalidomide and neurotrophism. *Skeletal Radiol.*, *48*, 517–525.

Stern, H. M., Brown, A. M. C., & Hauschka, S. D. (1995). Myogenesis in paraxial mesoderm: preferential induction by dorsal neural tube and by cells expressing Wnt-1. *Development*, *121*, 3675–3686.

Szymaniak, A. D., Mi, R., McCarthy, S. E., Gower, A. C., Reynolds, T. L., Mingueneau, M., et al. (2017). The Hippo pathway effector YAP is an essential regulator of ductal progenitor patterning in the mouse submandibular gland. *Elife*, *6*, e23499.

Taguchi, A., & Nishinakamura, R. (2017). Higher-order kidney organogenesis from pluripotent stem cells. *Cell Stem Cell*, *21*, P730–746.E6.

Taguchi, A., Kaku, Y., Ohmori, T., Sharmin, S., Ogawa, M., Sasaki, H., et al. (2014). Redefining the in vivo origin of metanephric nephron progenitors enables generation of complex kidney structures from pluripotent stem cells. *Cell Stem Cell*, *14*, 53–67.

Takeda, H., Nishimura, K., & Agata, K. (2009). Planarians maintain a constant ratio of different cell types during changes in body size by using the stem cell system. *Zoolog. Sci.*, *26*, 805–813.

Takuma, N., Sheng, H. Z., Furuta, Y., Ward, J. M., Sharma, K., & Hogan, B. L. M. (1998). Formation of Rathke's pouch requires dual induction from the diencephalon. *Development*, *125*, 4835–4840.

Tanaka, M., Tamura, K., & Ide, H. (1996). Citral, an inhibitor of retinoic acid synthesis, modifies chick limb development. *Dev. Biol.*, *175*, 239–247.

Tani-Matsuhana, S., Kusakabe, R., & Inoue, K. (2018). Developmental mechanisms of migratory muscle precursors in medaka pectoral fin formation. *Dev. Genes Evol.*, *228*(5), 189–196. https://doi.org/10.1007/s00427-018-0616-9.

Tapscott, S. J., Davis, R. L., Thayer, M. J., Cheng, P. F., Weintraub, H., & Lassar, A. B. (1988). MyoD1: a nuclear phosphoprotein requiring a Myc homology region to convert fibroblasts to myoblasts. *Science*, *242*(4877), 405–411.

Tata, P. R., & Rajagopal, J. (2017). Plasticity in the lung: making and breaking cell identity. *Development*, *144*(5), 755–766. 2017 Mar 1 https://doi.org/10.1242/dev.143784.

Teillet, M., Watanabe, Y., Jeffs, P., Duprez, D., Lapointe, F., & Le Douarin, N. M. (1998). Sonic hedgehog is required for survival of both myogenic and chondrogenic somitic lineages. *Development*, *125*, 2019–2030.

Terada, T. (2015). Ontogenic development of nerve fibers in human fetal livers: an immunohistochemical study using neural cell adhesion molecule (NCAM) and neuron-specific enolase (NSE). *Histochem. Cell Biol.*, *143*, 421–429.

Thorens, B. (2014). Neural regulation of pancreatic islet cell mass and function. *Diabetes Obes. Metab.*, *16*, 87–95.

Tiniakos, D. G., Mathew, J., Kittas, C., & Burt, A. D. (2008). Ontogeny of human intrahepatic innervation. *Virchows Arch.*, *452*, 435–442. https://doi.org/10.1007/s00428-007-0569-2.

Tuisku, F., & Hildebrand, C. (1994). Evidence for a neural influence on tooth germ generation in a polyphyodont species. *Dev. Biol.*, *165*, 1–9.

Ueishi, S., Shimizu, H., & Inoue, Y. H. (2009). Male germline stem cell division and spermatocyte growth require insulin signaling in *Drosophila. Cell Struct. Funct.*, *34*(1), 61–69.

Utsuyama, M., Kobayashi, S., & Hirokawa, K. (1997). Induction of thymic hyperplasia and suppression of splenic T cells by lesioning of the anterior hypothalamus in aging Wistar rats. *J. Neuroimmunol.*, *77*, 174–180.

Vainio, S. J., Itäranta, P. V., Peräsaari, J. P., & Uusitalo, M. S. (1999). Wnts as kidney tubule inducing factors. *Int. J. Dev. Biol.*, *43*, 419–423.

Voytik, S. L., Przyborski, M., Badylak, S. F., & Konieczny, S. F. (1993). Differential expression of muscle regulatory factor genes in normal and denervated adult rat hindlimb muscles. *Dev. Dyn.*, *198*, 214–224.

Wagner, D. E., Wang, I. E., & Reddien, P. W. (2011). Clonogenic neoblasts are pluripotent adult stem cells that underlie planarian regeneration. *Science*, *332*, 811–816.

Warburton, D., El-Hashash, A., Carraro, G., Tiozzo, C., Sala, F., Rogers, O., et al. (2010). Lung organogenesis. *Curr. Top. Dev. Biol.*, *90*, 73–158. https://doi.org/10.1016/S0070-2153(10)90003-3.

Wen, X., Huan, H., Wang, X., Chen, X., Wu, L., Zhang, Y., et al. (2016). Sympathetic neurotransmitters promote the process of recellularization in decellularized liver matrix via activating the IL-6/Stat3 pathway. *Biomed. Mater.*, *11*(6), 065007. 2016 Nov 4;.

Wenner, P., & O'Donovan, M. J. (2001). Mechanisms that initiate spontaneous network activity in the developing chick spinal cord. *J. Neurophysiol.*, *86*, 1481–1498.

White, I. A., Gordon, J., Balkan, W., & Hare, J. M. (2015). Sympathetic reinnervation is required for mammalian cardiac regeneration. *Circ. Res.*, *117*(12), 990–994. https://doi.org/10.1161/CIRCRESAHA.115.307465.

Whitsett, J. A., Kalin, T. V., Xu, Y., & Kalinichenko, V. V. (2019). Building and regenerating the lung cell by cell. *Physiol. Rev.*, *99*(1), 513–554. 163485.

Widelitz, R. B. (2008). Wnt signaling in skin organogenesis. *Organogenesis*, *4*, 123–133.

Wolpert, L., Beddington, R., Lawrence, P., & Jessell, T. M. (1998). *Principles of Development. Current Biology.* Oxford/New York/Tokyo: Ltd/Oxford University Press.

Youngblood, J. L., Coleman, T. F., & Davis, S. W. (2018). Regulation of pituitary progenitor differentiation by β-catenin. *Endocrinology*, *159*, 3287–3305.

Zakany, J., & Duboule, D. (2007). The role of *Hox* genes during vertebrate limb development. *Curr. Opin. Genet. Dev.*, *17*, 359–366.

Zhang, W., Behringer, R. R., & Olson, E. N. (1995). Inactivation of the myogenic bHLH gene MRF4 results in up-regulation of myogenin and rib anomalies. *Genes Dev.*, *9*, 1388–1399. https://doi.org/10.1101/gad.9.11.1388.

Brain involvement in phenotypic evolution

1 Environmental changes and animal evolution

It is generally observed that drastic changes in the environment almost always have been associated with great events in animal evolution, suggesting that a causal relationship may exist between them. Let us start with a succinct presentation of some notable facts on the correlation between sudden changes in the environment and great evolutionary transitions.

A drop in the sea level and rise of mountain ranges throughout the Earth led to large-scale climatic and ecological consequences that induced the global diversification of the Ordovician radiation, the Great Ordovician Biodiversification Event (GOBE). It followed the Cambrian explosion and took place between 485 and 443 mya (Drosser et al., 1996; Servais and Harpe, 2018). The major retreat of seas during the Devonian (380–360 mya) caused drastic climatic and ecological changes, leading to mass extinctions in the Earth's fauna and facilitated land colonization. The hyperoxic atmosphere during the Carboniferous period (360–300 mya) is thought to have favored tetrapod locomotion and advent of the flight in insects and birds (Dudley, 1998). During the Permian period (300–251 mya) occurred the largest recorded mass extinction in the Earth's history, also known as end-Permian mass extinction or Great Dying. It is explained by global warming as a result of increased volcanic activity (Shen et al., 2019), and the accompanying hypoxia that wiped out ~96% of all marine species and ~70% of terrestrial animals (Penn et al., 2018).

Another increase of volcanic activity and especially a meteoritic impact on the Earth that happened about 66 mya (Schulte et al., 2010) rapidly acidified the seawater, causing the mass extinction of the marine fauna and perturbation of the carbon cycle, which rebounded soon (in about 80,000 years) (Henehan et al., 2019). This period also marks the extinction of dinosaurs. Until this point, mammals were represented by small-body nocturnal insectivorous animals, but the end of the dinosaur era opened immense opportunities for geographic expansion and evolutionary adaptation of mammals, leading to the present dominant position of mammals in the hierarchy of the terrestrial vertebrate life. Even at smaller time scales of 10,000–100,000 years, it is observed that with the advance of glaciers the number of benthic ostracods decreased, whereas the retreat of glaciers was associated with increased species' diversity (Cronin and Raymo, 1997).

The Inductive Brain in Development and Evolution. https://doi.org/10.1016/B978-0-323-85154-1.00006-0

Even this glimpse of the above representative cases shows a clear correlation between drastic environmental changes and evolutionary events in the animal kingdom, which raises the question of the existence of a causal relationship between them. But, how can changing environmental conditions or stimuli be linked to evolutionary changes?

1.1 Stress response to sudden drastic changes in the environment

Extant animals have survived over long evolutionary periods of time by being adapted to their environmental niches. A sudden drastic change in the environment may present an existential risk and lead to the extinction of the species if it causes homeostatic disturbances that override its adaptational mechanisms or norm of reaction.

The disturbed homeostasis and the injurious environmental change are perceived as a threat in the central nervous system (Ulrich-Lai and Herman, 2009), which responds by activating the hypothalamic-pituitary-adrenal (limbic-hypothalamus-pituitary-adrenal, in humans) axis and other neurohormonal and physiological mechanisms of the stress response. The stress concept incorporates the environmental causal agent, the *stressor*, the disturbing effects of the stressor on the homeostasis, i.e., the *stress condition*, and the activation of the central stress circuit with the hypothalamus-pituitary-adrenal (HPA) axis and other physiological mechanisms for overcoming, or compensating for, injurious effects of the stressor, that is, the *stress response*.

According to the duration of the time of the action, biologists distinguish between

short-term (*acute*) *stressors* (e.g., detection of the approach of a predator, a temporary storm, seasonal depletion of food sources, etc.) inducing primarily the "fight-or-flight" response, which are irrelevant from the evolutionary viewpoint, and

long-term stressors, which persist for long periods of time, such as sudden transformation of a terrestrial into an aquatic habitat or the reverse, climatic changes, introduction of a new predator in the species' niche, etc., which, besides behavioral changes to avoid the stressor, may require other physiological and morphological adaptations.

According to the nature and origin of the stressor, biologists distinguish between

physical stressors (extremely warm and cold temperatures, physical exhaustion, injuries, illnesses, fatigue, etc.) and

psychological stressors, such as anger, grief, fear, panic, anxiety, etc.

Physical and psychological stressors are processed by different circuitries in the brain (Godoy et al., 2018).

All stressors act via the nervous system. An environmental agent will be perceived as a stressor after attaining a specific threshold of intensity or degree to which the animal responds by increased alertness, fear, and altered physiology. The threshold depends on the animal species and may be different not only in different species but even between conspecific individuals.

Stress-controlling mechanisms considered herein emerged early in the evolution of vertebrates. Expression of CRH (corticotropin-releasing hormone), the so-called stress hormone, is observed in all the vertebrate classes from fish to mammals (Yao et al., 2004; Crespi and Denver, 2005; Matsuda, 2013) as suggested by the identification of CRH-like and cortisone-like substances in invertebrate animals such as mollusks (Ottaviani et al., 1998).

A common characteristic of various forms of stress responses in vertebrates is activation of the neuroendocrine mechanism and the accompanying adaptive change of the behavior. Behavioral responses (fight, flight, or freeze) to stress arise from the brainstem and spinal neurons according to messages sent from evolutionarily newer brain structures.

The input of the stressful sensory and somatosensory stimuli is transmitted to the CNS via spinal and cranial sensory nerves. The short neuronal circuit determines the spinal stress response based on spinal reflexes, and the long neuronal circuit controlling the supraspinal stress response involves the prefrontal cortex, amygdala, hippocampus, and hypothalamus (Pacak and Palkovitz, 2001) (Fig. 4.1).

In humans, the affective and cognitive information from the limbic system (prefrontal cortex, amygdala, and hippocampus) is integrated into the nucleus of the solitary tract (NTS), which is essential for many behavioral and neuroendocrine processes (Rinaman, 2011; Myers et al., 2017). NTS also receives interoceptive input about the glucocorticoid level and afferent information about homeostasis (Fig. 4.2). The chemical output resulting from the integration and processing of the input in the brain determines the result that may be healthy or aberrant activation of the HPA axis.

1.2 The neuroendocrine control of the stress response

The neuroendocrine stress-controlling mechanism is part of the integrated control system in animals. The CNS detects the environmentally disturbed homeostasis by comparing it to the "anticipated perceptions of the external and internal environment" (Goldstein, 1990). As defined by McEwen (1999), stress is "a threat, real or implied, to the psychological or physiological integrity of an individual" (Greenberg et al., 2002). The definition of stress as a systemic response to both real or perceived stressors highlights the predominantly neural nature of the stress response. The stress response involves primarily the activation of the hypothalamus-pituitary-adrenal axis. The response begins with integration and processing of neural signals in the prefrontal cortex, amygdala, hippocampus and the paraventricular nucleus (PVN), resulting in secretion of the stress neurohormone, CRH (corticotropin-releasing hormone), from the PVN (Howland et al., 2017) (Fig. 4.3). The stress response may vary from dramatic neuroendocrine responses of the clinical level to very mild forms, and a continuum of intermediate forms in between.

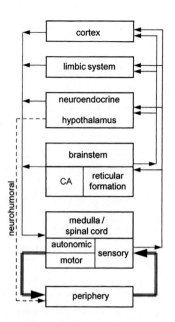

FIG. 4.1

Neuronal circuits in the organization of stress responses. *Horizontal thick and thin lines* indicate "short circuit": autonomic (sympathoadrenal and/or parasympathetic) and defense (withdrawal) spinal reflexes in response to stressful stimuli. Thin lines represent "long circuit": ascending (afferent) and descending (efferent) neuronal loops between the spinal cord/ medulla and "higher" brain centers. The *dashed line* indicates neurohumoral hypothalamic-pituitary outflow. *CA*, brainstem catecholaminergic neurons.

From Pacak, K., Palkovitz, M., 2001. Specificity of central neuroendocrine responses implications for stress-related disorders. Endocr. Rev. 22, 502–548.

CRH, which is a small neuropeptide of 41 amino acid residues, is released into the portal system and stimulates secretion in the anterior pituitary of the hormone adrenocorticotropic hormone (ACTH), endorphins, and melanocyte-stimulating hormone (MSH). ACTH and MSH "increase attention and motivation, also improve neuromuscular performance, all important components of the successful response to stress" (Strand, 1999).

The perception of stressors, and the stress condition they cause, trigger the stress response under the control of the nervous system. The stress response involves changes in the neuroendocrine system and the associating physiological changes aimed at normalizing the disturbed homeostasis. The stress response may be *immediate* or *delayed*.

Immediate response includes both changes in the HPA (hypothalamus-pituitary-adrenal) axis and related systemic changes, as well as changes in animal behavior intended to avoid, circumvent, or lessen the harmful effects of the stressor by settling into a new habitat, relocating, migrating, etc. The immediate stress response may

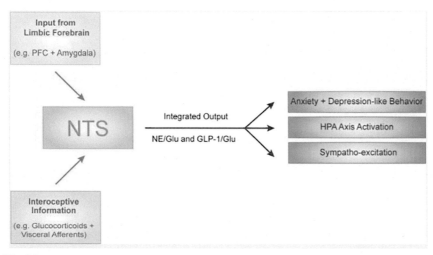

FIG. 4.2

Summary of NTS as nexus for integrating descending limbic input with ascending interoceptive information to generate behavioral, neuroendocrine, and autonomic stress responses. Regions of the limbic forebrain including the PFC and amygdala, among others, project to the NTS. The NTS also expresses a high density of glucocorticoid receptors and receives ascending visceral signals related to physiological status. At the level of the NTS, this information is integrated and multiple output circuits including NE/Glu and GLP-1/Glu coordinate organismal adaptation to adversity. *GLP-1*, glucagon-like peptide-1; *Glu*, glutamate; *NE*, norepinephrine; *NTS*, nucleus of the solitary tract; *PFC*, prefrontal cortex.

From Myers, B., Scheimann, J.R., Franco-Villanueva, A., Herman, J.P., 2017. Ascending mechanisms of stress integration: implications for brainstem regulation of neuroendocrine and behavioral stress responses. Neurosci. Biobehav. Rev. 74 (Pt. B), 366–375. https://doi.org/10.1016/j.neubiorev.2016.05.011.

also come in the form of the development in the animal or its offspring of qualitatively new traits, as is the case with intergenerational (see later in this chapter) and, less frequently, transgenerational developmental plasticity (see Chapter 5).

Delayed response to the environmental challenge also implies adaptation of the structure, size, and morphology of the existing parts or organs to the changing environment within the limits of the species-specific norm of reaction, under neural regulation as it is seen in cases of intergenerational and transgenerational developmental plasticity, or sporadically with the emergence of evolutionarily new traits.

1.3 Stress-induced developmental instability

Living in environments that are generally fluctuating and unpredictable exerted evolutionary pressure to optimize the balance between the potentiality to faithfully transmit phenotypic traits to the offspring and to maintain a significant degree of evolvability, with the balance clearly tilting on the side of developmental stability. Under normal

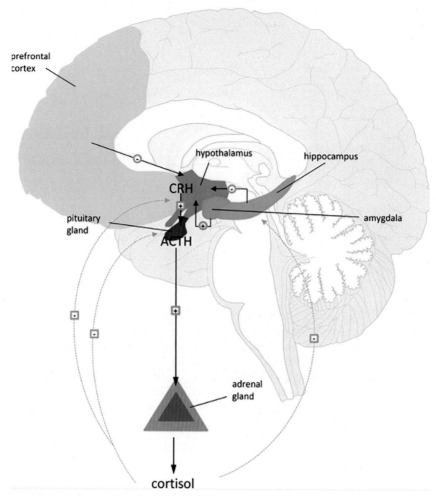

FIG. 4.3

Schematic representation of the hypothalamic-pituitary-adrenal (HPA) axis. In response to stress, corticotropin-releasing hormone (CRH) is synthesized in the paraventricular nucleus (PVN) of the hypothalamus and released into the hypophyseal portal blood. CRH binds to its receptors on pituitary corticotropes, stimulating the release of adrenocorticotrophic hormone (ACTH). Circulating ACTH binds to its receptors in the adrenal cortex and stimulates the release of cortisol, which mobilizes body systems to respond to the stressor. Elevated circulating cortisol inhibits further HPA axis activity (*blue squares*) by binding to its two receptor types, glucocorticoid receptors (GRs) and mineralocorticoid receptors (MRs), at the level of the hypothalamus, pituitary, and hippocampus. CRH-producing neurons in the PVN of the hypothalamus are innervated by afferent projections from multiple brain regions (*blue and green circles*), including the amygdala, which provides excitatory input, and the prefrontal cortex and hippocampus, which provide inhibitory input. *ACTH*, adrenocorticotropic hormone; *CRH*, corticotropin-releasing hormone; *GR*, glucocorticoid receptor; *MR*, mineralocorticoid receptor; *PVN*, paraventricular nucleus.

From Howland, M.A., Sandman, C.A., Glynn, L.M. 2017. Developmental origins of the human hypothalamic-pituitary-adrenal axis. Expert. Rev. Endocrinol. Metab. 12 (5), 321–339. https://doi.org/10.1080/17446651. 2017.1356222.

conditions, developmental mechanisms are highly resistant to random change or "noise," hence conserved across the animal taxa. However, chronic stress, under conditions of drastic changes in the environment, may prevail over developmental stability. The resulting developmental instability may increase the likelihood of uncontrolled changes in developmental pathways, leading to detrimental consequences, but very rarely to the emergence of new advantageous pathways or revival of the lost ancestral ones.

The stability of developmental pathways often is attributed to the existence in animals of the so-called "developmental constraints" that prevent evolution in certain directions. Regarding the origin, developmental constraints that ensure the stability of developmental pathways may be:

- *constitutive constraints* resulting from increased complexity and integration of structures and functions (cell differentiation, specialization, and division of labor) that make morphological change impossible because such changes would affect the whole embryo, or
- *functional constraints* determined by evolutionary inhibition of particular pathways in response to the pressure for higher stability in particular pathways. These constraints may be weakened or removed under the influence of environmental or internal factors.

Developmental stability is a relative stability because developmental constraints are not absolute or inalterable. Conservation and change are two sides of the same evolutionary coin. Under conditions of chronic environmental stress, advantageous changes in specific developmental pathways may occur that lead to the generation of new discrete morphological changes as it occurs in cases of intra-, inter-, and transgenerational developmental plasticity. A very interesting fact is that discrete morphological changes, both nonheritable (intragenerational) and heritable (inter- and transgenerational), occur not only in response to chronic stress but also under conditions that are nonstressful, but the presage approach of the cyclical seasonal stressful environmental changes, suggesting the existence of a transgenerational memory that the approach of the stressful condition is preceded by the presaging cue.

Evolutionary effects of the developmental instability resulting from stress conditions may emerge in the first offspring generation or in later generations. They may emerge abruptly or evolve stepwise; they may affect particular individuals, proportions of a population, or whole populations.

2 Evolutionary importance of behavioral changes

The existence of a relationship between behavior and evolution is not a new notion: "One cannot intelligently discuss behavior and structure separately. Behavior is what an animal does with its structure; structure is what an animal uses to behave" (Evans, 1966). "In animals, almost invariably, a change in behavior is the crucial factor initiating evolutionary innovation" (Mayr, 1988). If the behavior is a function of the structure, in a particular meaning, behavior then is the *raison d'être* of the structure, and the necessity to perform a specific behavior influences the evolution and perfection of an organ or part of the animal. "We hypothesize that the *same mechanisms* (emphasis mine—N.R.C.) act to alter both morphological and behavioral traits in response to the environment in which an organism develops" (Carvalho and Mirth, 2015).

An impressive example of the close relationship between morphology on the one hand and behavior on the other hand is the life history of holometabolous insects that, during metamorphosis, jointly change their morphology and behavior from that of a worm to that of adult insects. Similarly, most amphibians change, in parallel, their morphology and behavior from that of a fish to that of an adult amphibian.

2.1 Neural basis of animal behavior

Animal behavior comprises innate and learned behaviors. Animals perform most innate behaviors perfectly and invariably from their birth, without previous experience (instincts), whereas learned behaviors are acquired and mastered later through experience, but not transmitted to the progeny. The nervous system is organized in neuronal circuits, and both innate and learned behaviors "from simple reflex withdrawal away from a noxious to a complex mating dance" (Delcomyn, 1998) result from activation of specific neural circuits.

2.1.1 Innate behaviors

Innate behaviors result from activation of specific neural circuits that are relatively hardwired in the nervous system (Gillette, 1991) and are established during embryonic life, before birth, and hence experience independently, although experience sometimes can influence the innate behavior (Blumberg, 2017). Most of the knowledge on neural circuits comes from studies on invertebrates whose circuits have a smaller number of elements (neurons and synapses) compared with the larger number of components of vertebrate circuits (Golowasch, 2019).

According to Konrad Lorenz, a specific external stimulus triggers an "innate releasing mechanism" (German *angeborenes auslösendes Schema*—innate releasing schema) (Schleidt, 1962) that produces a specific behavior—a motor pattern known as fixed action patterns (FAPs). It is demonstrated that FAPs are wired in the embryonic nervous system. So, e.g., homotopic transplantation of parts of the Japanese quail (*Coturnix coturnix*) neural tube into chick embryos produces chimera domestic chicken (*Gallus gallus domesticus*) displaying quail crowing and head movements, subcomponents of the innate behavior originating from two different regions of their brain (Balaban, 1997). Also, the complex behavior of extrusion of the string of eggs from the reproductive tract to the gluing of the egg mass on a solid substrate in the marine gastropod mollusk, *Aplysia californica*, entails activation of several FAPs involving strictly determined movements of the head and neck. The egg deposition behavior is triggered by secretion of neuropeptide egg-laying hormone (ELH) by specific neurons (Purves et al., 2001b).

The hypothalamus, as a very ancient part of the vertebrate brain, is the site of several circuitries for innate behaviors. Most behaviors studied in animals involve hypothalamic circuits or are related with the hypothalamic-pituitary-terminal endocrine gland axes: "Specialized neuroendocrine circuits for innate behaviors thus seem to process sensory information relevant to ethological contexts and influence sensory perception and processing; integration by these circuits of multiple pathways of information relevant to different behaviors determines the behavioral state of the animal" (Manoli et al., 2006). A typical example of the role of the sensory perception in triggering innate behaviors is the milk let-down reflex in mammals; stimulation of the nipples by sucklings reaches the brain via afferent pathways and induces secretion of oxytocin by various brain centers, whereas the magnocellular oxytocin (OT) neurons that project their axons to the posterior pituitary stimulate the gland to release oxytocin in the blood (Knobloch et al., 2012) within a few seconds and squeeze milk down to the nipple in less than 1 min (Gorbman and Davey, 1991).

An example of complex innate behaviors is the spider web-constructing instinct. It is estimated that a cocoon-spinning spider has to perform 6400 individual movements to make its egg cocoon, based on a closed program, not modifiable by experience (Purves et al., 2001a). Annual migration and repatriation are other innate behaviors observed across the animal taxa, in invertebrates (insects, crabs, etc.) as well as vertebrates (fish, reptiles, birds, and mammals). How the details of the itinerary are wired in the brain and what sensory (visual, acoustic, olfactory, etc.) cues they use to reach the destination are not fully understood. According to one hypothesis, migratory birds use magnetite particles localized in the upper beak and the ethmoid region for geomagnetic orientation; the alternative hypothesis posits that geomagnetic orientation is a function of biradical reactions of macromolecules excited by light energy. It is believed that migrating birds that fly long distances to reach their destinations exploit changes in the light-dependent geomagnetic field (Muheim et al., 2002) to extract information for the navigational "map" that can be used for determining the position of the bird during migration (Beason, 2005).

The ability of birds to distinguish between the conspecific and heterospecific songs is another innate behavior established experience-independently during embryonic life (Whaling et al., 1997).

Types of behavior mentioned so far are neurally determined according to the sensory input from the periphery or the environment and can be modified by the sensory input. There are other innate behaviors that are centrally determined and are not modifiable by the sensory input, such as rhythmic behaviors (swimming, flight, walking, scratching, and chewing) that led to the concept of central pattern generators (CPGs) that produce rhythmic behaviors in the absence of rhythmic inputs and serve as the basis of FAPs (Croll et al., 1985; Suesswein and Byrne, 1988).

2.1.2 Neural circuits for innate behaviors

Some neural circuits for innate behaviors are characterized by a relatively high flexibility. "Single circuits can produce more than one behavior, and different circuits can combine to produce a single, new behavior" (Tierney, 1995). In some instances, the same neural circuit may generate different motor patterns by selectively activating specific neurons. For instance, the stomatogastric ganglion (STG) in the crab *Cancer borealis* is innervated by three modulatory projection neurons containing:

proctolin,

proctolin + neurotransmitter GABA, and

proctolin + GABA + CabTRP.

Activation of each of these modulating neurons elicits a distinct motor pattern, respectively, the MCN7 pyloric rhythm, MPN pyloric rhythm, and gastropyloric rhythm (Blitz, 1999) (Fig. 4.4).

In other instances, the activation of a neural circuit suppresses the activity of another circuit. So, e.g., in the predatory sea slug, *Pleurobranchaea californica*, activation of the swim escape circuit suppresses the activity of the feeding circuit (Jing and Gillette, 2000). Noteworthy, as an example of the flexibility of neural circuits, is the use in lampreys of the same central pattern generator to perform different behaviors, such as swimming, crawling, and burrowing (Ayers et al., 1983).

Neural circuits are dynamic information-processing networks that may be reconfigured by neuromodulators: "It is not possible to model the function of neural networks without including the actions of neuromodulators" (Harris-Warrick, 2011). They can be modulated to generate different outputs in response to external or internal stimuli. Secretion of specific neurohormones and neurotransmitters/neuromodulators can modify the processing of the sensory input in neural circuits and thus bias

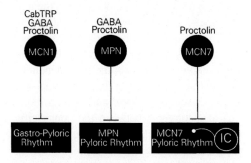

FIG. 4.4

Summary of the distinct motor patterns elicited from the crab STG network by selective activation of the proctolin neurons MCN1, MPN, and MCN7. MCN1 contains the transmitters CabTRP Ia, GABA, and proctolin, and it elicits a gastropyloric motor pattern. MPN contains proctolin and GABA but not CabTRP Ia. It elicits a particular pyloric rhythm and no gastric mill rhythm. MCN7 contains proctolin but not GABA and CabTRP Ia. MCN7 elicits a distinct pyloric motor pattern, which is dominated by IC neuron bursts. During this motor pattern, IC neuron activity regulates the pyloric frequency.

From Blitz, D.M., Christie, A.E., Coleman, M.J., Norris, B.J., Marder, E., Nusbaum, M.P., 1999. Different proctolin neurons elicit distinct motor patterns from a multifunctional neuronal network. J. Neurosci. 19, 5449–5463.

toward, or increase, the probability of specific behaviors (Kravitz, 1988). For instance, in lampreys, the central pattern generator for locomotion in the spinal cord can be activated in vitro by the excitatory neurotransmitter D-glutamate to produce swimming patterns, but addition of serotonin in vitro modulates the central pattern generator and the swimming pattern (Harris-Warrick and Cohen, 1985). The functional configuration of a neural circuit depends on the neuromodulators acting at any particular time (Diao et al., 2017).

Modulation of neural circuits for generating different outputs also emerges as a response to external stimuli, but evidence suggests that modulation may also be induced intrinsically, without external stimuli: "Extrinsic neuromodulation is known to be pervasive in nervous systems, but intrinsic neuromodulation is less recognized, even though it has now been demonstrated in sensory and neuromuscular circuits and in central pattern generators. By its nature, intrinsic neuromodulation produces local changes in neuronal computation, whereas extrinsic neuromodulation can cause global changes, often affecting many circuits simultaneously" (Katz and Frost, 1996).

Neuromodulators facilitate the switching of a neural circuit from an innate behavior to another. A case in point is the transition of the locust *Locusta migratoria* from the solitary to the gregarious phase characterized by crowding, crowd flying, and changes in morphology and morphometry as well as dark pigmentation of the body induced by injection of the neurohormone corazonin (Sugahara et al., 2015). Recall, the production of corazonin in the locust's nervous system is the result of a manipulative expression of the gene through activation of a circuit of the gene for the neurohormone. Similarly, injection of the adrenal steroid hormone, corticosteroid, in the eggs of the lizard *Lacerta vivipara* leads to production of offspring with altered antipredator behavior (Uller and Olsson, 2006).

Molting in insects is a function of the ecdysis circuit. It consists in execution of the ecdysis sequence, a series of several motor programs (White and Ewer, 2014), resulting in the shed of the cuticle in many invertebrates, including nematodes and arthropods (Roller et al., 2010). The process begins with secretion of ecdysis-triggering hormone (ETH) by endocrine Inka cells (Gammie and Truman, 1997), which in turn are induced by two brain neuropeptides, corazonin and eclosion hormone (EH) (Žitňan et al., 2007). EH is a neuromodulator of the ecdysis circuit that binds its receptor, ETHR (ETHR-A and ETHR-B), in the target neurons in the CNS. Neurons with ETH receptors stimulate neurons that have receptors for the neuropeptide bursicon. Secretion of bursicon activates neurons involved in the ecdysis central pattern generator (Marder, 2012).

Neural circuits are relatively stable and uncoupled from the evolution of genes. So, e.g., in the subphylum Crustacea, the circuitry of the stomatogastric ganglion (STG) and even the synaptic connections are conserved to a surprising degree, despite the evolution of genes during the last 350 million years. Although neural circuits are conserved in the evolution of metazoans, even small changes in neuronal circuits in various species can produce species-specific behaviors (Katz and Harris-Warrick, 1999).

When animals change their habitat and consequently the former sensory input ceases, the neural circuit, instead of being disassembled, may be used for another purpose. This seems to have been the case with the blind cave fish, in which, within the last ~10,000 years since they lost their eyes, the tectum shifted from visual processing to the processing of somatosensory input (Voneida and Fish, 1984).

In most mammals, including rodents and humans, most innate behaviors, especially those of the social nature (mating, maternal care, social receptivity, territorial behavior, fear, anxiety, reward, etc.), are determined by neuronal circuits in the limbic system centered around amygdala but including the hippocampus, solitary tract, prefrontal cortex, and hypothalamus. Afferent sensory stimuli processed in these neuronal circuits trigger execution of respective behaviors.

As for the ontogeny of neuronal circuits for innate behaviors, it is noteworthy that most in mice are established prenatally and experience independently, by the embryonic day 11–15 (E11–15) and migrate to their destinations by E18 (Sokolowski and Corbin, 2012). Evolution of the function of neural circuits results from evolution of the neurons involved in the circuits, the configuration of the circuit, and resulting computational properties, which may change by not involving any changes in genes.

Summarizing the brief evidence presented above, it can be said that innate behaviors in animals are programmed and executed by neuronal circuits: "Behavior is a readout of neural function" (Sato et al., 2020).

2.2 Neural basis of learned behavior

Environmental disasters such as permanently flooded habitats and introduction of new predators pose physiological-homeostatic or life-threatening problems that require adaptive changes in animal behavior, physiology, and morphology. In such cases, behavioral and physiological adaptations represent the first and immediate responses. Immediate behavioral response comes in the form of the flight instinct intended to escape the threatened habitat. Simultaneously, the animal responds by activating the stress response.

Animals that remain in a disaster-affected environment may experience chronic stress and learn new behaviors that diminish the harmful effect of the environment. Animals that live between the flooded and the land environment can learn to swim in shallow waters bordering the telluric habitat. For patterning and executing learned behaviors, animals use the existing species-specific fixed action patterns (FAPs) by modulating respective neural circuits.

The learning to swim in terrestrial animals may be facilitated by the fact that a single central pattern generator using a basic undulatory pattern to perform different forms of locomotion (swimming, crawling, and burrowing) evolved early in vertebrates such as lampreys (Ayers et al., 1983), which are able to switch between the above modes of locomotion. In view of the high conservancy of the neural circuits across the metazoan phyla, even species and other taxa that during their phylogeny lost the ability to swim may have conserved the relevant locomotion central pattern generator. The return to swimming may not be immediately perfect, but perfection and fine-tuning come later in the evolution; the learned behavior is not an "all-or-none" process.

Although it is not perfectly adaptive under the changed conditions of the environment, the learned behavior may be beneficial because of the following:

First, survival in a drastically changing environment does not require a perfect behavioral adaptation from the beginning; once the changed behavior makes survival possible, perfection of the behavior and morphophysiological changes may follow over time.

Second, drastic changes in the environment most of the time do not come up in the form of contrasting conditions of living such as, e.g., aquatic/terrestrial habitat, moist/dry terrain, cold/warm weather, short/long photoperiod, abundance/scarcity of food, herbivorous/carnivorous diet, presence/absence of predators, etc.; their intermediate states are more frequent in nature.

Third, often the same FAP can serve more than a single purpose. As J.L. Gould put it, "Learning involves a single set of basic processes which are shared throughout the animal kingdom from molluscs to man" (Gould, 1982).

Fourth, there is evidence showing that neural elements and connections for performing a behavior may be conserved after a species has lost the behavior. In such a case, the "learning" is reduced to activation of the ancestral circuit required for performing the new behavior.

Learned behavior is based on existing motor programs or motor program preadaptations. Through continued use, over time it may become inherited as an innate behavior and executed as a FAP. Both innate and learned behaviors are products of neural circuits. Although differences between these two types of behavior do exist, they do not amount to a "Chinese Wall" separating them. For a number of biological phenomena (e.g., imprinting, modification of song circuits and circuits of other behaviors to be dealt later in this chapter), it has been shown that both these forms of behavior are based on essentially similar neurobiological mechanisms. Like evolution of innate behaviors, the learning of new behaviors is associated with structural and functional changes in neurons and neural circuitries as it occurs in the song circuitries of zebra finches. In regard to the possibility of learned behaviors evolving into instincts, in his time, Darwin suggested: "If we suppose any habitual action to become inherited—and it can be shown that this does sometimes happen—then the resemblance between what originally was a habit and an instinct becomes so close as not to be distinguished" (Darwin, 1859, p. 209) and added: "Some intelligent actions, after being performed during several generations, become converted into instincts and are inherited" (Darwin, 1874). Learned behavior circuits in humans may be performed as effortless as spinning a web is to a spider or dead-reckoning is to a desert ant (Pinker, 1994).

Just like innate behaviors, learned behaviors are generated by activation of specific neural circuits. Simple behaviors are based on activation of FAPs; hence, it can be reasonably assumed that complex behaviors can be learned by integrating appropriate FAP circuits into neural circuits of the complex behaviors.

3 The relationship between behavior and morphology: Is it correlative or causal?

As mentioned earlier, environmental changes and stressors force animals to change their behavior. A vast body of empirical evidence shows that induction of behavioral changes as a result of stressors is sometimes associated with surprisingly rapid adaptive changes in morphology. This is typical for cases of predator-induced defenses (see Section 4.4 later in this chapter) in which adaptive changes in behavior and morphology occur almost simultaneously and both variables vary as an integrated unit (Van Buskirk and McCollum, 2000). Does this frequently observed correlation point to a causal relationship between the stress-induced behavioral change and the emergence of adaptive morphological defenses?

The systematic appearance of the correlation certainly proves that it cannot be a random occurrence. No systematic correlation would be possible in the absence of a direct or indirect causal relationship. The behavioral change precedes the development of the predator-induced morphological defense, and in a truly causal relationship, the cause precedes the effect.

In our causal inquiry of the development of the predator-induced defense, it is necessary to consider the predator as the ultimate element in the causal chain since:

– both the new behavior and development of the changed morphology appear only after the detection of the predator or predator risk, and
– there is no evidence that this type of defense may develop without, or separately from, predator-induced change in behavior.

As for the neural circuit(s) involved in the development of the morphological defenses, we are left with no alternative but to choose between two hypotheses: a single neural circuitry determines the rise of both the new behavior and the development of the adaptive defense or two separate, but causally related, neural circuits are involved in the correlation.

From the standpoint of the metabolic cost and the evolutionary parsimony, the first hypothesis looks more plausible, but the second is mechanistically preferable. If the latter were the case, then the temporal precedence of the behavior suggests that the behavioral circuit may be causal to, or triggers, the assemblage or activation of the neural circuit for the adaptive morphology.

Numerous cases of the correlated emergence of new behaviors and defensive morphologies are described in animals. To illustrate this, let us mention only several typical examples when upon detection of the predator or its kairomones in the environment, animals concurrently with the antipredator behavior develop defensive morphologies that reduce the likelihood of being eaten by the predator. Such is the case, e.g., with mayfly larvae, *Drunella coloradensis,* which on detecting the presence in the environment of brook trout *Salvelinus fontinalis* develop longer caudal filaments (Dahl and Peckarsky, 2002); the marine bryozoan *Membranipora membranacea,* exposed to extracts of its predator, the nudibranch *Doridella steinbergae,* within 2 days, develop spines of sizes that vary according to the concentration of the extract (Harvell, 1998); the blue mussel *Mytilus edulis* and the snail *Littorina obtusata*, outplanted into the habitat of their predator, the crab *Carcinus maenas*, develop defensive thicker shells (Trussell and Smith, 2000; Trussell, 2001); and tadpoles of the frog *Rana dalmatina* respond to the presence of the predatory fish threespine stickleback *Gasterosteus aculeatus* by developing a longer tail and more massive tail muscle, helping them to quickly hide in the mud or plants in the vicinity whereas in response to the presence of the dragonfly larvae develop deeper tail fin, enabling them to swim faster (Teplitsky et al., 2005a,b).

4 Inheritance of environmentally triggered changes: Evolutionary changes

Discovering the molecular mechanisms of the evolution of organs and parts of animals that occurred in the remote past is a task that leaves biologists with no option but to *draw inferences* from

- developmental mechanisms active in extant animals,
- intra- and transgenerational inheritance,
- the process of regeneration of the lost organs or parts, and
- paleontological evidence that may reveal the temporal order of the evolution of phenotypic traits.

Herein, evolutionary change is defined as any new phenotypic (behavioral, physiological, morphological, or life history) change, regardless of the inducing mechanism, arising in animals that is transmitted to the progeny. Any phenotypic change or any genotypic change (mutated gene) is first put to the test of ontogeny before it will be tested in the struggle for life during the postnatal life. Ontogeny is the workshop of evolutionary changes (Cabej, 2008). As sagaciously put by Walter Garstang, "Ontogeny does not recapitulate phylogeny: it creates it" (Garstang, 1922).

Evolutionary changes at the supracellular and behavioral levels are generated sometime during the ontogeny, in the process of individual development, between the zygote stage and adulthood. Hence, causally they are products of, and follow, changes in developmental pathways.

From an evolutionary standpoint, the emergence of a new trait or new organ commonly implies an evolutionary advantage that preserved it in the extant animals. The preservation of the new organ/trait in the extant animals also suggests that the molecular mechanisms of development have successfully passed the proof of natural selection. In this meaning, individual development is an archive of evolution and developmental mechanisms of extant animals represent a crucially important source of evolutionary knowledge. By revealing and critically reviewing developmental mechanisms, biologists hope to acquire valuable knowledge on the evolutionary events that took place in the animal phylogeny.

The evidence provided in Chapter 3 makes it clear that the nervous system/brain is crucially involved in the mechanisms of the development of organs. The question now is whether the nervous system has a substantial role in the emergence of evolutionary change and animal evolution in general: "Irrespective of their different nature, most animal adaptation mechanisms share the involvement of the central nervous system and often include endocrine activity" (Kolk et al., 2002).

Evolutionary change implies transmission of a new morphological/behavioral trait to the progeny, but this general definition is controversial. Intergenerational plasticity may be easily dismissed as a form of evolutionary change due to the loss of the "acquired trait" in the second generation. However, the problem is not so simple when one considers cases of transgenerational plasticity in which the new trait persists from two to many (Rechavi et al., 2011), or even an indefinite (Vastenhouw et al., 2003) number of consecutive generations. Moreover, observations on particular *Drosophila* strains show that they develop new "permanent" traits when reared for hundreds of generations under laboratory conditions, but switched back to the original wild traits within about 50 generations when reared in the ancestral environment (Teotónio and Rose, 2000; Teotónio et al., 2002). Thus, any limits set to the number of generations the new trait persists to be named evolutionary change will be arbitrary.

In this chapter, the role of the nervous system in determining the evolutionary changes will be illustrated primarily by evidence on the intergenerational predator-induced morphological defenses and the neurally induced speciation in animals. Substantial evidence on transgenerational developmental plasticity will be provided in Chapter 5.

4.1 Intergenerational plasticity

Intergenerational plasticity herein will be considered the appearance of a discrete new/modified trait in the offspring of animals that have experienced a specific environmental stimulus, although the offspring may have not been affected by the stimulus. The fact that in the absence of the triggering stimulus the new trait disappears after F1 makes it different from the evolutionary change. However, intergenerational plasticity and evolutionary change share an important attribute: the appearance in the progeny of a new, ancestrally not possessed trait. Given the appearance of new morphological traits involves developmental mechanisms, the considerable experimental accessibility offered by the intergenerational plasticity makes it an important source of information on the developmental mechanisms of the evolutionary change.

4.2 Predator-induced intergenerational plasticity

The first step in the process of induction of predatory-induced defensive morphologies in invertebrates is perception of the presence of the predator in the environment. The perception results from the processing of the visual, tactile, olfactory, and other cues in the central nervous system (CNS), with olfactory cues released by predators being predominant among them, especially in aquatic invertebrates.

4.3 Perception of kairomones in the CNS

Predator-induced plasticity requires the ability of preys to detect and identify the chemical cues, kairomones, released by predators. Olfactory perception has the advantage of being functional even when the predator is invisible because of physical barriers (for instance, when the water is too turbid for clear vision). The olfactory perception is also important for detecting the earlier presence of the predator in the prey's niche.

The olfactory system of crustaceans is not studied as well as that of vertebrates. Perception of kairomones in zebrafish takes place in the nostrils of the paired olfactory organ that lies on the dorsal side of the head, close to the eyes. The kairomone-containing water enters the nasal cavity through a funnel-shaped inlet. The olfactory organ is lined with a sensory epithelium consisting of four distinct sensory neurons: ciliated, microvillous, crypt, and kappe neurons (Hansen and Zeiske, 1998; Ahuja et al., 2015) (Fig. 4.5).

It is believed that the basic organization of the olfactory lobes is of glomerular structure and is the same in insects such as *Drosophila melanogaster* and crustaceans. Most olfactory neurons express one specific olfactory receptor (OR), and all ORNs expressing the same functional receptor project to one glomerulus of the antennal lobe. Chemoreceptor proteins in crustaceans are gustatory and olfactory receptor-like receptors, and gustatory receptor-like receptors are identified only in *Daphnia pulex*. Neurites from antennule extend to deuterocerebrum, but glomeruli in daphnids are not identified. In other crustaceans, olfaction is a function of the olfactory sensilla in the antennules that are lateral flagella of the first antennae, known as esthetascs, which are innervated by olfactory receptor neurons. The axons of the olfactory sensory neurons (OSNs) project to the neuropil compartment and to the bilateral olfactory lobes, which are organized into glomeruli. The inhibitory rim neurons also project in the neuropil. The local processing that takes place in glomeruli is followed by the central processing and the release of the behavioral output (Polanska et al., 2020) (Fig. 4.6).

Another form of olfaction is related to the so-called chemoreception that is mediated by the rest of chemosensilla distributed in various parts of the crustacean body, which are innervated by both chemoreceptor and mechanoreceptor neurons. Olfaction and distributed chemoreception share some functions, but the distant detection of sex pheromones, social cues, and alarm cues are functions of olfaction (Derby et al., 2016).

4.4 Evidence on predator-induced intergenerational morphological changes

Inducible defenses in cladocerans include the formation of spine neckteeth, crests, crowns of thorns, and helmets (Weiss et al., 2015a) (Fig. 4.7).

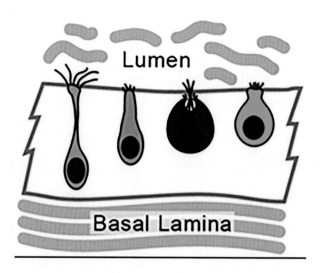

FIG. 4.5

Schematic representation of four types of olfactory sensory neurons with their laminar position. Ciliated neurons (*orange*) have round somata and slender dendrites that terminate in bundles of cilia on the epithelial surface. They constitute the most basal layer of olfactory sensory neurons. Microvillous neurons (*blue*) have bundles of microvilli on their apical surface. Crypt neurons (*red*) are globular-shaped and carry both microvilli and cilia on their apical surface. They are located more apical than microvillous neurons. Kappe neurons (*green*) are pear-shaped with an apical appendage resembling a cap (German: Kappe), have no cilia, and are located even more apical than crypt neurons. Kappe neurons (*green*) constitute a novel olfactory sensory neuron population.

From Ahuja, G., Nia, S.B., Zapilko, V., Shiriagin, V., Kowatschew, D., Oka, Y., et al., 2015. Kappe neurons, a novel population of olfactory sensory neurons. Sci. Rep. 4, 4037. https://doi.org/10.1038/srep04037.

The cladoceran *D. pulex* forms neckspine containing several teeth-like structures on the neck region on perceiving the presence of its insect predator, phantom midge larvae *Chaoborus* (Chaoboridae, Dipterae), but in the presence of the water bug *Notonecta glauca*, it develops a larger body by delaying the reproduction (Lüning, 1992), with both changes representing adaptive antipredation defenses. Moreover, *D. pulex* distinguishes between, and reacts specifically to, kairomones of different predators, the phantom midge larvae *Chaoborus* (Chaoboridae, Dipterae) and the three-spined stickleback *G. aculeatus* (Weiss et al., 2012).

Besides *D. pulex* and *Daphnia longispina*, neckteeth formation is also observed in *Daphnia curvirostris* complex in the field as well as under laboratory conditions, and in several juvenile instars (Juračka et al., 2011), as well as in *Daphnia sinevi* (Kotov et al., 2006), suggesting that neckteeth formation in daphnids might have occurred independently more than once.

Recently, biologists have identified the nature of chemical substances daphnids perceive to detect the presence of their *Chaoborus* larvae and fish predators. The *Chaoborus* kairomone is a cocktail of at least five different substances involved

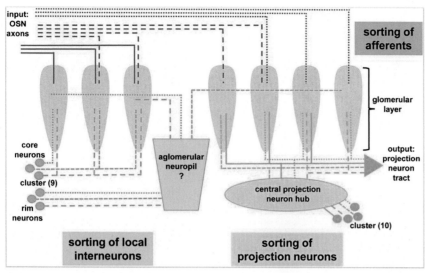

FIG. 4.6

Model summarizing the innervation of the olfactory glomeruli in *Coenobita clypeatus* with afferents, local olfactory interneurons, and olfactory projection neurons. The colored boxes indicate presumable sites where neurites are sorted toward their target structures.

From Polanska, M.A., Kirchhoff, T., Dircksen, H., Hansson, B.S., Harzsch, S., 2020. Functional morphology of the primary olfactory centers in the brain of the hermit crab Coenobita clypeatus (Anomala, Coenobitidae). Cell Tissue Res. 380 (3), 449–467. https://doi.org/10.1007/s00441-020-03199-5.

in the digestive processes of *Chaoborus* larvae (Weiss et al., 2018). The fish kairomone is a bile acid, 5α-cyprinol sulfate, which is released from the intestine, gills, and the urinary tract of the predator (Pohnert, 2019). In response to *Chaoborus* kairomones, *D. pulex* reacts by developing neckteeth, which prevent it from being eaten by the relatively small insect larvae, but on detecting the presence of the fish, it switches to a new life history: earlier reproduction at a smaller size (Stibor, 1992). Similarly, *Daphnia barbata*, upon detecting the presence of the predatory crustacean *Triops cancriformis*, forms a backward bending helmet, whereas in the presence of the insect backswimmer *N. glauca*, it forms a larger and longer helmet (Herzog and Laforsch, 2013; Herzog et al., 2016) and produces offspring of smaller body size that reproduce at a younger age (Lüning, 1992).

In all the above cases, adaptive defenses, which are specific for different predators, decrease the risk of predation.

Sometimes the adaptive response to the changed conditions is surprisingly delayed. This is the case with delayed effects of crowding in some parthenogenetic rotifer species (*Brachionus calycifloris, Brachionus angularis, Epiphanes senta*, etc.) that produce mictic female offspring, i.e., female and male individuals, only after fifth generation (Schröder and Gilbert, 2004). Similarly, the stressful rearing of the aphid *Dysaphis anthrisci majkopica* in the hostile plant *Chaerophyllum*

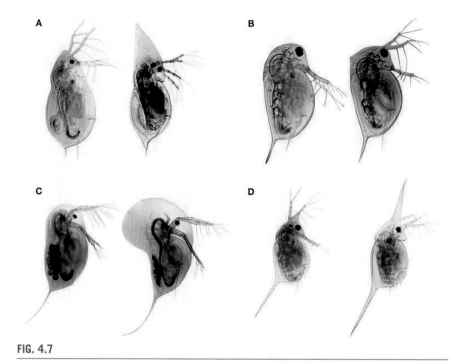

FIG. 4.7

Inducible morphological defenses are manifold in the genus *Daphnia*. The listed examples show helmet expression in *D. cucullata* (A), crest expression in *D. longicephala* (C), head- and tail-spine formation in *D. lumholtzi* (D), and neckteeth expression in *D. pulex* (B).
Undefended morphotypes are displayed on the *left side*, and the defended morphotype on the *right side*.

Images by Becker and Weiss. From Weiss, L.C., Albada, B., Becker, S.M., Meckelmann, S.W., Klein, J., Meyer, M., et al., 2018. Identification of Chaoborus kairomone chemicals that induce defences in Daphnia. Nat. Chem. Biol. 14, 1133–1139. https://doi.org/10.1038/s41589-018-0164-7.

maculatum after four to eight asexual reproduction leads to the heritable change of proboscis morphology and body size and adaptation to its often lethally hostile plant host (Shaposhnikov, 1965).

4.5 Neural mechanisms of intergenerational morphological and life history changes

Decades of studies on the predator-induced defenses and transgenerational changes in response to adversely changing environmental conditions have revealed that these inherited changes in morphology result from signal cascades starting in the central nervous system. This may be generalized in the following schema:

Environmental change → Reception by the sensory organs → Processing of the environmental stimuli in the CNS → Induction of a new developmental pathway → Development of the phenotypic change → Emplacement of parental factor(s) in the gamete(s) → The offspring CNS acquires the parental ability to develop the phenotypic change (for extended discussion, see Chapter 5).

In the following paragraphs, I present a few cases that illustrate the role of the nervous system as an inducer of the intergenerational phenotypic changes.

The water flea *Daphnia magna* is a small crustacean of the suborder Dafniidae (order Cladocera, class Branchiopoda) that under normal conditions is a parthenogenetic, female-only species, that produces only female offspring, but when the environmental conditions deteriorate (crowding, food source depletion, and shortening of the photoperiod), it switches to sexual reproduction by giving birth to a male + female generation and adaptively producing freezing- and desiccation-resistant eggs. The stressful stimuli are received by sensory neurons of the peripheral nervous system, which induce other neurons in the brain to secrete allatostatins that prevent the formation of methyl farnesoate (MF—crustacean juvenile hormone) and activate the cholinergic system (Eads et al., 2008) or secrete allatotropins that stimulate MF synthesis (LeBlanc and Medlock, 2015). MF binds its receptor MfR, which with MET (methoprene tolerant) and SRC (steroid receptor coactivator) forms a complex that acts as a transcription factor for the expression of male sex genes (LeBlanc et al., 2013; Olmstead and LeBlanc, 2002). Binding of MF by its receptor (MfR) induces expression of the *Dsx* (*doublesex*) gene, enabling switching to production of female + male generation. It is demonstrated that even naïve eggs exposed to MF aqueous solution produce male offspring (LeBlanc and Medlock, 2015) (Fig. 4.8).

D. pulex Leydig 1860, just like *D. magna*, forms female-only populations, but when conditions worsen (crowding and depletion of food sources), or it perceives sensory cues presaging deterioration of environmental conditions such as shortening of the day length and drop of temperature, it produces a sexually reproducing generation of male and female offspring. This is an adaptive life-history strategy that optimizes the fitness of the animal to the changing environment (Toyota et al., 2015a,b, 2017). In response, neurosecretory cells secrete allatotropins, which stimulate the synthesis of the hormone MF (Borst et al., 2002), which induces expression of the *dsx* gene that is responsible for the development of the male sex traits, but upstream it acts the brain glutamate system activated by sensing the abovementioned environmental factors. Rearing *D. pulex* under short photoperiod conditions leads to production of a mixed male + female generation and so does the administration of exogenous MF (Toyota et al., 2015a,b) (Fig. 4.9).

Brachionus plicatilis is a rotifer that induces sexual reproduction or diapause to the offspring in response to increased concentration of a kairomone, a protein (Snell et al., 2006) released by conspecific rotifers that is sensed by chemosensory neurons and perceived in the brain as an indicator of overcrowding and a forewarning of the consequent food depletion (Gilbert, 2003) (Fig. 4.10).

Upon detecting kairomones released by the predatory *Chaoborus* larvae, *D. pulex* forms defensive neckteeth on its head, which decrease the risk of being eaten by the predator. The kairomone is sensed by sensory neurons, and after being processed in the CNS, a signal cascade starts that activates an endocrine pathway leading to the

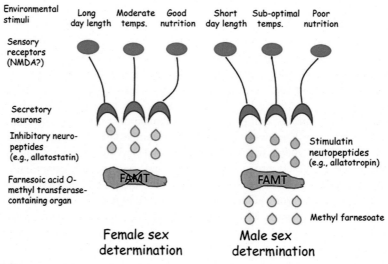

FIG. 4.8

Neuroendocrine model for the regulation of male sex determination in daphnid offspring. Specialized neurons in the maternal organisms transduce environmental signals via the NMDA receptor. In response, secretory neurons release peptide inhibitors or stimulators of methyl farnesoate synthesis through their regulatory action on the major contributing enzyme in methyl farnesoate synthesis: FAMT. Methyl farnesoate programs the egg/embryo to develop a male phenotype. In the absence of methyl farnesoate, embryos develop the female phenotype.

From LeBlanc, G.A., Medlock, E.K. 2015. Males on demand: the environmental–neuro-endocrine control of male sex determination in daphnids. FEBS J. 282, 4080–4093.

expression of morphogenetic factors that "result in defense morph formation" (Miyakawa et al., 2010). Moreover, the formation of neckteeth is induced by neuro-active substances, the cholinergic antagonist physostigmine and GABA antagonist picrotoxin and is inhibited by atropine (Barry, 2002; Weiss et al., 2012). The response by neckteeth formation is strictly specific: formation of defensive neckteeth in response to *Chaoborus* kairomones is associated with activation of the cholinergic neurotransmitters, whereas in response to fish kairomones is activated the GABAergic system (Weiss et al., 2012).

4.6 Predator-induced increase of winged offspring in insects

Upon visually perceiving their predators (ladybirds, hoverfly larvae, lacewing larvae), or even on detecting their kairomones, the pea aphid *Acyrthosiphon pisum* emits a volatile alarm pheromone, (E)-β-farnesene, whose olfactory perception by conspecific insects of the colony induces them to increase the proportion of winged offspring by several percentages (Dixon and Agarwala, 1999; Kunert and Weiser,

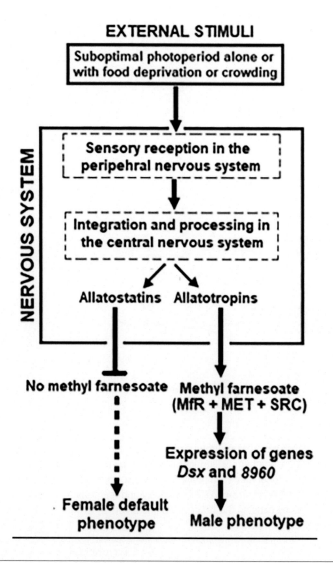

FIG. 4.9

Schematic of the mechanism of production of male and mictic female generations in *Daphnia pulex* in response to changing environmental conditions (see discussion in the text).

2003; Kunert et al., 2005). Administration of (E)-β-farnesene in the climate chamber increased production of winged aphids up to 124% compared with controls and in the field 600% compared with controls (Hatano et al., 2010). Cotton aphids, *Aphis gossypii* (Glover), under normal natural conditions produce offspring of four phenotypes: normal green apterae (unwinged), dark green apterae, "dwarf" yellow apterae, and alatae, but when they detect search track cues of their predator, ladybird

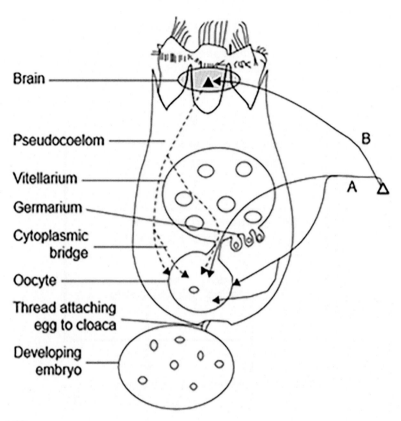

Brain

Pseudocoelom

Vitellarium

Germarium

Cytoplasmic
bridge

Oocyte

Thread attaching
egg to cloaca

Developing
embryo

B

A

FIG. 4.10

Mechanisms by which a crowding stimulus may induce oocytes of *Brachionus* to develop into mictic females. Proposed chemical inducer produced by rotifers is indicated by an *open triangle at the right*. Inducer may act directly on growing oocyte within maternal body cavity via various pathways (*solid-line arrows* A). Alternatively, inducer may act indirectly on maternal physiology, such as by causing the brain (*solid-line arrow* B) to secrete factor affecting the oocyte (*broken-line arrows*).

From Gilbert, J.J., 2003. Environmental and endogenous control of sexuality in a rotifer life cycle: developmental and population biology. Evol. Dev. 5, 19–24.

beetles, *Hippodamia convergens*, they increase the proportion of winged offspring for several sequential generations (Mondor et al., 2005). Predators also increased approximately threefold wing formation in the potato aphid, *Macrosiphum euphorbiae* (Kaplan and Thaler, 2012).

We do not know exactly how the alarm predatory signals are translated into neurohormonal mechanisms of insect wing development, but we have no visible reason to doubt that these mechanisms may be different from the normal pathways of the normal wing development, which are known to a considerable extent.

Development of wings in insects is a function of a gene regulatory network (GRN). The GRN is not self-regulatory but is ultimately activated by neurohormonal signals. Activation/inhibition of the insect wing GRN depends on the relative levels in the hemolymph of two hormones, ecdysone (Ec) and juvenile hormone (JH), with the former acting as an inducer and the latter as an inhibitor of the GRN. In turn, expression and suppression of Ec and JH expression are determined by neural signals originating in the insect brain in response to internal/external stimuli (Fig. 4.11).

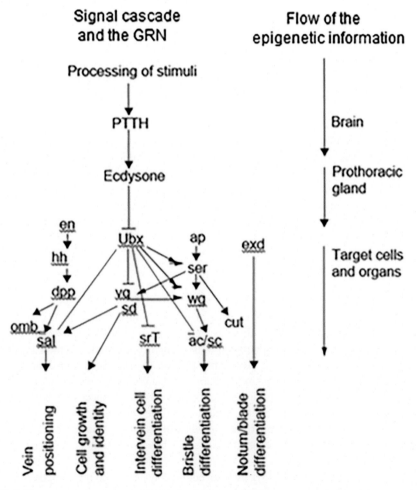

FIG. 4.11

Flow of epigenetic information along the ecdysone pathway for cell differentiation and growth in the process of wing development in insects.

Based on Abouheif, E., Wray, G.A., 2002. Evolution of the gene network underlying wing polyphenism in ants. Science 297, 249–252.

Ec production in the prothoracic gland is induced by secretion of the neurohormone prothoracicotropic hormone (PTTH) in brain neurons, and secretion of the neurohormone in turn is induced by reception via afferent interoceptors in the insect brain of information that the species-specific set point for body mass is attained. PTTH secretion is suppressed by another brain hormone, myoinhibitory peptide-prothoracicostatic peptide (MIP-PTSP) from specific midgut secretory neurons (Truman, 2006). Similarly, secretion of the juvenile hormone is induced by neuropeptides, allatotropins.

The elevation of Ec levels coincides with a drop in JH levels. At the onset of metamorphosis, Ec binds its nuclear receptor, EcR, which forms a dimer with another nuclear receptor, ultraspiracle (USP); the complex induces expression of a series of transcription factors to stimulate expression of the early response genes, which, in turn, induce expression of the late response effector genes, thus determining the ecdysone effect on wing development in a process that involves 468 genes (Li and White, 2003) or 3.4% of the 13,600 genes of the *Drosophila* genome (Adams, 2000). Similarly, high levels of JH in hemolymph and experimental application of JH and JH analogs inhibited the growth of wing discs in *Precis coenia* (Kremen and Nijhout, 1989; Miner et al., 2000).

Noteworthy in regard to the role of the CNS in the development of wings in insects is the case of fire ants, *Solenopsis invicta*. The mother queen of the colony releases a pheromone that prevents virgin queens from developing into functional queens. The pheromone via neural pathways decreases secretion of JH by corpora allata, but when the mother queen is removed from the colony, the JH level elevates, leading to dealation (wing shedding) and production of oocytes in the ovary (Fletcher and Blum, 1981) (Fig. 4.12).

5 Role of the brain in the process of speciation

Darwin was puzzled by the discreteness of species and absence (in his time) of slightly different forms between them as by his theory of common descent would predict; hence, his model of speciation was implicitly sympatric, which led him to the idea of sexual selection. According to him, "Sexual selection depends on the success of certain individuals over others of the same sex, in relation to the propagation of the species; whilst natural selection depends on the success of both sexes, at all ages, in relation to the general conditions of life" (Darwin, 1871b).

While developing the concept of sexual selection as a mechanism that leads to the evolution of secondary sexual characters, Darwin, however, did not consider it to be a mechanism of speciation.

In the early 1940s, Ernst Mayr introduced the concept of the reproductive isolation of a group of individuals from the rest of the population of a species as a precondition of speciation. This concept became the core of the biological definition of species. He wrote, "Sympatric speciation, if it occurs at all, must be an exceptional

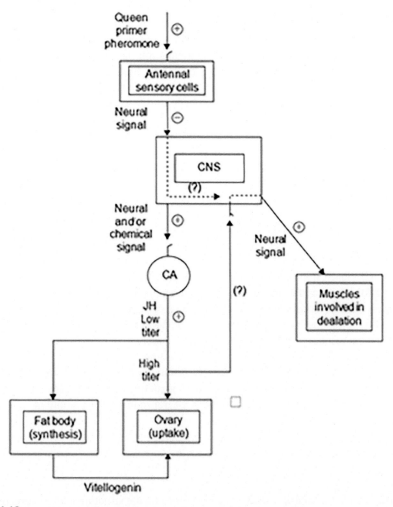

FIG. 4.12

Proposed general model for the mode of action of the primer pheromone of queen fire ants that inhibits dealation and ovary development in virgin queens. The pheromone triggers antennal receptors, which send inhibitory signals to the median neurosecretory cells in the brain. Largely inhibited, the median neurosecretory cells only weakly stimulate the corpora allata to synthesize JH, maintaining low titers of this hormone. At low levels, JH stimulates vitellogenin synthesis in the fat body. In the absence of the pheromone, the disinhibited neurosecretory cells send a stronger chemical and/or neural signal that triggers the corpora allata to produce larger quantities of JH. At higher titers, JH stimulates vitellogenin uptake by the ovaries and dealation. The latter process possibly involves an effect of JH on the nervous system. Dealation may result from a JH-independent pathway in the nervous system in lieu of, or in addition to, the JH-mediated pathway. These two possible pathways for control of dealation are flagged with question marks.

From Vargo, E.L., 1998. Primer pheromones in ants. In: Vander Meer, R.K., Breed, M.D., Espelie, K.E., Winston, M.L. (Eds.), Pheromone Communication in Social Insects. Ants, Wasps, Bees, and Termites. Westview Press, Boulder, CO, pp. 293–313.

process. The normal process of speciation in obligatorily sexual and cross-fertilizing organisms is that of geographical speciation" (Mayr, 1947). According to the neo-Darwinian doctrine, gene flow between populations would make it impossible the reproductive isolation within a panmictic population. Hence, the main form of speciation, if not the only one, is allopatric speciation, including the related peripatric and parapatric variants. With the irrefutable experimental evidence on the widespread occurrence of sympatric speciation, by the last quarter of the last century, gradually many neoDarwinians began recognizing the role of the sympatric speciation in evolution, while maintaining the primacy of the allopatric speciation.

The first and most important step in the process of speciation is emergence of the reproductive barrier between populations of a species, or emergence in particular individuals of the choosing sex of mating preference for individuals of the opposing sex that display some special morphological or behavioral characters. Over time, this change in mating behavior may lead to reproductive isolation of the group in sympatry, and evidence on such sudden shifts in mating preferences and sympatric speciation, especially in insects, fish, and birds, is accumulating.

The conceptual core of the sympatric evolution is reproductive isolation as a result and manifestation of changes in mate preferences and mate choice (Panhuis et al., 2001; Ryan and Rand, 2003). The mate choice is the function of the mate recognition system, consisting of male signaling or mating traits (Boughman et al., 2005) and female response to them (Butlin and Ritchie, 1991). The female preference and the resulting mate choice arise from the activation of specific neural circuits that integrate and process male signaling, including visual, auditory, olfactory, tactile, and behavioral traits. In rare cases, males may play the role of the choosy sex (Wong et al., 2005). Changes in mate choice and mating sensory signals represent the most frequently documented mechanism of reproductive isolation in sympatry (Cabej, 2019).

Mate choice is a behavior and, like any behavior, arises from the activation of specific neural circuits, in this case the male courtship circuitry. Like most innate behaviors, male mate courtship behavior is perfectly performed without previous experience.

Because of the simpler structure, a better idea on the structure and function of the circuitry may be obtained by a glimpse of the *D. melanogaster* male sexual behavior circuitry. In this insect, the male courtship behavior is a set of motor actions directed by FRU$^+$-specific circuit, whose name derives from the fact that several neurons of the circuit in males, but not in females, express the protein FRU^M encoded by the gene *fruitless*, along with the gene *doublesex* (*dsx*) (Pan et al., 2012).

The male courtship behavior circuitry is activated when the male fly via the foreleg gustatory sensory neurons detects cuticular pheromones and, in the form of neural signals, transmits this information to the first thoracic ganglion of the ventral nerve cord. The activation involves the perception of the neurally coded stimulus in the insect brain and integration of that information in higher brain centers that generate cognition and make the activation decision (Renoult and Mendelson, 2019). Two GABAergic neuronal populations, vAB3 and mAL (~30 neurons per hemisphere), convergently transmit, respectively, stimulatory and inhibitory gustatory information to the cluster of P1 protocerebral neurons representing the central driver of male mating behavior (Kallman et al., 2015). Whether to start or not the courting behavior depends on the balance between the two opposing actions of vAB3 and mAL neurons. The stimulatory role of vAB3 and inhibitory role of mAL neurons are indicated by the evidence that severance of the mAL axon tract increases substantially the response of P1 neurons to vAB3 stimulation. P1 also gets inhibitory inputs from olfactory neurons (Clowney et al., 2015).

D. melanogaster and *D. simulans* are two sibling species phenotypically quite similar (Shahandeh et al., 2018), living in sympatry in the Seychelles archipelago, Indian Ocean, but they are reproductively isolated (Billeter et al., 2009). They shared an ancestor between 3 and 5 million years ago (David et al., 2007). The neural circuit for the male courtship behavior in the sibling species *D. melanogaster* and *D. simulans* is a homolog circuit with constant anatomy (Shahandeh et al., 2020), wiring, and pheromone bouquet that receives similar neural input in both species; hence, it would be expected that the response to specific female pheromones by male individuals would be similar. However, contrary to the expectation, male individuals of these species behave in opposite ways to the same pheromone substance, 7-tricosene (7-T). *D. melanogaster* males respond to it by activating the courtship ritual, whereas *D. simulans* responds by suppressing the courtship (Sato et al., 2020). It is demonstrated that the opposing sexual responses of these sibling species to the same pheromone are not related to any genetic change or any change in the structure of the neural circuit. They result from a neural mechanism of alteration of the strength of the input from mAL and vAB3 to the neurons of the P1 cluster: "A change in the balance of excitation and inhibition onto courtship-promoting neurons transforms an excitatory pheromonal cue in *D. melanogaster* into an inhibitory cue in *D. simulans*" (Seeholzer et al., 2018). Both species use homologous peripheral sensory neurons to detect pheromone 7,11-HD but differ in the way they propagate the pheromone signal in the P1 neurons, leading to stimulation of the courtship in *D. melanogaster* and suppression in *D. simulans* (Fig. 4.13).

According to investigators, this valence change of the male response is mediated via a neural process involving central circuit modifications that manipulatively assign different courtship meanings to the same courtship pheromone (Seeholzer et al., 2018; Shahandeh et al., 2020).

Secretion by specific neurons of the male-specific neuropeptide NPFM, a homolog of the mammalian NPY (neuropeptide Y), also acts as an inhibitory input to the circuit suppressing the courtship behavior in *Drosophila*: activation of P1 increases NPFM activity, which in turn via NPF receptor neurons suppresses male courtship (Liu et al., 2019).

In most cases, insects use the same trait for mate recognition/choice and heterospecific recognition/reproductive isolation (Ryan and Rand, 1993, 2003). So, e.g., sympatric butterflies *Pieris protodice* and *Pieris occidentalis*, under natural conditions, do not hybridize, although there is no postmating mechanism to prevent their hybridization. To distinguish between heterospecific and conspecific mates, they use

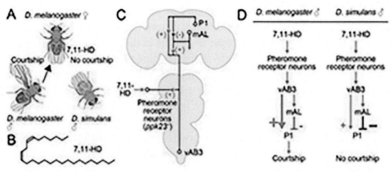

FIG. 4.13

Circuit basis for the conspecific preference in mate choice. (A) *D. melanogaster* female-specific pheromone 7,11-HD attracts and repels male *D. melanogaster* and male *D. simulans*, respectively. (B) Structure of 7,11-HD. (C) A pathway through which the 7,11-HD information reaches P1 neurons. + and – denote excitatory and inhibitory connections, respectively. (D) Proposed species differences in the pheromone processing pathway. *Thick lines* indicate predominant pathways.

From Sato, K., Tanaka, R., Ishikawa, Y., Yamamoto, D., 2020. Behavioral evolution of Drosophila: unraveling the circuit basis. Genes (Basel). 11 (2), 157. Published 2020 Feb 1. https://doi.org/10.3390/genes11020157.

a visual signal, the wing melanization pattern. In a study, experimental modification of wing melanization was carried out in male *P. protodice* so that they resembled males of *P. occidentalis*, and females of the latter could not discriminate and hybridized with *P. protodice* males (Wiernasz and Kingsolver, 1992). *Drosophila montana* and *D. lummei* females use the male courtship song both for recognizing conspecifics and mate choice (Saarikettu et al., 2005). Frogs use the advertisement call for species recognition, mate choice, and reproductive isolation (Ryan and Rand, 2003). Males of the replete group of *Drosophila* species use auditive signals for discriminating against heterospecific flies, but they use the A-type courtship song to distinguish conspecific from heterospecific flies and B type to sexually stimulate conspecific females (Ewing and Miyan, 1986).

An example of the use of olfactory signals for mate choice and reproductive isolation is the use of cuticular hydrocarbons in *Drosophila serrata* and *D. birchii* (Blows and Allan, 1998).

An evolutionary explosion of cases of sympatric speciation occurred among cichlid fish in lakes Victoria, Malawi, and Tanganyika, in East Africa (despite a significant gene flow) (Kocher, 2004; Haesler and Seehausen, 2005), from single panmictic populations, mostly within several to a dozen thousand years.

Studies on the continental European bird subspecies showed that morphological evolution occurred independently of changes in genes, and no differences in mitochondrial DNA were observed in a survey of 41 named bird species (Zink, 2004).

Most of the time, the sympatric speciation in the presence of the gene flow and recombination in a population is based on the ability of the sexual courtship behavior to respond differently to the same cues. This exceptional property of neural circuits relies on their ability of "differential interpretation of the same peripheral cue" (Khallaf et al., 2020). The structure and the function of neural circuits are modified by neuropeptides released by secretory neurons and neuromodulators that, by transforming the firing properties of circuit neurons and modifying the synaptic strength, reconfigure neuronal circuits and change their computational properties, often altering their chemical output (Getting, 1989; Montag-Sallaz et al., 2003; Marder, 2012; Flint, 2019).

There is abundant evidence that the courtship behavior circuitry can induce neurocognitive segregation and reproductive isolation within a panmictic population. No changes in genes or GRNs are necessary for this type of sympatric speciation.

5.1 Mate preference and mate choice

Mate preference is an esthetic property, an attraction that emerges in the nervous system of the choosy sex toward certain forms and sensory signals or patterns of a conspecific individual of the opposing sex, received through animal senses; it is the source of what Darwin called admiration of the females (or the choosing sex) for "the beauty of their partners" (Darwin, 1871a,b). Mate choice additionally implies intention and attempt to mate.

How do mate preferences evolve? Ronald Fisher's "runaway" hypothesis posits that female preference for particular male ornamentation traits gives males selective advantage to leave more progeny for the same trait that will be exaggerated through generations until the cost of developing it will cause negative selection of the trait. Observational evidence against the hypothesis has been presented. So, e.g., the poeciliid fish *Xiphophorus helleri* has evolved a swordtail after the genus *Xiphophorus* diverged from *Priapella*, but contrary to what would be predicted by the runaway hypothesis, *Priapella olmecae* displays a stronger preference for the swordtail (Basolo, 1998). Now, 90 years after being first proposed, the runaway hypothesis is still waiting validation.

According to the "good genes" hypothesis, male attractiveness is an indicator of "good genes" in males, but convincing evidence in support of the hypothesis lacks and cases that reject this possibility are described. As an example to illustrate the hypothesis is presented the fact that female guppies of *Poecilia reticulata* display mate preference for males with orange spots, but this is a consequence of their general sensory preference for orange-colored objects, including orange-colored food, rather than an indicator of "good genes" or higher fitness of these males (Rodd et al., 2002). Selection experiments for female mate choice and male attractiveness with *P. reticulata* for three generations also failed to show results that would be expected according to the previous hypotheses (Hall et al., 2004).

According to the sensory exploitation hypothesis, males evolve their signaling traits to match or exploit preexisting female preferences; hence, no genetic correlation exists between female mating biases and the evolution of the male signaling: "a signal and a receiver in animal communication—can evolve out of concert, with the evolution of one component lagging behind that of the other" (Ryan, 1998), implying that the natural selection is not involved in their evolution and the sender evolves mating signals to exploit mate preferences of the choosing sex but is hard to causally explain how the male is induced to evolve mating signals to satisfy female mating biases.

While there is no evidence on any genetic correlation between the evolution of male signaling and female preferences, we are certain that the evolution of mate choice is essentially related to the evolution of the neural circuits for mate choice. Mate choice is a neurobiologically made decision.

5.2 Reproductive isolation in sympatry

For a long time, the idea of the sympatric speciation as a mechanism of evolution of new species was dismissed because of the apparent incompatibility with the biological concept of species and the idea of the absence of the gene flow as an indispensable condition of speciation.

Reproductive isolation is a condition of speciation and the primary criterion of the biological species concept. In the well-known Ernst Mayr's definition, "species are groups of actually or potentially interbreeding natural populations, which are reproductively isolated from other such groups" (Mayr, 1942). Now, eight decades after the formulation, the definition seems too restrictive by implying the absence of the gene flow and the reproductive isolation as conditions of speciation: as shown earlier, gene flow is demonstrated to occur between sister species and sensu stricto sister species are not reproductively isolated but can hybridize and produce viable offspring.

Now we know, as will be shown, that many species, while being potentially interbreeding, are nevertheless reproductively isolated by neurocognitive mechanisms. They have evolved interspecific neurocognitive mechanisms to establish and maintain reproductive isolation of populations or species in the same geographic area while remaining potentially capable of hybridizing and producing viable offspring.

Reproductive isolation is the first stage in the process of sympatric speciation. It takes place in random-mating populations of an area as a result of changes in mating preferences in groups of individuals whose mating preference shifts toward conspecifics of the opposite sex displaying specific phenotypic characters. When changes are heritable, they lead to the formation of a separate mating group within the original population. Formation of separate mating groups spatially may result from

- emergence in a group of individuals of sensory-determined preferences for individuals of the opposite sex displaying specific phenotypic traits, or
- migration of a particular group to a new niche because of a sensory preference.

In the first case, reproductive isolation of populations in sympatry requires a shift in mating preferences in a population, whereas in the second case, it is an automatic result of the population that carves a new niche.

Now reproductive isolation in sympatry is a scientifically demonstrated fact, accepted by most biologists. It is a behavioral trait based on neurobiological mechanisms; it seems the most likely way of separating and isolating reproductively populations in sympatry.

5.3　Evolution of receiver biases

Female sensory biases are neurocognitive outputs of the integration and processing of the male signaling traits in the female courtship behavior circuits or her extended circuit. As a rule, receiver biases precede the evolution of mate preferences in conspecific individuals of the opposing sex (West-Eberhard, 1984), with the latter evolving in directions of matching the female preferences; in a particular case, females of the fish *P. olmecae* have evolved preference for the swordtail, even though its male conspecifics have not evolved the swordtail yet (Basolo, 1998).

The signaling traits of male individuals are communicated to the receiver of the opposite sex through the visual, auditory, olfactory, and tactile senses, which in the sensory neurons and in the receiver's CNS are converted into electrical impulses, the universal currency of the nervous system. The receiver is the driver of selection for the target of selection, i.e., the signaling traits (Ryan and Cummings, 2013). The processing of the sensory signals in the female's mate preference circuit may lead to attractivity (pleasurable sensation) that is manifested or not in the form of mate choice. The female mating preference provides the male a selective advantage that ultimately leads to further strengthening or exaggeration of the signaling traits.

5.4　Evolution of sender signaling traits

Male courtship in *Drosophila* is the function of a neural circuit comprising a group of ~20 *fru-*, *dsx-*, and *tru*-expressing neurons (Baker et al., 2001). The fact that its activation is also associated with *fru* expression in the olfactory, gustatory, and auditory systems led to the idea that all *fru*-expressing neurons in the fly's sensory organs and CNS may be part of an extended male courtship circuit in *Drosophila* (Stockinger et al., 2005).

The female preference, or aversion, plays selective roles in the evolution of male signaling traits. Male display traits are sometimes lost in reptiles and mammals as a result of the loss of the corresponding female biases, but often they are lost even in the presence of the respective female preferences. While a general explanation of the evolution of male signaling traits in such cases is difficult, a well-grounded idea supported by vast experimental and observational evidence is that both processes of evolution of the receiver biases and male signaling traits are determined by neurocognitive processes.

Although mating preferences are inherited innate behaviors that vary between individual animals and even may change within the lifetime of the individual, as exemplified by the case of females of the satin bowerbirds (*Ptilonorhynchus violaceus*), which display different mating preferences during different stages of life (Coleman et al., 2004). They can also change through experience, as demonstrated in the case of the freshwater fish *X. helleri* that have an innate preference for long swordtail, but they respond by shifting to the preference for swordless fish after exposure to predators (Johnson and Basolo, 2003). Experimental administration of hormones in female túngara frogs increases their receptivity and permissiveness (Lynch et al., 2006). And finally, the female guppy reverts to the mate preference for brighter males from its innate preference for black males when exposed to its predator *Cichlasoma biocellatum* (Gong and Gibson, 1996).

In some cases, the increased mate preference is related to changes in the number and activity of neurons involved in the processing of signals in neural circuits (Phelps et al., 2006). This is the case with the increase by 50%–100% in the size and number of the hypothalamic GnRH neurons in adults compared with juveniles in the fish *Porichthys notatus* (Grober et al., 1994). The female Lincoln's sparrow (*Melospiza lincolnii*) and European starlings (*Sturnus vulgaris*) increase song preference as a result of increased attractivity of the conspecific song environment, which leads to increased activity of particular neurons in the forebrain (Sockman and Lyons, 2017; Heimovics et al., 2011).

Observations on two sister species *Xiphophorus multilineatus* and *X. nigrensis* indicated that no genetic correlation exists between the evolution of male signaling traits and the female preferences (Morris and Ryan, 1996).

5.5 Mate recognition system

Receiver preferences and male signaling traits in dioecious animals form a communication entity that is known as the mate recognition system. An archetypal case of this system is the lek ritual, where males of some bird species come together at particular sites (leks) to display their mating traits to conspecific females.

The differences observed in the display score of male marine iguanas by different females (Wikelski et al., 2001), along with a degree of habituation of females to male songs of field cricket, *Gryllus lineaticeps* (females exposed to more attractive songs respond less than usual to the normal male songs) attest the presence of a "subjectivity" in their mate recognition system (Wagner et al., 2001).

The mate recognition system is based on perception in the CNS of visual, olfactory, auditory, tactile, or electrical stimuli. The process of the reproductive isolation/speciation may initiate with perception of one of these stimuli types. Three endemic pupfish morphospecies (*Cyprinodont beltrani*, *C. labiosus*, and *C. maya*) in Lake Chichancanab (Mexico) that diverged only 2000–12,000 years ago show "very little genetic change" that is within the limits of the intraspecific genetic variation, hence displaying "no genetic divergence" (Strecker et al., 1996), are partially reproductively isolated, and still continue to interbreed (Strecker and Kodric-Brown, 1999), suggesting that no genetic changes are responsible for the evolution of morphological differences and mate recognition systems between them. Only their mate recognition system has differently evolved in each of them by using perception of visual or olfactory cues for mate recognition and preference. Two first species have diverged only in the olfactory part of the recognition system (production and perception of chemical signals) but very little in the visual part (their nuptial colors are almost identical) and hence are still interbreeding. By contrast, *C. maya* uses both visual and olfactory cues for mate recognition and, consequently, evolved a complete mate preference for conspecific males.

In the mate recognition system, not only do receivers influence the development and evolution of mate signaling traits, but the latter also can influence the behavioral plasticity of receivers. This has been demonstrated among others in recent experiments with Lincoln's sparrows (*M. lincolnii*). The treatment of females with high-performance recordings of the same song in the wild elevated the threshold for releasing the female phonotaxis (body movements induced by hearing the song) toward male song compared with the birds treated with low-performance song recordings. This behavioral change in females was related to changes in the level of neurotransmitter concentrations in the auditory forebrain (caudomedial mesopallium and caudomedial nidopallium), the brain region where the song processing takes place (Sockman and Lyons, 2017).

6 Neurocognitive mechanisms of reproductive isolation

6.1 Visual-cognitive mechanisms of reproductive isolation and mate choice

In vertebrates, the visual perception of the signaling traits begins in retinal neurons that convert the received visual data into electrical signals and, via the optic nerve, relay these further to the optic chiasm and the thalamic lateral geniculate nucleus for processing and creating the image in the higher brain centers.

Visual cues may be the most widespread vertebrate signals for discriminating not only between conspecific and heterospecific individuals but also for mate choice. They have played an important role in evolution of the mate signaling traits in fish, reptilians, and birds.

The most widely used cues for recognition of conspecifics and mate choice in fish are color and body size. Some of the most impressive and unequivocal cases of sympatric speciation are observed in the relatively young volcano crater lakes that are isolated from other bodies of water. Three sympatric cichlid sister species, *Amphilophus amarillo*, *A. sagittae*, and *A. xiloaensis*, inhabit the small young volcanic crater lake ~3.7 km^2 Lake Xiloa, Nicaragua. These sympatric sister species use body color to distinguish heterospecific from conspecific individuals and for mate choice and, at least for two of them, the body color must have been used for the initial divergence of their populations from the ancestral population of *A. citrinellus* of Lake Managua. Not only these three cichlid sister species but also the normal gray and the gold forms of *A. xiloaensis* now represent incipient sympatric species that have differentiated to a considerable extent (Elmer et al., 2009) (Fig. 4.14).

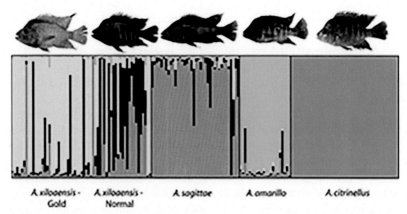

FIG. 4.14

Model-based cluster analyses distinguish the genetic differentiation of Lake Managua *A. citrinellus* from the gold and normal *A. xiloaensis* and the sister taxa *A. sagittae*, and *A. amarillo* in Lake Xiloá. *Amphilophus xiloaensis* gold is shown in *yellow*, *A. xiloaensis* normal in *gray*, *A. sagittae* in *beige*, *A. amarillo* in *green*, and the ancestral population *A. citrinellus* from Lake Managua in *blue*, along with photograph exemplars of each species/morph.

From Elmer, K.R., Lehtonen, T.K., Meyer, A., 2009. Color assortative mating contributes to sympatric divergence of neotropical cichlid fish. Evolution 63 (10), 2750–2757. https://doi.org/10.1111/j.1558-5646.2009.00736.

Another example of more recent evolution of new cichlid species was described a decade ago in another volcanic lake, again in Nicaragua. It occurred in the ~1800-year-old Lake Apoyeque, and the genetic analysis shows that the lake was colonized only ~100 years ago or is populated by fish for the same number of generations. It is suggested that both *Amphilophus* cf. *citrinellus* morphs of thin- and thick-lipped fishes may be sympatric incipient species (Elmer et al., 2009).

Lake Victoria was formed about 12,400 years ago but in it have evolved more than 500 distinct cichlid fish species. They remain reproductively isolated, although they can produce viable hybrids. The reproductive isolation of these hundreds of morphologically similar species of cichlid fish in Lake Victoria is enabled by their ability to discriminate between homo- and heterospecific fish and mate choice of females based on the male body color and pattern (Seehausen et al., 1997a,b; Seehausen and van Alphen, 1998).

Bioluminescent flash signals are essential species-specific and sex-specific components of the courtship behavior in many nocturnal insects. Experimental evidence from studies on *Photinus* firefly suggested that the nerves from terminal abdominal ganglion (A8), which innervates the paired lanterns (light organs) (Christensen et al., 1983), are octopaminergic nerves (Robertson and Carlson, 1976; Christensen et al., 1983). Both males and females of the *Photinus* group of fireflies have lanterns. Octopamine released by nerves innervating these organs induces production of nitric oxide synthase (NOS), which via a chain of reactions induces synthesis of NO (nitric oxide) and O_2, leading to the formation of oxyluciferin, which produces light by emitting photons (Aprille et al., 2004) (Fig. 4.15).

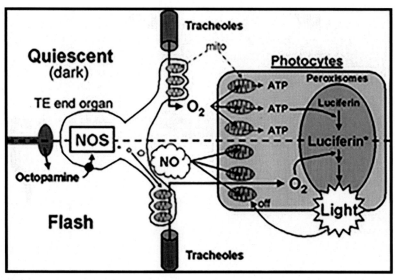

FIG. 4.15

Illustration of the scheme for NO interaction with mitochondria in the mechanism for oxygen gating for on/off switching of firefly flashing. See text for explanation.

From Aprille, J.R., Lagace, C.J., Modica-Napolitano, J., Trimmer, B.A., 2004. Role of nitric oxide and mitochondria in control of firefly flash. Integr. Comp. Biol. 44, 213–219. Modified and used with permission.

There are no noticeable morphological differences between firefly males; hence, the most important element determining courtship behavior in these flies is the flashing pattern, which in turn is determined by neural pulses of the release of the neurotransmitter octopamine (Greenfield, 2001; Trimmer et al., 2001), regulated by a central pattern generator. Generally, females prefer males that have higher flash pattern rates (Demary et al., 2006). To prevent heterospecific mating, the firefly *Photinus ignitus* has evolved a mechanism of reproductive isolation that sharply decreases the mate preference when the duration of the flash exceeds the species-specific range (Cratsley and Lewis, 2003).

Recently, experimental data are presented from two luminous species of firefly, *Luciola cruciata* and *L. lateralis*, and one nonluminous species, *Lucidina biplagiata*, contradicting the above mechanism; nevertheless, the investigators admit that "there might be a possibility that neurotransmitter nitric oxide (NO) functions to induce rapid pulses of bioluminescence" (Ohtsuki et al., 2014).

6.2 Olfactory-cognitive mechanisms of reproductive isolation and mate choice

Olfactory recognition mechanisms used for discriminating against heterospecific individuals and for mate choice are based on identification of kairomones, species-specific chemical substances released by animals. They belong to a large group of more than 3500 substances, known as semiochemicals. Besides hormones, the group also comprises substances that influence the behavior of heterospecific animals (kairomones, allomones, and synomones). In distinction from the mate signaling traits, whose receivers are female conspecifics, pheromones are mostly released by females and males act as receivers of these mating female signals.

In *Drosophila*, secretion of pheromones is regulated by brain neuropeptides, pheromonotropic hormones, such as pheromone biosynthesis activating neuropeptide (PBAN), and PBAN-like and other neuropeptides stimulated by a "neural input from the ventral nerve cord" on the pheromone gland (Altstein et al., 1996; Iglesias et al., 1998).

In *Drosophila*, the courtship behavior is induced by perception of cuticular hydrocarbons (CHCs) (Smadja and Butlin, 2009), a group of more than 100 compounds (Drijfhout et al., 2009). Closely related groups of *Drosophila* species differ in the predominant types of their cuticular CHCs: 7-tricosene (7-T), 7-T+6T, and 7,11-heptacosadiene (7,11-HD). Even each individual animal within a species produces a specific blend of CHCs, depending on the age, sex, diet, and geographical origin. Regarding the unique role of the structure and function of the central courtship circuit in *Drosophila* in generating different and even opposing results in response to the same pheromone, noteworthy is the case of two *Drosophila* sibling species, *D. melanogaster* and *D. simulans* (Seeholzer et al., 2018). Both species produce and detect the female courtship pheromone 7,11-HD (7,11-heptacosadiene) using homologous sensory neurons, but while in *D. melanogaster* males, the pheromone acts as an aphrodisiac stimulating the courtship behavior and a song by wing vibrations, males of the sibling species *D. simulans* do not execute the courtship behavior and even respond to the pheromone as to a repellant. The difference results from differences in propagation of the signal differently to the courtship-promoting P1 neurons. It is a modification in the central courtship circuit that assigns different courtship meanings to the same courtship pheromone (Seeholzer et al., 2018). Unlike *D. melanogaster*, mAL interneurons in *D. simulans* develop stronger courtship inhibitory activity by antagonizing vAB3 ascending neurons leading to suppression of the courtship function of P1 neurons (Seeholzer et al., 2018; Sato et al., 2020) (Fig. 4.13).

Four allopatric *Drosophila mojavensis* populations inhabiting four regions of the south-western part of the United States (Baja California peninsula, mainland Sonoran Desert, Mojave Desert, and Santa Catalina Island) (Matzkin, 2014) are classified as subspecies (Pfeiler et al., 2009) that display differences in morphological, neurophysiological, and behavioral traits, but interpopulational crosses produce viable flies, indicating that their evolution is a result of reproductive isolation in sympatry rather than changes in genes. They diverged ~250,000 years ago and represent a case of incipient speciation (Khallaf et al., 2020). All the subspecies display similar courtship rituals.

A complete reproductive isolation barrier exists between the northern subspecies, on the one hand, and the southern subspecies on the other hand, as well as between the two southern subspecies, but not between the northern subspecies, indicating that only northern subspecies are not completely reproductively isolated. Pheromones that determine con-subspecific recognition and mate choice in four *D. mojavensis* subspecies are four male-specific acetates [R and S enantiomers of (Z)-10-heptade-cen-2-yl acetate (R&S-HDEA); heptadec-2-yl acetate (HDA); (Z,Z)-19,22-octaco-sadien-1-yl acetate (OCDA)]. R&S-HDEA and HDA are volatile pheromones, whereas the courtship-suppressing OCDA is a nonvolatile pheromone expressed in mated females but not in virgin females; hence, it suppresses the male courtship behavior of mated females only. Females of the four subspecies have similar chemical profiles, which may explain why males of all the subspecies cannot discriminate between females of other subspecies.

Of five courtship pheromone receptors (Or47b1, Or47b2, Or65a, Or67d, and Or88a) expressed in *Drosophila* olfactory neurons, OR65a is responsible for the detection of R-HDEA and HDA pheromones and increased female receptivity in *D. moj. wrigleyi* toward con-subspecific males. Along with the pheromones, auditory cues are also involved in mate recognition in *D. moj. wrigleyi*. Differences induced by R-HDEA in courtship behaviors among *D. mojavensis* subspecies result from differences in the neural processing of the pheromonal stimulus in each of the subspecies (Khallaf et al., 2020).

The use of new pheromones or changes in the contents of the pheromone blends induce changes in mating preferences. So, e.g., when caterpillars and adult African butterfly *Bicyclus anynana* butterflies were fed on corn leaves (their usual laboratory food), coated with artificial banana, mango, coffee, and almond odors, as adult butterflies they acquired a learned preference for banana and mango odors but not for coffee and almond. The trait was inherited in the next generation (Gowri et al., 2019). Females of *B. anynana* exposed to a pheromone blend containing an increased amount of the male sex pheromone hexadecanal (MSP2) shifted mating preference toward males emanating this blend compared with the wild-type males (Dion et al., 2020).

Exposure of *B. anynana* butterfly females to manipulated pheromone blends released by conspecific males induces learning and acquisition of mating preference/choice for individuals releasing the new pheromone blends and the acquired mate choice, and the trait is inherited in the female offspring. Learning to prefer new pheromone blends of conspecific males and its transmission to the offspring may have played a role in evolution of *B. anynana* (Dion et al., 2020).

Two sister cichlid species in Lake Malawi, *Pseudotropheus emmiltos* and *Pseudotropheus fainzilberi*, are reproductively isolated mainly by olfactory cues, and investigators believe that "divergence of olfactory signals may have been an important influence on the explosive radiation of the East African species flock" (Plenderleith et al., 2005).

About 5 million years ago, in northwestern America a rapid process of speciation of 35 salamander species of the *Plethodon* genus took place. The fact that this coincided with the shift from the male abrading the female skin with the premaxillary teeth to the delivery of the male mental gland pheromones into the female nares indicates that the change in the sexual behavior may have induced the accelerated speciation process in this salamander group (Wiens et al., 2007).

Smadja and Butlin provide a table of 40 cases of olfactory reproductive isolation in species of different animal groups (insects, annelids, fish, squamates, and mammals), but there are likely to be many other examples of chemosensory speciation in insects (Smadja and Butlin, 2009). There is no doubt that selective pressures have played an important role in the divergence of the sender's signaling traits and pheromones (Higgie et al., 2000; Gries et al., 2001) for distinguishing between homospecific/heterospecific traits and mate choice.

A few studies that suggest the involvement of genetic changes in cuticular hydrocarbons in *Drosophila* (Gleason et al., 2005; McBride, 2007) do not determine whether these changes were accumulated before or after the beginning of the process of speciation. Studies on the genetic changes identified in various species of the butterfly *B. anynana* and the evolution of the pheromonal components or pheromonal blends showed that evolution of "pheromonal dialects" or composition of putative MSPs (male sex pheromones) in the butterfly evolved before the speciation process and preceded the genetic divergence between *B. anynana* species, which still "share a similar mitochondrial haplotype" and gene flow is still occurring between their populations (Bacquet et al., 2016), indicating that they are still potentially interbreeding and have not evolved postmating isolating mechanisms.

6.3 Auditory neurocognitive mechanisms of reproductive isolation and mate choice

Acoustic signals may also have played a role in the reproductive isolation and speciation in sympatry, at least in insects, amphibians, and birds.

Acoustic signals of the male song serve as mating signals by female crickets (Hedwig, 2005). Male singing in crickets is regulated by an auditory center in the anterior protocerebrum, which is activated by cholinergic neurotransmitters and suppressed by GABA (gamma-aminobutyric acid). Male songs function as cues for mate choice and discrimination of conspecific from heterospecific individuals when a morphological distinction is impossible.

Hearing in many insects is a function of special ears consisting of a tracheal sac with a tympanum on the front and a chordotonal organ with a group of sensillae, each of which contains several sensory cells that receive sounds and transmit their input to the CNS. Development of the chordotonal organ in the moth *Lymnatria dispar* is induced by signals from the auditory nerve (IIIN1b1) (Lewis and Fullard, 1996).

Neural circuits for auditory recognition and mate choice are remarkably plastic. How fast insects can adapt neural circuits responsible for conspecific call recognition in response to selective pressures for interspecific discrimination can be illustrated by the case of two sibling cricket species that have evolved two different temporal filters after diverging from their common ancestor. Females of the cricket *Teleogryllus commodus* have evolved two temporal filters, one for chirps and one for the trill of the male song, whereas females of the closely related species, *Teleogryllus oceanicus*, develop a single filter for the chirp part of the male song (Hennig and Weber, 1997). Similarly, two North American cryptic crickets that are indistinguishable morphologically inhabit in sympatry are reproductively isolated from each other because of the evolution of two different male songs and female preferences, even though they could interbreed and produce fertile hybrid offspring. Reproductive isolation has also been demonstrated between two other sister cricket species, *Gryllus texensis* and *Gryllus rubens* in the south-eastern part of North America, which are morphologically indistinct from each other in nature but can produce viable hybrids experimentally (Gray and Cade, 2000). Also, a group of *Laupala* crickets in Hawaii Island underwent, and are still experiencing, an exceptionally high rate of speciation of 4.7 species per million years, that is, 26 times greater than the average speciation time of arthropods in general. The reproductive isolation and consequent speciation in this group are results of divergence of the courtship song's pulse rate in particular populations, arising through changes in the function of song circuits rather than genetic changes (Mendelson and Shaw, 2005).

Three closely related cichlid species in Lake Malawi, East Africa, *Pseudotropheus zebra*, *P. callainos*, and *P.* "zebra gold" live in sympatry and are prematingly isolated due to differences in sounds they generate during the courtship ritual (Amorim et al., 2004).

Male auditory signals (mating calls) in the túngara frogs, *Physalaemus pustulosus*, are processed in the female midbrain *torus semicircularis* and hypothalamus as deduced by expression of the *egr-1* gene in five torus nuclei and three hypothalamic nuclei that are also involved in the processing of the auditory signals. In response to these conspecific male calls, túngara frogs display phonotaxis, movements of the body related to the calls; they show only a little or no response to heterospecific calls of the distant relative frog, *Physalaemus enesefae* (Hoke et al., 2005) (Fig. 4.16).

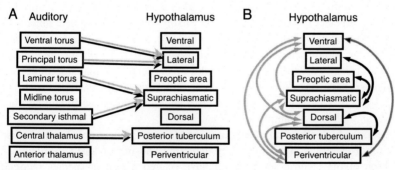

FIG. 4.16

Functional connectivity of the túngara frog hypothalamus. *Gray arrows* show significant relationships ($P < 0.05$) in frogs that heard irrelevant acoustic stimuli (*P. enesefae* whine and chuck-only), and *black arrows* indicate relationships in frogs exposed to behaviorally relevant stimuli (conspecific whine and whine-chuck). (A) egr-1 levels in the midbrain and thalamic nuclei implicated in auditory processing are significant predictors of hypothalamic expression patterns. Relationships between auditory and hypothalamic regions do not vary with the relevance of stimulus. (B) egr-1 correlations between hypothalamic regions differ based on the behavioral relevance of acoustic stimulus.

From Hoke, K.L., Ryan, M.J., Wilczynski, W., 2005. Social cues shift functional connectivity in the hypothalamus. Proc. Natl. Acad. Sci. U. S. A. 102 (30), 10712–10717. https://doi.org/10.1073/pnas.0502361102.

In songbirds (order Passeriformes), songs serve as auditory mating signals, products of the male song system, which comprises two inborn prenatally wired song circuitries:

- The *song production circuitry*, which is responsible for both song acquisition and production of the learned songs. It comprises the high vocal center (HVC), robust nucleus of the arcopallium (RA), tracheosyringeal motor nucleus (nXIIts), and muscles of the syrinx, the song organ.
- The *song learning* circuitry, which is responsible for song acquisition only and comprises HVC, Area X, dorsolateral anterior thalamic nucleus (DLM), lateral magnocellular nucleus of the nidopallium (LMAN), and RA (robust nucleus of the arcopallium), which is necessary for song acquisition only (Nottebohm, 2005) (Fig. 4.17).

The high plasticity of the neural circuits determining male auditory signals in amphibians, birds, and fish and the ensuing (or preceding) female preferences for changed auditory signals suggest that such changes may have played an important role in the evolution and speciation in these groups.

6.4 Electrocognitive speciation

By the middle of the last century, Hans W. Lissmann (1909–1995) observed that a fish species inhabiting river Nile and other East African rivers avoided hitting obstacles when swimming backward, an observation that he proposed to explain with the fish having an electric organ, which "may enable the animal to detect objects in the vicinity of its body" (Lissmann, 1951). Ever since, a number of mammalian species of the order Monotremata, including platypus (*Ornithorhynchus anatinus*) and echidnas of the Tachyglossidae family and most recently, Guiana dolphin (*Sotalia guianensis*), have been identified to possess the bioelectric sense (Czech-Damal et al., 2013). Electric organs have evolved at least six times in cartilaginous elasmobranchs and teleosts (Crampton, 2019).

Due to the high electrical conductivity of the water, fish evolved a mechanism of emission of electrical signals for prey detection and intraspecific communication. Evolution of the electric sense might have played a role in species recognition and mate choice in sympatric speciation (Feulner et al., 2009). A group of ~800 elasmobranch fish use electrical signals for prey detection and intraspecific communication. Electrogenesis is also used by ~200 mormyriform freshwater fish species in 18 genera of the family Mormyridae in Central Africa and about 115 gymnotiform species comprising 28 genera and 7 families in neotropical freshwaters of South and Central America (Hopkins, 1999).

The electric sense can be considered a key innovation not only regarding its primary function, i.e., electrolocation, but also as an efficient trait for "electric" mate recognition.

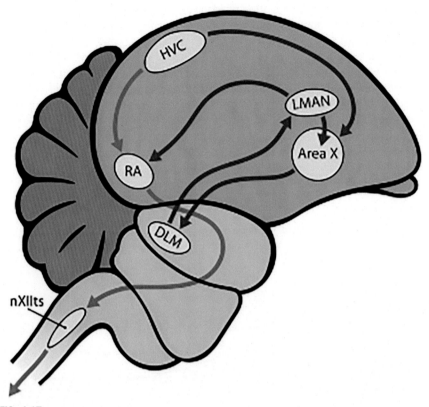

FIG. 4.17

The song system of songbirds. Nucleus HVC feeds information into two pathways that ultimately lead to the neurons in the tracheosyringeal half of the hypoglossal nucleus (nXIIts) that project to vocal muscles. HVC projects to nucleus RA directly (PDP), and indirectly via Area X, the dorsolateral anterior thalamic nucleus (DLM), and LMAN (AFP) in a manner that shares similarities with the mammalian pathway cortex-basal ganglia-thalamus-cortex. *AFP*, anterior forebrain pathway; *DLM*, dorsolateral anterior thalamic nucleus; *LMAN*, lateral magnocellular nucleus of the nidopallium; *nXIIts*, the tracheosyringeal motor nucleus in the brain stem; *PDP*, posterior descending pathway; *RA*, robust nucleus of the arcopallium.

From Nottebohm, F., 2005. The neural basis of birdsong. PLoS Biol. 3(5), e164.

6.4.1 Electrogenesis

Generation of electrical signals in fish is a function of the electric organ consisting of muscle-derived cells or electrocytes. The emitted electric signals are used for both prey detection and intraspecific communication. In the latter case, fish use a regular constant waveform of electrical discharge pulse (EOD), suggested to serve species recognition and a variable sequence of pulse intervals (100–300 ms) to express an individual's identity and motivation (Carlson and Hopkins, 2004). EOD's duration may change in the course of ontogeny; in adults generally it becomes three times longer than in juveniles (Nguyen et al., 2020).

Emission of EOD pulses is the output of a neural circuit starting with the knollenorgan, whose sensory cells send the input of stimuli in a coded form to the postsynaptic neurons of the ipsilateral nucleus of the electrosensory lateral line lobe (nELL) in the fish's brain. nELL neurons project to the anterior EL anterior (Ela) subdivision exterolateral nucleus (EL), where the analysis of the EOD waveform is performed, whereas the temporal processing or analysis of the interpulse intervals (IPI) is performed in the EL posterior (ELp) subdivision. The latter, along the nELL, sends axons to the medioventral nucleus (MV) (Fig. 4.18).

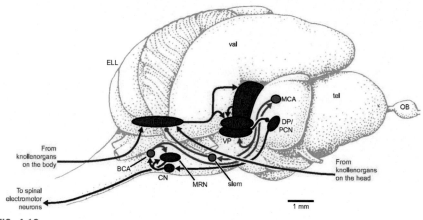

FIG. 4.18

Neuroanatomy of the knollenorgan electrosensory (*red*), electromotor (*blue*), and corollary discharge (*purple*) pathways. Excitatory connections are indicated by *arrows* and inhibitory connections by *punctate terminals*. Knollenorgan electroreceptors project somatotopically to the hindbrain nucleus of the electrosensory lateral line lobe (nELL), which projects to two nuclei of the torus semicircularis: the exterolateral nucleus (EL) and the medioventral nucleus (MV). Each EOD command originates in the command nucleus (CN). The CN projects to the medullary relay nucleus (MRN), which in turn innervates the electromotor neurons in the spinal cord that innervate the electric organ. The CN output is influenced by excitation from the precommand nucleus (PCN) and the dorsal posterior thalamic nucleus (DP), both of which receive inhibition from the ventroposterior nucleus (VP). When the CN initiates an EOD, it sends a copy of this signal to the bulbar command-associated nucleus (BCA), which projects to the MRN and to the mesencephalic command-associated nucleus (MCA), which mediates inhibition of the motor pathway through the VP, and inhibition of the KO sensory pathway through the sublemniscal nucleus (slem). *ELL*, electrosensory lateral line lobe; *OB*, olfactory bulb; *tel*, telencephalon; *val*, valvula of the cerebellum.

From Baker, C.A., Kohashi, T., Lyons-Warren, A.M., Ma, X., Carlson, B.A., 2013. Multiplexed temporal coding of electric communication signals in mormyrid fishes. J. Exp. Biol. 216, 2365–2379. https://doi.org/10.1242/jeb. 082289.

6.4.2 Electroreception

Fish that produce an electric current in the form of EODs also have electroreceptors for sensing such currents (Sawtell et al., 2005). Generally, South American gymnotiform fish emit continuous quasisinusoidal EOD waves, whereas African mormyriform fish produce EOD pulses that are separated from each other by intervals that are longer than the pulse duration (Sawtell et al., 2005). As mentioned earlier, electroreceptors synapse on the primary afferent nerve fibers. The electrical current coming through the skin depolarizes receptor cells, which terminate in the electrosensory lobe (ELL), but higher brain centers are also involved in the processing of the electrical stimuli. It is suggested that the afferent input from the electroreceptors in the ELL pyramidal cells is transformed into a more abstract electrical representation or "image" (Metzner et al., 1998).

6.5 Electrosensory communication in reproductive isolation and speciation

Most electric fish display sexual dimorphism in the EOD waveform, suggesting that they may use electrical signaling for social communication, including electric mate recognition. The suggestion is validated in experiments, indicating that EOD signals are used for mate choice in fish (Arnegard et al., 2006; Feulner et al., 2009; Nagel et al., 2018).

Fourteen described *Campylomormyrus* fish species in the freshwater system of the Congo Basin, Central Africa, live in sympatry, and it is believed that species of this genus evolved in sympatry (Feulner et al., 2007). They discriminate between con- and heterospecific fish based on EOD patterns, but cases are described when some of them cannot recognize all the species of the group as heterospecific. Their divergent EOD waveforms coincide with differences in their rostrum shape (Feulner et al., 2009). For instance, because of similarities in EOD patterns, females of the sympatric weakly electric fish species *Campylomormyrus compressirostris* in the African freshwater system cannot distinguish their own from heterospecific males of distantly related *C. tamandua*, while they distinguish as heterospecific males of the more closely related species *C. rhynchophorus* (Feulner et al., 2009). The reproductive isolation and divergence in sympatry of *C. compressirostris* and *C. rhynchophorus* have been related to assortative mating as a result of the evolution of longer EODs in *C. rhynchophorus* (Feulner et al., 2009).

It is suggested that "the EOD waveform provides sufficient information for species recognition in some but not all African weakly electric fishes" (Nagel et al., 2018). EOD waveform and sequence of pulse intervals (SPIs) were sufficient for species recognition in *C. compressirostris* males (Nagel et al., 2018), but the interval between pulses (temporal pattern) may also help in species recognition (Kramer and Kuhn, 1994). There is evidence suggesting that in many of these cases, species recognition may be supported by the perception of visual and olfactory cues.

Three East African sympatric morphotypes (sibling species), the so-called Magnostipes complex of the Ogooué River system, Gabon, have diverged recently and

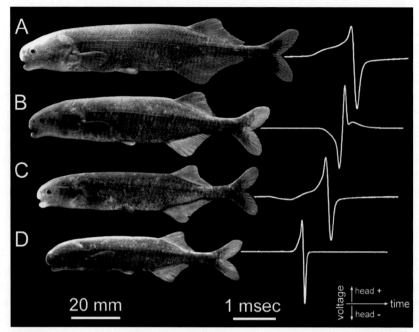

FIG. 4.19

Example of a sympatric assemblage of morphologically similar mormyrids from the *Brienomyrus* species flock of Gabon. Photographs (scaled to specimen standard length) of the only four *Brienomyrus* signal forms occurring in Mouvanga Creek (2819923.50S, 11841918.10E) are shown next to voltage traces of electric organ discharges (EODs) recorded from the same individuals: (A) bp1, specimen 3016, adult male, CU80355; (B) type I morph, specimen 3025, adult female or nonbreeding male, CU80231; (C) type II morph, specimen 3014, adult female or nonbreeding male, CU80358; and (D) sn3, specimen 3027, adult female or nonbreeding male, CU80356. Scale bars of 20 mm and 1 ms are indicated.

From Arnegard, M.E., Bogdanowicz, S.M., Hopkins, C.D., 2005. Multiple cases of striking genetic similarity between alternative electric fish signal morphs in sympatry. Evolution 59, 324–343.

are still indistinguishable genetically (their gene pools are similar) and morphologically but are plainly distinct in regard to the EOD rate and waveform (Arnegard et al., 2005) (Fig. 4.19).

Remarkable differences in electrical signals between the sympatric morphotypes, in the absence of genetic relevant differences, suggest that a neural mechanism via EODs has promoted species isolation of *Brienomyrus* flock in the Gabon river system (Arnegard et al., 2005).

Abundant evidence from studies on African freshwater fish of the genus *Brienomyrus* has shown that changes in EOD patterns represent neural innovations that have led to reproductive isolation and formation of new species: "neural innovations can drive the diversification of signals and promote speciation … evolutionary change in the functional organization of sensory pathways can establish new perceptual abilities that trigger explosive diversification" (Carlson et al., 2011).

6.6 Nonsexual neurocognitive speciation in insects: Speciation via host plant shift

Shifts of particular insect populations to new host plants may isolate them reproductively from the original stock and initiate the formation of new species. The host shift speciation in sympatry is based on the emergence in a particular group of individuals of a new preference for odors emitted by a plant that is different from the "normal" host plant. The new preference implies a "neural innovation" in the processing of the new plant odor in the respective olfactory circuits of the insect. Female insects perceive conspecific male pheromones and host plant odors as a single chemosensory landscape. The olfactory information is received through the olfactory receptor neurons, which project to the olfactory center (AL) in the deuterocerebrum, consisting of a number of glomeruli, and further to the higher brain centers (mushroom body and lateral protocerebrum) for olfactory processing.

It has been argued that the host shift speciation may be a form of ecological speciation resulting from mate choice in different environments of populations of an originally single species; it has also been argued that host shift speciation is but another form of allopatric speciation, determined by the interruption of gene flow between two populations that diverged in preference for the mating host. From the standpoint of causality, however, the issue boils down to a cause-and-effect issue, whereby the cause precedes the effect. In this case, the emergence of the preference for mating in a new host precedes and conditions the reproductive isolation and divergent selection of populations; hence, syllogistically, the emergence of the new neurally determined preference is the ultimate cause of the speciation process. It is the emergence of the preference for the new host that brings about the interruption of the gene flow between the two populations and the consequent "allopatry," not the other way around.

In the following, a few well-documented examples on the host shift speciation will be briefly presented.

Rhagoletis pomonella is a fruit fly now comprising six sibling species. No domestic apple trees existed in the United States until the 1700s, but suddenly in 1867, in ~150 generations or years, the fly was reported to have shifted from its original native host, the hawthorn (*Crataegus* spp.), to domestic apple trees (*Malus pumila* P. Mill) to evolve a new host race (Olsson et al., 2006a). Now, these host races that are interfertile in the laboratory show considerable allele frequency differences, indicating that the gene flow between the two races is prevented, even though flies of both hosts are randomly dispersed. From the beginning, investigators suspected that some "intrinsic biological factors" rather than any spatial separations prevented the gene flow from occurring (Feder et al., 1988).

No changes in any relevant genes have been identified, or assumed, as the cause of the host shift. Flies identify the new host based on the olfactory perception of the host volatiles, but no differences are found in the number or class of the antennal olfactory receptor neurons (ORNs) that may have induced the differential host preference between the new host race and the original population. Instead, in the new host race, the existing ORNs have only tuned to the new host volatiles (Olsson et al., 2006a). A correlation is observed to exist between the differences in host volatile preferences on the one hand, and the differences in the temporal firing patterns and sensitivity (patterns of receptor expression) in the peripheral (antennal) olfactory receptor neurons on the other hand. The latter contribute to deciphering the volatile signal in the central neural processing (Olsson et al., 2006b), and it is suggested that only a few neurons are involved in host shifting (Tait et al., 2016).

Now, most investigators in the field agree that "the apple race of *R. pomonella* has evolved an increased preference for a specific blend of volatiles from apple fruit, and decreased response to hawthorn volatiles, during ~150 years of the host shift. Because mate choice in *R. pomonella* is directly tied to host choice, the difference in host odor preference led to the premating reproductive isolation between apple and hawthorn sibling species" (Linn et al., 2003).

The European acorn moth *Cydia splendana* (Lepidoptera, Tortricidae) females in Sweden use only the oak *Quercus robur* as their host, but in Central and Southern Europe, they use the chestnut *Castanea sativa* and several oak species. Males of each of these populations respond to *different* sex pheromone blends of the isomers of 8,10-dodecadien-1-yl acetate, indicating that they represent two pheromone races of the moth in the initial stage of speciation. These populations of the acorn moth are reproductively isolated under natural conditions, and the underlying cause of isolation is the mate choice resulting from the divergence in the two populations of preference for pheromone blends (Bengtsson et al., 2014).

Noctuid moth species of the African cotton leafworm, *Spodoptera littoralis*, in Africa and Mediterranean countries and the tobacco cutworm, *Spodoptera litura*, in Asia and Oceania are sibling species reproductively isolated under natural conditions but produce viable hybrids under no-choice situations, indicating that the genetic divergence and the process of speciation are not yet completed. Males of both species are attracted by heterospecific females in a no-choice situation, but in a choice situation, males of *S. littoralis* are preferentially attracted by conspecific females due to some additional pheromone components that lack the sibling species. By contrast, males of *S. litura* do not discriminate between conspecific and heterospecific females because their pheromone components overlap with these of *S. littoralis* (Saveer et al., 2015).

Enchenopa binotata is a species complex of treehoppers comprising 11 sap-feeding sister species in eastern North America (Cocroft et al., 2008) that are morphologically similar; hence, they are identified according to their host plants. Each species of the complex specializes on a different host plant species. Male mating signals consist of two or more successive bouts of "whines" (tones with declining

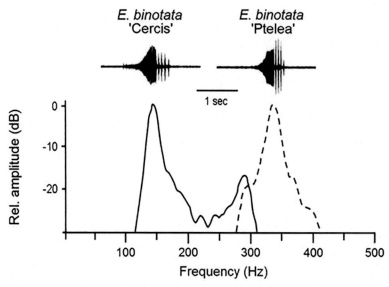

FIG. 4.20

Variation in male mating signal frequency for *Enchenopa binotata* "Cercis" and *E. binotata* "Ptelea." Waveforms of each species' signal with the corresponding amplitude spectra showing the frequency difference between species.

From McNett, G.D., Cocroft, R.B., 2008. From host shifts favor vibrational signal divergence in Enchenopa binotata treehoppers. Behav. Ecol. 19 (3), 650–656. https://doi.org/10.1093/beheco/arn017.

frequency). The signals are qualitatively similar across the species, but they differ quantitatively, especially in respect to the frequency and the number of pulses that follow the whine (Fig. 4.20).

Differences in the frequency of female response and female preferences for particular mating signals may have played a role in the divergence of male mating signals among species of the complex (Cocroft et al., 2008). The result of this interaction between male mating signals and respective female preferences represents the initial and primary cause of speciation in sympatry. Host shift then promoted the divergence in different ways, including the differences in the physiology of the host that determines differences in the timing of the hatch of the insect eggs and consequently the reproductive isolation.

References

Adams, M. D. (2000). The genome sequence of *Drosophila melanogaster*. *Science, 287,* 2185–2195.

Ahuja, G., Nia, S. B., Zapilko, V., Shiriagin, V., Kowatschew, D., Oka, Y., et al. (2015). Kappe neurons, a novel population of olfactory sensory neurons. *Sci. Rep., 4,* 4037. https://doi.org/10.1038/srep04037.

Altstein, M., Gazit, Y., Aziz, O. B., Gabay, T., Marcus, R., Vogel, Z., et al. (1996). Induction of cuticular melanization in Spodoptera littoralis larvae by PBAN/MRCH: development of a quantitative bioassay and structure function analysis. *Arch. Insect Biochem. Physiol., 31*, 355–370.

Amorim, M. C. P., Knight, M. E., Stratoudakis, Y., & Turner, G. F. (2004). Differences in sounds made by courting males of three closely related Lake Malawi cichlid species. *J. Fish Biol., 65*, 1358–1371.

Aprille, J. R., Lagace, C. J., Modica-Napolitano, J., & Trimmer, B. A. (2004). Role of nitric oxide and mitochondria in control of firefly flash. *Integr. Comp. Biol., 44*, 213–219.

Arnegard, M. E., Bogdanowicz, S. M., & Hopkins, C. D. (2005). Multiple cases of striking genetic similarity between alternative electric fish signal morphs in sympatry. *Evolution, 59*, 324–343.

Arnegard, M. E., Jackson, B. S., & Hopkins, C. D. (2006). Time-domain signal divergence and discrimination without receptor modification in sympatric morphs of electric fishes. *J. Exp. Biol., 209*, 2182–2198.

Ayers, J., Carpenter, G. A., Currie, S., & Kinch, J. (1983). Which behavior does the lamprey central motor program mediate? *Science, 221*, 1312–1314.

Bacquet, P. M., de Jong, M. A., Brattström, O., Wang, H.-. L., Molleman, F., Heuskin, S., et al. (2016). Differentiation in putative male sex pheromone components across and within populations of the African butterfly *Bicyclus anynana* as a potential driver of reproductive isolation. *Ecol. Evol., 6*(17), 6064–6084. Published 2016 Jul 29 https://doi.org/10.1002/ece3.2298.

Baker, B. S., Taylor, B. J., & Hall, J. C. (2001). Are complex behaviors specified by dedicated regulatory genes? Reasoning from *Drosophila. Cell, 105*, 13–24.

Balaban, E. (1997). Changes in multiple brain regions underlie species differences in a complex, congenital behavior. *Proc. Natl. Acad. Sci. U. S. A., 94*, 2001–2006.

Barry, M. J. (2002). Progress toward understanding the neurophysiological basis of predator-induced morphology in *Daphnia pulex. Physiol. Biochem. Zool., 75*, 179–186.

Basolo, A. (1998). Evolutionary change in a receiver bias: a comparison of female preference functions. *Proc. R. Soc. B Biol. Sci., 265*, 2223–2228.

Beason, R. C. (2005). Mechanisms of magnetic orientation in birds. *Integr. Comp. Biol., 45*, 563–573.

Bengtsson, M., Boutitie, A., Jósvai, J., Toth, M., Andreadis, S., Rauscher, S., et al. (2014). Pheromone races of *Cydia splendana* (Lepidoptera, Tortricidae) overlap in host plant association and geographic distribution. *Front. Ecol. Evol.* https://doi.org/10.3389/fevo.2014.00046. 06 August 2014.

Billeter, J. C., Atallah, J., Krupp, J. J., Millar, J. G., & Levine, J. D. (2009). Specialized cells tag sexual and species identity in *Drosophila melanogaster. Nature, 461*(7266), 987–991.

Blitz, D. M. (1999). Different proctolin neurons elicit distinct motor patterns from a multifunctional neuronal network. *J. Neurosci., 19*, 5449–5463.

Blows, M. W., & Allan, R. A. (1998). Levels of mate recognition within and between two *Drosophila* species and their hybrids. *Am. Nat., 152*, 834–837.

Blumberg, M. S. (2017). Development evolving: the origins and meanings of instinct. *Wiley Interdiscip. Rev. Cogn. Sci., 8*(1–2). https://doi.org/10.1002/wcs.1371.

Borst, D. W., Wainwright, G., & Rees, H. H. (2002). In vivo regulation of the mandibular organ in the edible crab, *Cancer pagurus. Proc. Biol. Sci., 269*, 483–490.

Boughman, J. W., Rundle, H. D., & Schluter, D. (2005). Parallel evolution of sexual isolation in sticklebacks. *Evolution, 59*, 361–373.

Butlin, R. K., & Ritchie, M. G. (1991). Variation in female mate preference across grasshopper hybrid zone. *J. Evol. Biol., 4*, 227–240.

Cabej, N. R. (2008). *Epigenetic Principles of Evolution* (p. 263). Dumont, NJ: Albanet.

Cabej, N. R. (2019). *Epigenetic Principles of Evolution* (p. 622). Dumont, NJ: Albanet.

Carlson, B. A., & Hopkins, C. D. (2004). Central control of electric signaling behavior in the mormyrid *Brienomyrus brachyistius*: segregation of behavior-specific inputs and the role of modifiable recurrent inhibition. *J. Exp. Biol., 207*, 1073–1084.

Carlson, B. A., Hasan, S. M., Hollmann, M., Miller, D. B., Harmon, L. J., & Arnegard, M. E. (2011). Brain evolution triggers increased diversification of electric fishes. *Science, 332*(6029), 583–586. https://doi.org/10.1126/science.1201524.

Carvalho, M. J. A., & Mirth, C. K. (2015). Coordinating morphology with behavior during development: an integrative approach from a fly perspective. *Front. Ecol. Evol.*. https://doi.org/10.3389/fevo.2015.00005.

Christensen, T. A., Sherman, T. G., McCaman, R. E., & Carlson, A. D. (1983). Presence of octopamine in firefly photomotor neurons. *Neuroscience, 9*, 183–189.

Clowney, E. J., Iguchi, S., Bussell, J. J., Scheer, E., & Ruta, V. (2015). Multimodal chemosensory circuits controlling Male courtship in *Drosophila*. *Neuron, 87*(5), 1036–1049. https://doi.org/10.1016/j.neuron.2015.07.025.

Cocroft, R. B., Rodriguez, R. L., & Hunt, R. E. (2008). Host shifts, the evolution of communication, and speciation in the *Enchenopa binotata* species complex of treehoppers. In K. Tilmon (Ed.), *Specialization, Speciation, and Radiation*. Berkeley: California University Press. https://doi.org/10.1525/9780520933828-009.

Coleman, S. W., Patricelli, G. L., & Borgia, G. (2004). Variable female preferences drive complex male displays. *Nature, 428*, 742–745.

Crampton, W. G. R. (2019). Electroreception, electrogenesis and electric signal evolution. *J. Fish Biol., 95*(1), 92–134.

Cratsley, C. K., & Lewis, S. M. (2003). Female preference for male courtship flashes in *Photinus ignitus* fireflies. *Behav. Ecol., 14*, 135–140.

Crespi, E. J., & Denver, R. J. (2005). Ancient origins of human developmental plasticity. *Am. J. Hum. Biol., 17*, 44–54.

Croll, R. P., Davis, W. J., & Kovac, M. P. (1985). Neural mechanisms of motor program switching in the mollusc *Pleurobranchaea*. I. Central motor programs underlying ingestion, egestion, and the "neutral" rhythm(s). *J. Neurosci., 5*, 48–55.

Cronin, T. M., & Raymo, M. E. (1997). Orbital forcing of deep-sea benthic species diversity. *Nature, 385*, 624–627.

Czech-Damal, N. U., Dehnhardt, G., Manger, P., & Hanke, W. (2013). Passive electroreception in aquatic mammals. *J. Comp. Physiol. A Neuroethol. Sens. Neural Behav. Physiol., 199*(6), 555–563. https://doi.org/10.1007/s00359-012-0780-8.

Dahl, J., & Peckarsky, B. L. (2002). Induced morphological defenses in the wild: predator effects on a mayfly, *Drunella coloradensis*. *Ecology, 83*, 1620–1634.

Darwin, C. R. (1859). *The Origin of Species by Means of Natural Selection or the Preservation of Favored Races in the Struggle for Life* (1st ed., p. 209). London: John Murray.

Darwin, C. R. (1871a). *The Descent of Man and Selection in Relation to Sex. vol. 1* (1st ed., p. 63). London: John Murray.

Darwin, C. R. (1871b). *The Descent of Man, and Selection in Relation to Sex. vol. 2* (1st ed., p. 398). London: John Murray.

Darwin, C. R. (1874). *The Descent of Man and Selection in Relation to Sex* (2nd, p. 67). London: Revised and augmented. Murray.

David, J. R., Lemeunier, F., Tsacas, L., & Yassin, A. (2007). The historical discovery of the nine species in the *Drosophila melanogaster* species subgroup. *Genetics, 177*, 1969–1973.

Delcomyn, F. (1998). *Foundations of Neurobiology* (p. 602). New York: W.H. Freeman and Co.

Demary, K., Michaelidis, C. I., & Lewis, S. M. (2006). Firefly courtship: behavioral and morphological predictors of male mating success in *Photinus greeni*. *Ethology, 112*, 485–492.

Derby, C. D., Kozma, M. T., Senatore, A., & Schmidt, M. (2016). Molecular mechanisms of reception and perireception in crustacean chemoreception: a comparative review. *Chem. Senses, 41*, 381–398. https://doi.org/10.1093/chemse/bjw057.

Diao, F., Elliott, A. D., Diao, F., Shah, S., & White, B. H. (2017). Neuromodulatory connectivity defines the structure of a behavioral neural network. *Elife, 6*. https://doi.org/10.7554/eLife.29797, e29797.

Dion, E., Pui, L. X., Weber, K., & Monteiro, A. (2020). Early-exposure to new sex pheromone blends alters mate preference in female butterflies and in their offspring. *Nat. Commun., 11*(1). https://doi.org/10.1038/s41467-019-13801-2.

Dixon, A. F. G., & Agarwala, B. K. (1999). Ladybird-induced life-history changes in aphids. *Proc. R. Soc. B Biol. Sci., 266*, 1549–1553.

Drijfhout, F. P., Kather, R., & Martin, S. J. (2009). The role of cuticular hydrocarbons in insects. In W. Zhang, & H. Liu (Eds.), *Behavioral and Chemical Ecology*. New York: Nova Science Publishers, Inc.

Drosser, M. L., Fortey, R. A., & Li, X. (1996). The ordovician radiation. *Am. Sci., 84*, 122–131.

Dudley, R. (1998). Atmospheric oxygen, giant Paleozoic insects and the evolution of aerial locomotor performance. *J. Exp. Biol., 201*, 1043–1050.

Eads, B. D., Andrews, J., & Colbourne, J. K. (2008). Ecological genomics in *Daphnia*: stress responses and environmental sex determination. *Heredity, 100*, 184–190.

Elmer, K. R., Lehtonen, T. K., & Meyer, A. (2009). Color assortative mating contributes to sympatric divergence of neotropical cichlid fish. *Evolution, 63*(10), 2750–2757. https://doi.org/10.1111/j.1558-5646.2009.00736.

Evans, H. E. (1966). *The Comparative Ethology and Evolution of the Sand Wasps* (p. 2). Cambridge: Harvard Univ. Press. Quoted by Wcislo, W.T., 1989. Behavioral environments and evolutionary change. Annu. Rev. Ecol. Syst. 20, 137–169.

Ewing, A. W., & Miyan, J. A. (1986). Sexual selection, sexual isolation and the evolution of song in the *Drosophila repleta* group of species. *Anim. Behav., 34*, 421–429.

Feder, J. L., Chilcote, C. A., & Bush, G. L. (1988). Genetic differentiation between sympatric host races of the apple maggot fly *Rhagoletis pomonella*. *Nature, 336*, 61–64.

Feulner, P. G. D., Kirschbaum, F., Mamonekene, V., Ketmaier, V., & Tiedemann, R. (2007). Adaptive radiation in African weakly electric fish (Teleostei: Mormyridae: *Campylomormyrus*): a combined molecular and morphological approach. *J. Evol. Biol., 20*, 403–414. https://doi.org/10.1111/j.1420-9101.2006.01181.x.

Feulner, P. G., Plath, M., Engelmann, J., Kirschbaum, F., & Tiedemann, R. (2009). Electrifying love: electric fish use species-specific discharge for mate recognition. *Biol. Lett., 5*(2), 225–228. https://doi.org/10.1098/rsbl.2008.0566.

Fletcher, D., & Blum, M. (1981). Pheromonal control of dealation and oogenesis in virgin queen ants. *Science, 212*, 73–75.

Flint, J. (2019). Outside in. *PLoS Genet., 15*(2), e1008014. Published 2019 Feb 28 https://doi.org/10.1371/journal.pgen.1008014.

Gammie, S. C., & Truman, J. W. (1997). Neuropeptide hierarchies and the activation of sequential motor behaviors in the Hawkmoth, *Manduca sexta*. *J Neurosci., 17*, 4389–4397.

Garstang, W. (1922). The theory of recapitulation: a critical re-statement of the biogenetic law. *J. Linn. Soc. London, Zool.*, *35*, 81–101 (98).

Getting, P. A. (1989). Emerging principles governing the operation of neural networks. *Annu. Rev. Neurosci.*, *12*, 185–204.

Gilbert, J. J. (2003). Environmental and endogenous control of sexuality in a rotifer life cycle: developmental and population biology. *Evol. Dev.*, *5*, 19–24.

Gillette, R. (1991). Central nervous system. Section D. In C. L. Prosser (Ed.), *Neural and Integrative Animal Physiology* (4th ed., pp. 574–611). Wiley-Liss (595).

Gleason, J. M., Jallon, J. M., Rouault, J. D., & Ritchie, M. G. (2005). Quantitative trait loci for cuticular hydrocarbons associated with sexual isolation between *Drosophila simulans* and *D. sechellia*. *Genetics*, *171*(4), 1789–1798. https://doi.org/10.1534/genetics.104.037937.

Godoy, L. D., Rossignoli, M. T., Delfino-Pereira, P., Garcia-Cairasco, N., & de Lima Umeoka, E. H. (2018). A comprehensive overview on stress neurobiology: basic concepts and clinical implications. *Front. Behav. Neurosci.*. https://doi.org/10.3389/fnbeh.2018.00127. 03 July 2018.

Goldstein, D. S. (1990). Neurotransmitters and stress. *Biofeedback Self Regul.*, *3*, 243–271.

Golowasch, J. (2019). Neuromodulation of central pattern generators and its role in the functional recovery of central pattern generator activity. *J. Neurophysiol.*, *122*(1), 300–315. https://doi.org/10.1152/jn.00784.2018.

Gong, A., & Gibson, R. M. (1996). Reversal of a female preference after visual exposure to a predator in the guppy, *Poecilia reticulata*. *Anim. Behav.*, *52*, 1007–1015.

Gorbman, A., & Davey, K. (1991). Endocrines. In C. L. Prosser (Ed.), (vols. 693–654). *Neural and Integrative Animal Physiology* (4th ed., pp. 709–711). Wiley-Liss.

Gould, J. L. (1982). *Ethology: The Mechanisms and Evolution of Behavior* (p. 177). New York, NY: W.W. Norton & Co.

Gowri, V., Dion, E., Viswanath, A., Piel, F. M., & Monteiro, A. (2019). Transgenerational inheritance of learned preferences for novel host plant odors in *Bicyclus anynana* butterflies. *Evolution*, *73*(12), 2401. https://doi.org/10.1111/evo.13861.

Gray, D. A., & Cade, W. H. (2000). Sexual selection and speciation in field crickets. *Proc. Natl. Acad. Sci. U. S. A.*, *97*, 14449–14454.

Greenberg, N., Carr, J. A., & Summers, C. (2002). Causes and consequences of stress. *Integr. Comp. Biol.*, *42*, 508–516.

Greenfield, M. D. (2001). Missing link in firefly bioluminescence revealed: NO regulation of photocyte respiration. *Bioessays*, *23*, 992–995.

Gries, G., Schaefer, P. W., Gries, R., Liska, J., & Gotoh, T. (2001). Reproductive character displacement in *Lymantria monacha* from northern Japan? *J. Chem. Ecol.*, *27*, 1163–1176.

Grober, M. S., Fox, S. H., Laughlin, C., & Bass, A. H. (1994). GnRH cell size and number in a teleost fish with two male reproductive morphs: sexual maturation, final sexual status and body size allometry. *Brain Behav. Evol.*, *43*, 61–78.

Haesler, M. P., & Seehausen, O. (2005). Inheritance of female mating preference in a sympatric sibling species pair of Lake Victoria cichlids and implications for speciation. *Proc. R. Soc. B Biol. Sci.*, *272*, 237–245.

Hall, M., Lindholm, A. K., & Brooks, R. (2004). Direct selection on male attractiveness and female preference fails to produce a response. *BMC Evol. Biol.*, *4*, 1. Published 2004 Jan 14 https://doi.org/10.1186/1471-2148-4-1.

Hansen, A., & Zeiske, E. (1998). The peripheral olfactory organ of the zebrafish, Danio rerio: an ultrastructural study. *Chem. Senses*, *23*, 39–48.

Harris-Warrick, R. M. (2011). Neuromodulation and flexibility in Central Pattern Generator networks. *Curr. Opin. Neurobiol.*, *21*(5), 685–692. https://doi.org/10.1016/j.conb.2011.05.011.

Harris-Warrick, R. M., & Cohen, A. H. (1985). Serotonin modulates the central pattern generator for locomotion in the isolated lamprey spinal cord. *J. Exp. Biol.*, *116*, 27–46.

Harvell, C. D. (1998). Genetic variation and polymorphism in the inducible spines of a bryozoan. *Evolution*, *52*, 80–86.

Hatano, E., Kunert, G., & Weisser, W. W. (2010). Aphid wing induction and ecological costs of alarm pheromone emission under field conditions. *PLoS One*, *5*(6), e11188. Published 2010 Jun 23 https://doi.org/10.1371/journal.pone.0011188.

Hedwig, B. (2005). Pulses, patterns and paths: neurobiology of acoustic behavior in crickets. *J. Comp. Physiol. A Neuroethol. Sens. Neural Behav. Physiol.*, *192*, 677–689.

Heimovics, S. A., Salvante, K. G., Sockman, K. W., & Riters, L. V. (2011). Individual differences in the motivation to communicate relate to levels of midbrain and striatal catecholamine markers in male European starlings. *Horm. Behav.*, *60*, 529–539.

Henehan, M. J., Ridgwell, A., Thomas, E., Zhang, S., Alegret, L., Schmidt, D. N., et al. (2019). Rapid ocean acidification and protracted Earth system recovery followed the end-Cretaceous Chicxulub impact. *Proc. Natl. Acad. Sci. U. S. A.*, *116*(45), 22500–22504.

Hennig, R. M., & Weber, T. (1997). Filtering of temporal parameters of the calling song by cricket females of two closely related species: a behavioral analysis. *J. Comp. Physiol. A*, *180*, 621–630.

Herzog, Q., & Laforsch, C. (2013). Modality matters for the expression of inducible defenses: introducing a concept of predator modality. *BMC Biol.*, *11*, 113.

Herzog, Q., Tittgen, C., & Laforsch, C. (2016). Predator-specific reversibility of morphological defenses in *Daphnia barbata*. *J. Plankton Res.*, *38*, 771–780.

Higgie, M., Chenoweth, S., & Blows, M. W. (2000). Natural selection and the reinforcement of mate recognition. *Science*, *290*, 519–521.

Hoke, K. L., Ryan, M. J., & Wilczynski, W. (2005). Social cues shift functional connectivity in the hypothalamus. *Proc. Natl. Acad. Sci. U. S. A.*, *102*(30), 10712–10717. https://doi.org/10.1073/pnas.0502361102.

Hopkins, C. D. (1999). Signal evolution in electric communication. In M. D. Hauser, & M. Konishi (Eds.), *The Design of Animal Communication* (pp. 461–491). Cambridge, MA: M.I.T. Press.

Howland, M. A., Sandman, C. A., & Glynn, L. M. (2017). Developmental origins of the human hypothalamic-pituitary-adrenal axis. *Expert. Rev. Endocrinol. Metab.*, *12*(5), 321–339. https://doi.org/10.1080/17446651.2017.1356222.

Iglesias, F., Pilar-Marco, M., Jacquin-Joly, E., Camps, F., & Fabrias, G. (1998). Regulation of sex pheromone biosynthesis in two noctuid species, *S. littoralis* and *M. brassicae*, may involve both PBAN and the ventral nerve cord. *Arch. Insect Biochem. Physiol.*, *37*, 295–304.

Jing, J., & Gillette, R. (2000). Escape swim network interneurons have diverse roles in behavioral switching and putative arousal in *Pleurobranchaea californica*. *J. Neurophysiol.*, *81*, 654–667.

Johnson, J. B., & Basolo, A. L. (2003). Predator exposure alters female mate choice in the green swordtail. *Behav. Ecol.*, *14*, 619–625.

Katz, P. S., & Frost, W. N. (1996). Intrinsic neuromodulation. Altering neural circuits from within. *Trends Neurosci.*, *19*, 54–61.

Juračka, P. J., Laforsch, C., & Petrusek, A. (2011). Neckteeth formation in two species of the *Daphnia curvirostris* complex (Crustacea: Cladocera). *J. Limnol.*, *70*(2), 359–368.

Kallman, B. R., Kim, H., & Scott, K. (2015). Excitation and inhibition onto central courtship neurons biases *Drosophila* mate choice. *Elife*, *4*, e11188. Published 2015 Nov 14 https://doi.org/10.7554/eLife.11188.

Kaplan, I., & Thaler, J. S. (2012). Phytohormone-mediated plant resistance and predation risk act independently on the population growth and wing formation of potato aphids, *Macrosiphum euphorbiae*. *Arthropod Plant Interact.*, *6*(2). https://doi.org/10.1007/s11829-011-9175-y.

Katz, P. S., & Harris-Warrick, R. M. (1999). The evolution of neuronal circuits underlying species-specific behavior. *Curr. Opin. Neurobiol.*, *2*, 628–633.

Khallaf, M. A., Auer, T. O., Grabe, V., Depetris-Chauvin, A., Ammagarahalli, B., Zhang, D.-D., et al. (2020). Mate discrimination among subspecies through a conserved olfactory pathway. *Sci. Adv.*, *6*(25), eaba5279. Published 2020 Jun 17 https://doi.org/10.1126/sciadv.aba5279.

Knobloch, H. S., Charlet, A., Hoffmann, L. C., Eliava, M., Khrulev, S., Cetin, A. H., et al. (2012). Evoked axonal oxytocin release in the central amygdala attenuates fear response. *Neuron*, *73*, 553–566.

Kolk, S. M., Kramer, B. M. R., Cornelisse, L. N., Scheenen, W. J. J. M., Jenks, B. G., & Roubos, E. W. (2002). Multiple control and dynamic response of the *Xenopus* melanotrope cell. *Comp. Biochem. Physiol. B*, *132*, 257–268.

Kocher, T. D. (2004). Adaptive evolution and explosive speciation: the cichlid fish model. *Nat. Rev. Genet.*, *5*, 288–298.

Kotov, A. A., Ishida, S., & Taylor, D. J. (2006). A new species in the *Daphnia curvirostris* (Crustacea: Cladocera) complex from the eastern Palearctic with molecular phylogenetic evidence for the independent origin of neckteeth. *J. Plankton Res.*, *28*, 1067–1079.

Kramer, B., & Kuhn, B. (1994). Species recognition by the sequence of discharge intervals in weakly electric fishes of the genus *Campylomormyrus* (Mormyridae, Teleostei). *Anim. Behav.*, *48*, 435–445. https://doi.org/10.1006/anbe.1994.1257.

Kremen, C., & Nijhout, H. F. (1989). Juvenile hormone controls the onset of pupal commitment in the imaginal disks and epidermis of *Precis coenia* (Lepidoptera: Nymphalidae). *J. Insect Physiol.*, *35*, 603–612.

Kravitz, E. A. (1988). Hormonal control of behavior. *Science*, *241*, 1775–1781.

Kunert, G., & Weiser, W. W. (2003). The interplay between density- and trait-mediated effects in predatory-prey interactions: a case study in aphid wing polymorphism. *Oecologia*, *135*, 304–312.

Kunert, G., Otto, S., Roese, U. S. R., Gershenzon, J., & Weisser, W. W. (2005). Alarm pheromone mediates production of winged dispersal morphs in aphids. *Ecol. Lett.*, *8*, 596–603.

LeBlanc, G. A., Wang, Y. H., Holmes, C. N., Kwon, G., & Medlock, E. K. (2013). A transgenerational endocrine signaling pathway in crustacea. *PLoS One*, *8*(4), e61715.

LeBlanc, G. A., & Medlock, E. K. (2015). Males on demand: the environmental–neuroendocrine control of male sex determination in daphnids. *FEBS J.*, *282*, 4080–4093.

Lewis, F. P., & Fullard, J. H. (1996). Neurometamorphosis of the ear in the gypsy moth, *Lymnatria dispar*, and its homologue in the earless forest tent caterpillar moth, *Malacosoma disstria*. *J. Neurobiol.*, *31*, 245–262.

Li, T.-R., & White, K. P. (2003). Tissue-specific gene expression and ecdysone-regulated genomic networks in *Drosophila*. *Dev. Cell*, *5*, 59–72.

Linn, C., Feder, J. L., Nojima, S., Dambroski, H. R., Berlocher, S. H., & Roelofs, W. (2003). Fruit odor discrimination and sympatric host race formation in *Rhagoletis*. *Proc. Natl. Acad. Sci. USA*, *100*, 11490–11493.

Lissmann, H. W. (1951). Continuous electrical signals from the tail of a fish, *Gymnarchus niloticus* Cuv. *Nature*, *167*, 201–202. https://doi.org/10.1038/167201a0.

Liu, W., Ganguly, A., Huang, J., Wang, Y., Ni, J. D., Gurav, A. S., et al. (2019). Neuropeptide F regulates courtship in *Drosophila* through a male-specific neuronal circuit. *Elife*, *8*, e49574. Published 2019 Aug 12 https://doi.org/10.7554/eLife.49574.

Lüning, J. (1992). Phenotypic plasticity of *Daphnia pulex* in the presence of invertebrate predators; morphological and life history responses. *Oecologia*, *92*, 383–390.

Lynch, K. S., Crews, D., Ryan, M. J., & Wilczynski, W. (2006). Hormonal state influences aspects of female mate choice in the Túngara Frog (*Physalaemus pustulosus*). *Horm. Behav.*, *49*(4), 450–457. https://doi.org/10.1016/j.yhbeh.2005.10.001.

Manoli, D. S., Meissner, G. W., & Baker, B. S. (2006). Blueprints for behavior: genetic specification of neural circuitry for innate behaviors. *Trends Neurosci.*, *29*, 444–451.

Marder, E. (2012). Neuromodulation of neuronal circuits: back to the future. *Neuron*, *76*, 1–11. https://doi.org/10.1016/j.neuron.2012.09.010.

Matsuda, K. (2013). Regulation of feeding behavior and psychomotor activity by corticotropin-releasing hormone (CRH) in fish. *Front. Neurosci.*, *7*, 91. https://doi.org/10.3389/fnins.2013.00091.

Matzkin, L. M. (2014). Ecological genomics of host shifts in *Drosophila mojavensis*. *Adv. Exp. Med. Biol.*, *781*, 233–247. https://doi.org/10.1007/978-94-007-7347-9_12.

Mayr, E. (1942). *Systematics and the Origin of Species*. New York: Columbia University Press.

Mayr, E. (1988). *Toward a New Philosophy of Biology* (p. 408). Cambridge MA: Harvard University Press.

Metzner, W., Koch, C., Wessel, R., & Gabbiani, F. (1998). Feature extraction by burst-like spike patterns in multiple sensory maps. *J. Neurosci.*, *18*, 2283–2300.

Miner, A. L., Rosenberg, A. J., & Nijhout, H. F. (2000). Control of growth and differentiation of the wing imaginal disk of *Precis coenia* (Lepidoptera: Nymphalidae). *J. Insect Physiol.*, *46*, 251–258.

Miyakawa, H., Imai, M., Sugimoto, N., Ishikawa, Y., Ishikawa, A., Ishigaki, H., et al. (2010). Gene up-regulation in response to predator kairomones in the water flea, *Daphnia pulex*. *BMC Dev. Biol.*, *10*, 45. https://doi.org/10.1186/1471-213X-10-45.

Montag-Sallaz, M., Baarke, A., & Montag, D. (2003). Aberrant neuronal connectivity in CHL1-deficient mice is associated with altered information processing-related immediate early gene expression. *J. Neurobiol.*, *57*, 67–80.

Morris, M. R., & Ryan, M. J. (1996). Sexual difference in signal-receiver coevolution. *Anim. Behav.*, *52*, 1017–1024.

Mayr, E. (1947). Ecological factors in speciation. *Evolution*, *1*, 263–288.

McBride, C. S. (2007). Rapid evolution of smell and taste receptor genes during host specialization in *Drosophila sechellia*. *Proc. Natl. Acad. Sci. U. S. A.*, *104*, 4996–5001.

McEwen, B. S. (1999). Stress. In R. A. Wilson, & F. Keil (Eds.), *The MIT Encyclopedia of the Cognitive Sciences*. Cambridge: MIT Press.

Mendelson, T. C., & Shaw, K. L. (2005). Rapid speciation in an arthropod. *Nature*, *433*, 375–376.

Mondor, E. B., Rosenheim, J. A., & Addicott, J. F. (2005). Predator-induced transgenerational phenotypic plasticity in the cotton aphid. *Oecologia*, *142*, 104–108.

Muheim, R., Beckman, J., & Akesson, S. (2002). Magnetic compass orientation in European robins is dependent on both wavelength and the intensity of light. *J. Exp. Biol.*, *205*, 3845–3856.

Myers, B., Scheimann, J. R., Franco-Villanueva, A., & Herman, J. P. (2017). Ascending mechanisms of stress integration: Implications for brainstem regulation of neuroendocrine and behavioral stress responses. *Neurosci. Biobehav. Rev.*, *74*(Pt. B), 366–375. https://doi.org/10.1016/j.neubiorev.2016.05.011.

Nagel, R., Kirschbaum, F., Engelmann, J., Hofmann, V., Pawelzik, F., & Tiedemann, R. (2018). Male-mediated species recognition among African weakly electric fishes. *R. Soc. Open Sci.*, *5*(2), 170443. https://doi.org/10.1098/rsos.170443.

Purves, D., Augustine, G. J., Fitzpatrick, D., Katz, L. C., LaMantia, A.-S., McNamara, J. O., et al (Eds.). (2001a). *Neuroscience* (2nd ed., p. 984). Sunderland MA: Sinauer Associates, Inc. Publishers.

Purves, D., Augustine, G. J., Fitzpatrick, D., Katz, L. C., LaMantia, A.-S., McNamara, J. O., et al (Eds.). (2001b). *Neuroscience* (2nd ed., p. 991). Sunderland MA: Sinauer Associates, Inc. Publishers.

Nguyen, L., Mamonekene, V., Vater, M., Bartsch, P., Tiedemann, R., & Kirschbaum, F. (2020). Ontogeny of electric organ and electric organ discharge in *Campylomormyrus rhynchophorus* (Teleostei: Mormyridae). *J. Comp. Physiol. A*, *206*, 453–466. https://doi.org/10.1007/s00359-020-01411-z.

Nottebohm, F. (2005). The neural basis of birdsong. *PLoS Biol.*, *3*(5), e164.

Ohtsuki, H., Yokoyama, J., Ohba, N., Ohmiya, Y., & Kawata, M. (2014). Expression of the *nos* gene and firefly flashing: a test of the nitric-oxide-mediated flash control model. *J. Insect Sci.*, *14*, 56. https://doi.org/10.1093/jis/14.1.56.

Olmstead, A. W., & LeBlanc, G. A. (2002). Juvenoid hormone methyl farnesoate is a sex determinant in the crustacean *Daphnia magna*. *J. Exp. Zool.*, *293*, 736–739.

Olsson, S. B., Linn, C. E., Jr., & Roelofs, W. L. (2006a). The chemosensory basis for behavioral divergence involved in sympatric host shifts. I. Characterizing olfactory receptor neuron classes responding to key host volatiles. *J. Comp. Physiol. A Neuroethol. Sens. Neural Behav. Physiol.*, *192*, 279–288.

Olsson, S. B., Linn, C. E., Jr., & Roelofs, W. L. (2006b). The chemosensory basis for behavioral divergence involved in sympatric host shifts II: olfactory receptor neuron sensitivity and temporal firing pattern to individual key host volatiles. *J. Comp. Physiol. A Neuroethol. Sens. Neural Behav. Physiol.*, *192*, 289–300.

Ottaviani, E., Franchini, A., & Franceschi, C. (1998). Presence of immunoreactive corticotropin-releasing hormone and cortical molecules in invertebrate gametocytes and lower and higher vertebrate thymus. *Histochem. J.*, *30*, 61–67.

Pacak, K., & Palkovitz, M. (2001). Specificity of central neuroendocrine responses implications for stress-related disorders. *Endocr. Rev.*, *22*, 502–548.

Pan, Y., Meissner, G. W., & Baker, B. S. (2012). Joint control of Drosophila male courtship behavior by motion cues and activation of male-specific P1 neurons. *Proc. Natl. Acad. Sci. U. S. A.*, *109*(25), 10065–10070. https://doi.org/10.1073/pnas.1207107109.

Panhuis, T. M., Butlin, R., Zuk, M., & Tregenza, M. (2001). Sexual selection and speciation. *Trends Ecol. Evol.*, *16*, 364–372.

Penn, J. L., Deutsch, C., Payne, J. L., & Sperling, E. A. (2018). Temperature-dependent hypoxia explains biogeography and severity of end-Permian marine mass extinction. *Science*, *362*(6419). https://doi.org/10.1126/science.aat1327, eaat1327.

Pfeiler, E., Castrezana, S., Reed, L. K., & Markow, T. A. (2009). Genetic, ecological and morphological differences among populations of the cactophilic *Drosophila mojavensis* from southwestern USA and northwestern Mexico, with descriptions of two new subspecies. *J. Nat. Hist.*, *43*(15–16), 923–938.

Phelps, S. M., Rand, A. S., & Ryan, M. J. (2006). A cognitive framework for mate choice and species recognition. *Am. Nat.*, *167*, 28–42.

Pinker, S. (1994). *The Language Instinct*. New York: William Morrow and Co.

Plenderleith, M., van Oosterhout, C., Robinson, R. L., & Turner, G. F. (2005). Female preference for conspecific males based on olfactory cues in a Lake Malawi cichlid fish. *Biol. Lett.*, *1*, 411–414.

Pohnert, G. (2019). Kairomones: finding the fish factor. *Elife*, *8*. https://doi.org/10.7554/eLife.48459, e48459.

Polanska, M. A., Kirchhoff, T., Dircksen, H., Hansson, B. S., & Harzsch, S. (2020). Functional morphology of the primary olfactory centers in the brain of the hermit crab *Coenobita clypeatus* (Anomala, Coenobitidae). *Cell Tissue Res.*, *380*(3), 449–467. https://doi.org/10.1007/s00441-020-03199-5.

Rechavi, O., Minevich, G., & Hobert, O. (2011). Transgenerational inheritance of an acquired small RNA-based antiviral response in *C. elegans*. *Cell*, *147*(6), 1248–1256. https://doi.org/10.1016/j.cell.2011.10.042.

Renoult, J. P., & Mendelson, T. C. (2019). Processing bias: extending sensory drive to include efficacy and efficiency in information processing. *Proc. Biol. Sci.*, *286*(1900), 20190165. https://doi.org/10.1098/rspb.2019.0165.

Rinaman, L. (2011). Hindbrain noradrenergic A2 neurons: diverse roles in autonomic, endocrine, cognitive, and behavioral functions. *Am. J. Physiol. Regul. Integr. Comp. Physiol.*, *300*, R222–R235. https://doi.org/10.1152/ajpregu.00556.2010.

Robertson, H. A., & Carlson, A. D. (1976). Octopamine: presence in firefly lantern suggests a transmitter role. *J. Exp. Zool.*, *195*(1), 159–164. https://doi.org/10.1002/jez.1401950116.

Rodd, F. H., Hughes, K. A., Grether, G. F., & Baril, C. T. (2002). A possible non-sexual origin of mate preference: are male guppies mimicking fruit? *Proc. R. Soc. B Biol. Sci.*, *269*, 475–481.

Roller, L., Žitňanová, I., Dai, L., Šimo, L., Park, Y., Satake, H., et al. (2010). The ecdysis triggering hormone signaling in arthropods. *Peptides*, *31*(3), 429–441.

Ryan, M. J. (1998). Sexual selection, receiver biases, and the evolution of sex differences. *Science*, *281*, 1999–2003.

Ryan, M. J., & Rand, A. S. (1993). Species recognition and sexual selection as unitary problem in animal communication. *Evolution*, *47*(2), 647–657.

Ryan, M. J., & Rand, A. S. (2003). Mate recognition in túngara frogs: a review of some studies of brain, behavior, and evolution. *Acta Zool. Sin.*, *49*, 713–726.

Ryan, M. J., & Cummings, M. E. (2013). Perceptual biases and mate choice. *Annu. Rev. Ecol. Evol. Syst.*, *44*, 437–459.

Saarikettu, M., Liimatainen, J. O., & Hoikkala, A. (2005). The role of male courtship song in species recognition in *Drosophila montana*. *Behav. Genet.*, *35*, 257–263.

Sato, K., Tanaka, R., Ishikawa, Y., & Yamamoto, D. (2020). Behavioral evolution of *Drosophila*: unraveling the circuit basis. *Genes (Basel)*, *11*(2), 157. Published 2020 Feb 1 https://doi.org/10.3390/genes11020157.

Saveer, A. M., Becher, P. G., Birgersson, G., Hansson, B. S., Witzgall, P., & Bengtsson, M. (2015). Mate recognition and reproductive isolation in the sibling species *Spodoptera littoralis* and *Spodoptera litura*. *Front. Ecol. Evol.*. https://doi.org/10.3389/fevo.2014.00018.

Sawtell, N. B., Williams, A., & Bell, C. C. (2005). From sparks to spikes: information processing in the electrosensory systems of fish. *Curr. Opin. Neurobiol.*, *15*, 437–443.

Schleidt, W. M. (1962). Die historische Entwicklung der Begriffe "Angeborenes auslösendes Schema" und "Angeborener Auslösemechanismus" in der Ethologie. *Z. Tierpsychol.*, *19*, 697–722.

Schröder, T., & Gilbert, J. J. (2004). Transgenerational plasticity for sexual reproduction and diapause in the life cycle of monogonont rotifers: intraclonal, intraspecific and interspecific variation in response to crowding. *Funct. Ecol.*, *18*, 458–466.

Schulte, P., Alegret, L., Arenillas, I., Arz, J. A., Barton, P. J., Bown, P. R., et al. (2010). The Chicxulub asteroid impact and mass extinction at the Cretaceous-Paleogene boundary. *Science*, *327*, 1214–1218.

Seehausen, O., van Alphen, J. J. M., & Witte, F. (1997a). Cichlid fish diversity threatened by eutrophication that curbs sexual selection. *Science*, *277*, 1808–1811.

Seehausen, O., Witte, F., Katunzi, E. F. B., Smits, J., & Bouton, N. (1997b). Patterns of the remnant cichlid fauna in southern Lake Victoria. *Conserv. Biol.*, *11*, 890–905.

Seehausen, O., & van Alphen, J. J. M. (1998). The effect of male coloration on female mate choice in closely related Lake Victoria cichlids (Haplochromis nyererei complex). *Behav. Ecol. Sociobiol.*, *42*, 1–8.

Seeholzer, L. F., Seppo, M., Stern, D. L., & Ruta, V. (2018). Evolution of a central neural circuit underlies *Drosophila* mate preferences. *Nature*, *559*(7715), 564–569. https://doi.org/10.1038/s41586-018-0322-9.

Servais, T., & Harpe, D. A. T. (2018). The Great Ordovician Biodiversification Event (GOBE): definition, concept and duration. *Lethaia*, *51*(2), 151–164.

Shahandeh, M. P., Pischedda, A., & Turner, T. L. (2018). Male mate choice via cuticular hydrocarbon pheromones drives reproductive isolation between *Drosophila* species. *Evolution*, *72*(1), 123–135. https://doi.org/10.1111/evo.13389.

Shahandeh, M. P., Pischedda, A., Rodriguez, J. M., & Turner, T. L. (2020). The genetics of male pheromone preference difference between *Drosophila melanogaster* and *Drosophila simulans*. *G3 (Bethesda)*, *10*(1), 401–415. https://doi.org/10.1534/g3.119.400780.

Shaposhnikov, G. K. (1965). Morphological divergence and convergence in an experiment with aphids (Homoptera, Aphidinae). *Entomol. Rev.*, *44*, 1–12 (according to Görür, G., 2000. The role of phenotypic plasticity in host race formation and sympatric speciation in phytophagous insects, particularly in aphids. Turk. J. Zool. 24, 63–68).

Shen, J., Chen, J., Algeo, T. J., Yuan, S., Feng, Q., Yu, J., et al. (2019). Evidence for a prolonged Permian–Triassic extinction interval from global marine mercury records. *Nat. Commun.*, *10*, 1563. https://doi.org/10.1038/s41467-019-09620-0.

Smadja, C., & Butlin, R. (2009). On the scent of speciation: the chemosensory system and its role in premating isolation. *Heredity*, *102*, 77–97. https://doi.org/10.1038/hdy.2008.55.

Snell, T. W., Kubanek, J., Carter, W., Payne, A. B., Kim, J., Hicks, M. K., et al. (2006). A protein signal triggers sexual reproduction in *Brachionus plicatilis* (Rotifera). *Mar. Biol.*, *149*, 763–773.

Sockman, K. W., & Lyons, S. M. (2017). How song experience affects female mate-choice, male song, and monoaminergic activity in the auditory telencephalon in Lincoln's sparrows. *Integr. Comp. Biol.*, *57*, 891–901.

Sokolowski, K., & Corbin, J. G. (2012). Wired for behaviors: from development to function of innate limbic system circuitry. *Front. Mol. Neurosci.*. https://doi.org/10.3389/fnmol.2012.00055.

Stibor, H. (1992). Predator induced life-history shifts in a freshwater cladoceran. *Oecologia*, *92*(2), 162–165. https://doi.org/10.1007/BF00317358.

Stockinger, P., Kvitsiani, D., Rotkopf, S., Tirián, L., & Dickson, B. J. (2005). Neural circuitry that governs *Drosophila* male courtship behavior. *Cell*, *121*, 795–807.

Strand, F. L. (1999). *Neuropeptides* (p. 144). MIT Press.

Strecker, U., Meyer, C. G., Sturmbauer, C., & Wilkens, H. (1996). Genetic divergence and speciation in an extremely young species flock in Mexico formed by the genus *Cyprinodon* (Cyprinodontidae, Teleostei). *Mol. Phylogenet. Evol.*, *6*, 143–149.

Strecker, U., & Kodric-Brown, A. (1999). Mate recognition systems in a species flock of Mexican pupfish. *J. Evol. Biol.*, *12*, 927–935.

Suesswein, A. J., & Byrne, J. H. (1988). Identification and characterization of neurons initializing patterned neural activity in the buccal ganglia in *Aplysia*. *J. Neurosci.*, *5*, 48–55.

Sugahara, R., Saeki, S., Jouraku, A., Shiotsuki, T., & Tanaka, S. (2015). Knockdown of the corazonin gene reveals its critical role in the control of gregarious characteristics in the desert locust. *J. Insect Physiol.*, *79*, 80–87. https://doi.org/10.1016/j.jinsphys.2015.06.009.

Tait, C., Batra, S., Ramaswamy, S. S., Feder, J. L., & Olsson, S. B. (2016). Sensory specificity and speciation: a potential neuronal pathway for host fruit odour discrimination in *Rhagoletis pomonella*. *Proc. Biol. Sci.*, *283*(1845). https://doi.org/10.1098/rspb.2016.2101.

Teotónio, H., & Rose, M. R. (2000). Variation in the reversibility of evolution. *Nature*, *408*, 463–466.

Teotónio, H., Matos, M., & Rose, M. R. (2002). Reverse evolution of fitness in *Drosophila melanogaster*. *J. Evol. Biol.*, *15*(4), 608–617.

Teplitsky, C., Plenet, S., & Joly, P. (2005a). Costs and limits of dosage response to predation risk: to what extent can tadpoles invest in anti-predator morphology. *Oecologia*, *145*, 364–370.

Teplitsky, C., Plenet, S., Léna, J.-. P., Mermet, N., Malet, E., & Joly, P. (2005b). Escape behaviour and ultimate causes of specific induced defences in an anuran tadpole. *J. Evol. Biol.*, *18*, 180–190.

Tierney, A. J. (1995). Evolutionary implications of neural circuit structure and function. *Behav. Processes*, *35*(1–3), 173–182. https://doi.org/10.1016/0376-6357(95)00041-0.

Toyota, K., Miyakawa, H., Hiruta, C., Furuta, K., Ogino, Y., Shinoda, T., et al. (2015a). Methyl farnesoate synthesis is necessary for the environmental sex determination in the water flea *Daphnia pulex*. *J. Insect Physiol.*, *80*, 22–30. https://doi.org/10.1016/j.jinsphys.2015.02.002.

Toyota, K., Miyakawa, H., Yamaguchi, K., Shigenobu, S., Ogino, Y., Tatarazako, N., et al. (2015b). NMDA receptor activation upstream of methyl farnesoate signaling for short day-induced male offspring production in the water flea, *Daphnia pulex*. *BMC Genomics*, *16*(1), 1392. https://doi.org/10.1186/s12864-015-1392-9.

Toyota, K., Sato, T., Tatarazako, N., & Iguchi, T. (2017). Protein kinase C is involved with upstream signaling of methyl farnesoate for photoperiod-dependent sex determination in the water flea *Daphnia pulex*. *Biol. Open*, *6*(2), 161–164. https://doi.org/10.1242/bio.021857.

Trimmer, B. A., Aprille, J. R., Dudzinski, D. M., Lagace, C. J., Lewis, S. M., Michel, T., et al. (2001). Nitric oxide and the control of firefly flashing. *Science*, *292*, 2486–2488.

Truman, J. W. (2006). Steroid hormone secretion in insects comes of age. *Proc. Natl. Acad. Sci. U. S. A.*, *103*, 8909–8910.

Trussell, G. C., & Smith, L. D. (2000). Induced defenses in response to an invading crab predator. *Proc. Natl. Acad. Sci. U. S. A.*, *97*, 2123–2127.

Trussell, G. C. (2001). Phenotypic clines, plasticity and morphological trade-offs in an intertidal snail. *Evolution*, *54*, 151–166.

Uller, T., & Olsson, M. (2006). Corticosterone during embryonic development influences behaviour in an ovoviviparous lizard. *Ethology*, *112*, 390–397.

Ulrich-Lai, Y. M., & Herman, J. P. (2009). Neural regulation of endocrine and autonomic stress responses. *Nat. Rev. Neurosci.*, *10*, 397–409. https://doi.org/10.1038/nrn2647.

Van Buskirk, J., & McCollum, S. A. (2000). Functional mechanisms of an inducible defense in tadpoles: morphology and behavior influence mortality risk from predation. *J. Evol. Biol.*, *13*, 336–347.

Vastenhouw, N. L., Fischer, S. E., Robert, V. J., Thijssen, K. L., Fraser, A. G., Kamath, R. S., et al. (2003). A genome-wide screen identifies 27 genes involved in transposon silencing in *C. elegans. Curr. Biol.*, *13*, 1311–1316. https://doi.org/10.1016/S0960-9822(03)00539-6.

Voneida, T. J., & Fish, S. E. (1984). Central nervous system changes related the reduction of visual input in in a naturally blind fish (*Astyanax hubbsi*). *Am. Zool.*, *24*, 773–782.

Wagner, W. E., Jr., Smeds, M. R., & Wiegman, D. D. (2001). Experience affects female responses to male song in variable field cricket *Gryllus lineaticeps* (Orthoptera, Gryllidae). *Ethology*, *107*, 769–776.

Weiss, L. C., Kruppert, S., Laforsch, C., & Tollrian, R. (2012). *Chaoborus* and gasterosteus anti-predator responses in *Daphnia pulex* are mediated by independent cholinergic and gabaergic neuronal signals. *PLoS One*, *7*(5). https://doi.org/10.1371/journal.pone.0036879, e36879.

Weiss, L. C., Leimann, J., & Tollrian, R. (2015a). Predator-induced defences in *Daphnia longicephala*: location of kairomone receptors and timeline of sensitive phases to trait formation. *J. Exp. Biol.*, *218*, 2918–2926.

Weiss, L. C., Albada, B., Becker, S. M., Meckelmann, S. W., Klein, J., Meyer, M., et al. (2018). Identification of *Chaoborus* kairomone chemicals that induce defences in *Daphnia. Nat. Chem. Biol.*, *14*, 1133–1139. https://doi.org/10.1038/s41589-018-0164-7.

West-Eberhard, M. J. (1984). Sexual selection, competitive communication and species-specific signals in insects. In T. Lewis (Ed.), *Insect Communication* (pp. 283–324). Toronto: Academic Press.

Whaling, C. S., Solis, M. M., Doupe, A. J., Soha, J. A., & Marler, P. (1997). Acoustic and neural bases for innate recognition of song. *Proc. Natl. Acad. Sci. U. S. A.*, *94*, 12694–12698.

White, B. H., & Ewer, J. (2014). Neural and hormonal control of postecdysial behaviors in insects. *Annu. Rev. Entomol.*, *59*, 363–381. https://doi.org/10.1146/annurev-ento-011613-162028.

Wiens, J. J., Kuczynski, C. A., Duellman, W. E., & Reeder, T. W. (2007). Loss and re-evolution of complex life cycles in marsupial frogs: does ancestral trait reconstruction mislead? *Evolution*, *61*(8), 1886–1899.

Wiernasz, D. C., & Kingsolver, J. G. (1992). Wing melanin pattern mediates species recognition in *Pieris occidentalis. Anim. Behav.*, *43*, 89–94.

Wikelski, M., Carbone, C., Bednekoff, P. A., Choudhury, S., & Tebbich, S. (2001). Why is female choice not unanimous? Insights from costly mate sampling in marine iguanas. *Ethology*, *107*, 623–638.

Wong, B. B. M., Fisher, H. S., & Rosenthal, G. G. (2005). Species recognition by male swordtails via chemical cues. *Behav. Ecol.*, *16*, 818–822.

Yao, M., Westphal, N. J., & Denver, R. J. (2004). Distribution and acute stress-induced activation of corticotropin-releasing hormone neurones in central nervous system of *Xenopus laevis*. *J. Neuroendocrinol.*, *16*, 880–893.

Zink, R. M. (2004). The role of subspecies in obscuring avian biological diversity and misleading conservation policy. *Proc. R. Soc. B Biol. Sci.*, *271*, 561–564.

Žitňan, D., Kim, Y. J., Žitňanová, I., Roller, L., & Adams, M. E. (2007). Complex steroid-peptide-receptor cascade controls insect ecdysis. *Gen. Comp. Endocrinol.*, *153*(1–3), 88–96. https://doi.org/10.1016/j.ygcen.2007.04.002.

Maintenance and change of phenotype: Inheritance of acquired traits

Animals maintain their adult structure for considerable long periods of time in a relative steady state, or dynamic equilibrium (Fliessgleichgewicht sensu von Bertalanffy), despite the continual losses they experience at the molecular, cellular, and supracellular levels. The array of phenotypes an animal can express within life-compatible environmental conditions is known as reaction norm (Reaktionsnorm) in a term coined more than a century ago by the German zoologist Richard Woltereck (1877–1944) based on his observations in experiments on the water flea *Daphnia* grown on varying laboratory conditions (Woltereck, 1909).

The reaction norm is an inherited species' character; it describes the adaptational potential of species to environmental changes and consists of all the phenotypes the animal can express in response to different environmental conditions or stimuli but are not inherited by the offspring in the absence of the triggering stimuli. The adaptation may involve physiological, behavioral, morphological, and life-history changes. The species-specific "upper and lower limits of change" represent the extreme life-compatible phenotypes beyond which animals cannot survive.

Performance of vital functions results in a continual loss of matter and energy at the molecular level as well as cells at the supracellular and organismic levels. Despite these unavoidable losses, animals maintain for considerable periods of time their normal functions and structure and their physiology and anatomy. This fact indicates that they have mechanisms for detecting and assessing losses and restoring the normal state at all the above levels. The normal function of animal cells requires their chemical and physical environment, the interstitial fluid that connects the cells with the blood, be kept in a steady state. The state and the ability of animals to maintain their internal environment in a relatively steady state within relatively narrow limits is known as homeostasis (Claude Bernard's milieu intérieur), and it is a primary function of the central nervous system.

Given that animal behavior is also controlled and regulated by the nervous system, in the following we will briefly review some well-known examples of the neural control of the physiology and morphology/morphometry of animals, although the neural control of the morphological development (organogenesis) is discussed in Chapter 3.

The Inductive Brain in Development and Evolution. https://doi.org/10.1016/B978-0-323-85154-1.00002-3

1 Maintenance of the normal physiological state

For a long time, it is known that the central nervous system plays a crucial role in controlling and regulating functions of all the organs and organ systems, digestive, circulatory, and excretory systems, etc. Herein we will only consider a few examples of the role of the nervous system in maintaining physicochemical properties of body fluids in a steady state:

1. the neural control of the chemical contents of the body fluids (blood, lymph, and interstitial fluid),
2. the maintenance of the water contents and glucose "within the narrow limits" even under abnormal, reduced or abundant, intake, and
3. the maintenance of constant body temperature even under biologically extreme (cold or warm) environmental temperatures.

1.1 Maintenance of the constant chemical contents of extracellular fluids

A relatively constant chemical and physical content of the body fluids is a prerequisite for the normal functioning of cells throughout the organism and the animal's survival. The content of body fluids is species-specific, but it may vary within certain limits in individuals of the same species. The most complex fluid is the blood plasma, which, besides cations (Na^+, K^+, Ca^{2+}) and anions such as Cl^-, contains numerous other products of metabolism and many secreted proteins, hormones, neuromodulators, and nutritional substances, as well as red, white, and other types of blood cells, all within a relatively narrow range of limits.

1.1.1 Neural regulation of water and electrolyte balance

Maintenance of the species-specific concentration of electrolytes in the body and the intercellular fluids is a necessary condition for the normal functioning of cells. The elimination of electrolytes and water with urine, perspiration, excretion, and exhalation requires a special subsystem of control that continuously monitors the state of the osmolality that in mammalians is chiefly regulated by osmoreceptor neurons of the circumventricular organ OVLT (from Latin *organum vasculosum of the lamina terminalis*) and SFO (subfornical nucleus). The drop in the water content of the body increases the concentration of electrolytes (mainly Na^+ and K^+) in fluids, causes a drop in the blood pressure, stimulating activation of the renin-angiotensin system (RAS) osmoreceptor neurons in OVLT, and increases production of angiotensin II (ANG II) in the subfornical organ and other parts of the brain and body. The subfornical organ neurons sense this because of the high number of ANG II receptors in their neurons that bind ANG II. OVLT and SFO send dense projections to the hypothalamic POA that integrates their input and, both via the median preoptic nucleus and directly, these osmoreceptors send signals to the "osmometers," magnocellular neuroendocrine cells (MNCs), in the hypothalamic paraventricular nucleus (PVN) and supraoptic nucleus (Bourque, 2008) (Fig. 5.1).

The role of SFO in regulating of water and electrolyte balance is also supported by the fact that drinking is induced by experimental administration of ANGII in the SFO, while this effect can be blocked by lesions of the SFA (Coble et al., 2014).

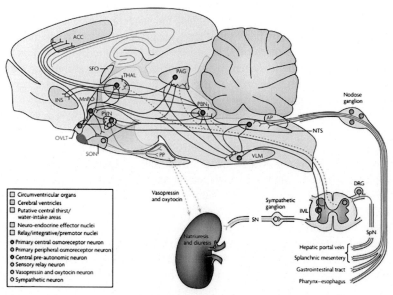

FIG. 5.1

Osmoregulatory circuits in the mammalian brain and the periphery. Sagittal illustration of the rat brain, in which the relative positions of relevant structures and nuclei have been compressed into a single plane. Only structures that have been directly implicated in the osmotic control of osmoregulatory responses are shown. Neurons and pathways are color-coded to distinguish osmosensory, integrative, and effector areas. Although visceral sensory pathways that relay information from dorsal root ganglion neurons are known to ascend through the spinal cord, specific evidence that peripheral osmosensory information ascends through this route is only partial (Vallet and Baertschi, 1982); this tract is therefore illustrated as a dashed line. Abbreviations: ACC, anterior cingulate cortex; AP, area postrema; DRG, dorsal root ganglion; IML, intermediolateral nucleus; INS, insula; MnPO, median preoptic nucleus; NTS, nucleus tractus solitarius; OVLT, organum vasculosum laminae terminalis; PAG, periaqueductal gray; PBN, parabrachial nucleus; PP, posterior pituitary; PVN, paraventricular nucleus; SFO, subfornical organ; SN, sympathetic nerve; SON, supraoptic nucleus; SpN, splanchnic nerve; THAL, thalamus; VLM, ventrolateral medulla.

From Bourque, C.W., 2008. Central mechanisms of osmosensation and systemic osmoregulation. Nat. Rev. Neurosci. 9, 519–531.

1.1.2 Regulation of glucose levels

Despite the higher blood glucose levels after meals, glucose is maintained at a relatively constant level by virtue of a neurohormonal system involving interactions between the pancreas, liver, and other organ systems under CNS coordination. A drop in the blood glucose level is sensed both peripherally by the pancreatic Langerhans islet β-cells, glucose-sensing neurons in the mouth, and neurons in the hepatoportal region (Hevener et al., 1997; Hevener et al., 2001; Fujita et al., 2007) and centrally by glucose sensors in the forebrain and the brainstem, especially in the VMH (ventromedial hypothalamus). In response to the drop of the blood glucose level, the sympathetic nervous system signals the pancreatic α-cells to increase glucagon production, which by converting the liver glycogen into glucose restores the normal glucose level in the blood. In response to high blood glucose levels after meals, the parasympathetic system by releasing acetylcholine stimulates β-cells to secrete insulin, but noncholinergic stimuli from sympathetic adrenergic nerves are also involved in the process (Thorens, 2014) (Fig. 5.2).

The parasympathetic system stimulates β-cell proliferation, whereas sympathetic signals inhibit the proliferation. This is corroborated by the experimental evidence that β-cell proliferation is stimulated by infusion of nutrients in the brain and periphery. Infusion of glucose + NEFA (nonesterified fatty acids) in the brain stimulates β-cell proliferation via vagus nerve whereas peripheral infusions of glucose + NEFA induce proliferation of β-cells by inhibiting the sympathetic system (Moullé et al., 2019).

Interesting regarding the role of the nervous system in the maintenance of blood glucose homeostasis is the phenomenon of the cephalic phase insulin release (CPIR). The term defines the anticipatory insulin secretion stimulated by the food visual (sight), olfactory (aroma), and gustatory (savor) stimuli, 10 min before the meal is absorbed and the elevated level of glucose is detected in β-cells. This nonglucose insulin secretion is again neurally determined (Eliasson et al., 2017).

1.2 Maintenance of normal body temperature

In the warm-blooded animals, the normal function at the organismal and cell levels requires maintenance of a constant core body temperature (temperature of the inner part of the body), the Tcore (core body temperature), within a narrow range, even under conditions of challenging temperature fluctuations in the environment. This implies the existence in the organism of a central temperature regulator, which for more than half a century is known to be in the hypothalamic neurons (Hammel et al., 1963). The maintenance of a constant Tcore is a function of a central regulatory circuit. The organism receives sensory input on the environmental temperature from nerve endings via cool and warm thermoreceptors in the skin and on the body temperature via thermoreceptors in internal organs and the brain.

In high environmental temperature, the organism decreases thermogenesis, heat production, by inhibiting thyroid hormone secretion and by stimulating the skin heat loss via vasodilation, cholinergic stimulation of sweating, and panting in some

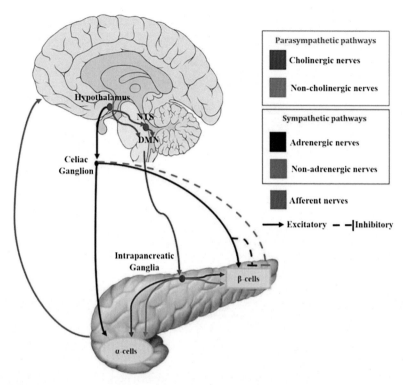

FIG. 5.2

Schematic of the most significant neural connections between the brain and the pancreas. The neural pathways to the α- and β-cells include postganglionic parasympathetic (*green color range*) and sympathetic (*red color range*) nerves. Afferent connections from the pancreas to the brain are also depicted (*gray*). Abbreviations: NTS, nucleus of solitary tract; DMN, dorsal motor nucleus of the vagus.

From Güemes, A., Georgiou, P. Review of the role of the nervous system in glucose homoeostasis and future perspectives towards the management of diabetes. Bioelectron Med. 4, 9 (2018). https://doi.org/10.1186/s42234-018-0009-4.

animals. When temperature drops, the organism stimulates thermogenesis via the opposite processes of induction of thyroid secretion, shivering, and adrenergic constriction of skin blood vessels.

All the above start with the processing of the environmental and internal body and brain temperatures in a neural circuit comprising several other brain regions with the hypothalamic POA (preoptic area) at its center (Zhao et al., 2017) (Fig. 5.3).

Depending on the output of the processing, instructions for thermogenesis or heat loss signals are sent to thermoeffectors to maintain the homeostatic Tcore. Cool and warm skin thermoreceptors parallelly transmit their information in the form of electrical signals to respective sensory neurons of the dorsal root ganglia (DRG), which

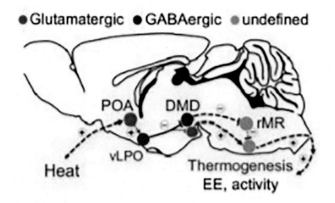

FIG. 5.3

Model for heat-induced suppression of thermogenesis. The *solid line* represents the connection verified in the current study. *Dashed lines* represent proposed connections based on our data and other reports. (+), activation. (−), inhibition. Abbreviations: DMD, dorsomedial hypothalamus; POA, hypothalamic preoptic area; rMR, rostral medullary region; vLPO, ventral part of the lateral preoptic nucleus; EE, energy expenditure.

From Zhao, Z.D., Yang, W.Z., Gao, C., Fu, X., Zhang, W., Zhou, Q. et al., 2017. A hypothalamic circuit that controls body temperature [published correction appears in proc. Natl. Acad. Sci. USA 114(9), E1755]. Proc. Natl. Acad. Sci. U. S. A. 114(8), 2042–2047. https://doi.org/10.1073/pnas.1616255114.

relay it to the second-order sensory neurons in the dorsal horn (DH) and from there to the lateral parabrachial nucleus (LPB) (Fig. 5.4). The skin cool and warm sensory pathways in the median preoptic area (MnPO) subnucleus, respectively, inhibit and excite the warm neurons (W-S). Signals from MnPO, via the dorsomedial hypothalamus (DMH), reach synaptically BAT (brown adipose tissue) cells or project to the rostral raphe pallidus (rRPa) to excite the shiver and BAT premotor neurons and excite/inhibit cutaneous vasoconstriction (CVC) sympathetic premotor neurons. The premotor neurons, via respective motor neurons, project to CVC sympathetic pre-ganglionic neurons (SPNs) and BAT SPNs in the intermediolateral nucleus (IML), whereas shiver premotor neurons project to alpha (α) and gamma (γ) motoneurons in the ventral horn (VH) of the spinal cord (Morrison, 2016).

The fact that animals have species-specific normal temperatures suggests that they have set points for maintaining the normal temperature by comparing the body temperature with the respective set points. Hammel et al. (1963) discovered that in dogs this set point was in the hypothalamus. The temperature set points are species-specific but not immutable or permanent. They are "neurally adjustable" under experimental conditions. So, e.g., rabbits reared in lower temperatures develop lower set points as a result of an increase in the number of warm-sensitive neurons (Tzschenke and Nichelmann, 1997), and from chicken eggs incubated at higher than optimal temperatures hatch birds of higher temperature set points (Tzschenke and Basta, 2002).

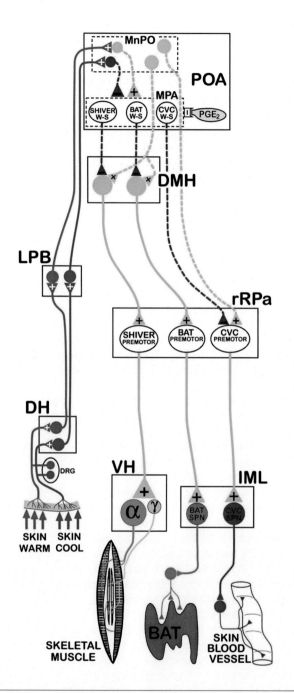

FIG. 5.4

Functional neuroanatomical model for the fundamental pathways providing the
thermoregulatory control and pyrogenic activation of cutaneous vasoconstriction (CVC) and
brown adipose tissue (BAT) and shivering thermogenesis. Cool and warm cutaneous
thermoreceptors transmit signals to respective primary sensory neurons in the dorsal root

(Continued)

2 Maintenance of animal morphometry

In Chapter 3, we have shown that the CNS is the ultimate source of information for the development of many organs and other structures in animals. From this, we could not reasonably extrapolate that the CNS is responsible for maintaining in a relative steady state the continuously eroding adult animal structure, although this would be plausible from an evolutionary viewpoint. Three facts seem to point in that direction:

First, the empirical evidence that the CNS induces the development of many organs (reviewed in Chapter 3) strongly suggests that the maintenance of the adult structure may be a function of the nervous system rather than another organ, organ system, or tissue.

Second, the evidence of the existence in the CNS of set points for body mass, and the number of cells (see below).

FIG. 5.4, cont'd ganglia (DRG), which relay this information to second-order thermal sensory neurons in the dorsal horn (DH). Cool sensory DH neurons glutamatergically activate third-order sensory neurons in the external lateral subnucleus of the lateral parabrachial nucleus (LPB), while warm sensory DH neurons project to third-order sensory neurons in the dorsal subnucleus of the LPB. Thermosensory signals driving thermoregulatory responses are transmitted from the LPB to the preoptic area (POA), which contains the microcircuitry through which cutaneous and core thermal signals are integrated to regulate the balance of POA outputs that are excitatory (*dashed green*) and inhibitory (*dashed red*) to thermogenesis-promoting neurons in the dorsomedial hypothalamus (DMH) and to CVC sympathetic premotor neurons in the rostral raphe pallidus (rRPa). Within the POA, GABAergic interneurons (*red*) in the median preoptic (MnPO) subnucleus are postulated to receive a glutamatergic input from skin cooling-activated neurons in LPB and inhibit each of the distinct populations of warm-sensitive (W-S) neurons in the medial preoptic area (MPA) that control CVC, BAT, and shivering. In contrast, glutamatergic interneurons (*dark green*) in the MnPO are postulated to be excited by glutamatergic inputs from skin warming-activated neurons in LPB and, in turn, excite the populations of W-S neurons in MPA. Prostaglandin E 2 (PGE 2) binds to EP3 receptors, which are postulated to inhibit the activity of each of the classes of W-S neurons in the POA. Preoptic W-S neurons may provide inhibitory control of CVC by inhibiting CVC sympathetic premotor neurons in the rostral ventromedial medulla, including the rRPa, which project to CVC sympathetic preganglionic neurons (SPNs) in the intermediolateral nucleus (IML). Preoptic W-S neurons may provide inhibitory thermoregulatory control of BAT and shivering thermogenesis by inhibiting BAT sympathoexcitatory neurons and shivering-promoting neurons, respectively, in the DMH, which, when disinhibited during skin and core cooling, provide respective excitatory drives to BAT sympathetic premotor neurons and to skeletal muscle shivering premotor neurons in the rRPa. These, in turn, project, respectively, to BAT SPNs in the IML and to alpha (α) and gamma (γ) motoneurons in the ventral horn (VH) of the spinal cord.

From Morrison, S.F., 2016. Central control of body temperature. F1000Res. 5, F1000 Faculty Rev-880. Published 2016 May 12. https://doi.org/10.12688/f1000research.7958.1.

Third, biological regeneration whereby the innervation is essentially involved in restoring the lost organ or part.

2.1 Maintenance of the body mass, the number of cells, and cell-type proportions

Evidence from experiments with the deer mouse (*Peromyscus maniculatus*) led investigators to the idea of the existence in the CNS of a set point determining and regulating the species-specific body mass. They observed that intraperitoneal implantation of an artificial load equal to 7.5% and 10% of the animal's weight was associated with a decreased food intake and compensatory equal loss of weight. After removal of the load, animals showed increased food consumption and regained the lost weight. The perception of the increased body mass as a result of the implant led to reduced food intake to reduce the "excess" body weight and, the reverse, the lower weight after the removal of the implant stimulated food intake and regaining of the lost weight. In their interpretation, the set point is sensitive to changes in the perception of body mass that emerges as a result of the information sent to multiple loci in the cerebral cortex by mechanoreceptors located within muscles and tendons (Adams et al., 2001).

Another neural pathway of the maintenance of constant body mass was revealed quite recently from experiments of implantation in rats, suggesting the existence of a sensor for body mass in osteocytes of the long bones of extremities, which sends afferent information on the increased body mass in a "gravitostat" in animal's CNS, which in turn suppresses food intake and reduces body fat (Jansson et al., 2018).

The adult weight in insects is determined by three main factors: the *growth rate*, *critical weight*, and the *PTTH* (prothoracicotropic hormone) *delay time*, which in turn are neurally determined by the CNS.

– Growth rate depends on the secretion of brain neuropeptides, PTTH (prothoracicotropic hormone) and ILPs (insulin-like peptides) (Ikeya et al., 2002) by specific neurons in the insect brain. Ablation of these neurons causes growth retardation and developmental delay (Rulifson et al., 2002).
– Critical weight coincides with suppression of juvenile (JH) hormone as a result of processing in the insect brain of the stretch stimuli related to the increased body mass (Gorbman and Davey, 1991), to which the brain responds by secreting neuropeptide allatostatins.
– PTTH delay time, which is the interval of time between suppression of JH secretion and beginning of the PTTH secretion during which the insect continues to grow.

And finally, let us consider an example of the neural control of the number and proportions of cells of various types in a lower invertebrate, the planaria. After experimental amputations, Takeda et al. (2009) observed that planarians were "maintaining a constant ratio of different cell types" in the regenerating body via apoptosis (programed cell death), proliferation, and differentiation of neoblasts. In their interpretation, the maintenance of constant proportions of different cell types during growth and degrowth indicates the presence of a "counting mechanism" that regulates the absolute and relative numbers of various cell types, a noncell-autonomous mechanism. To the self-addressed question, "Do any organs have the capacity to determine the size of other organs?" they respond assenting, "the brain may have the capacity to determine the size of other organs in planarians" (Takeda et al., 2009). It is also noteworthy that in planarians brain amputation is followed by regression of testes (Wang et al., 2007), while the neuropeptide Y-8 (NPY-8) is responsible for cell differentiation (Collins III et al., 2010) and the maintenance of male germ cells (Chong et al., 2013).

Furthermore, production and maintenance of germline stem cells (GSCs) in *Drosophila* are regulated by brain insulin-like peptides (Dilps) secreted by insulin-producing neurons (Ueishi et al., 2009), and activation of insulin signaling suppresses the stem cell loss and repopulates the GSC niche in starved *Drosophila* testes (McLeod et al., 2010).

Hair regrowth after the telogen (shedding phase), that is, in the anagen (growth phase), is induced and regulated by the sympathetic nervous system that increases secretion of norepinephrine and Shh (Brownell et al., 2011), which stimulates the proliferation of hair follicle stem cells (HFSCs) (Fan et al., 2018; Schwartz et al., 2020). The same nervous mechanism of hair regrowth and production of new hair are activated in mice in response to cold temperatures (Schwartz et al., 2020).

The neurotransmitter serotonin regulates embryonic development and maintains the cardiac structure in adult humans (Nebigil et al., 2000; Nebigil and Maroteaux, 2001).

3 The process of regeneration

Regeneration or restoration of a lost biological supracellular structure requires a huge amount of biological information, and the empirical evidence during more than half a century indicates that innervation, local products of the neural activity, and neuro-hormonal factors are essentially involved in the process of the restoration of the lost structure.

Animals may accidentally lose parts of their body, but they have evolved mechanisms of replacing or regenerating them wholly, *restitutio ad integrum*, or partly/defectively, *restitutio* cum *defectum*. Regeneration is observed at widely different degrees in the overwhelming majority of animal taxa, from sponges and planarians to vertebrates. Among vertebrate classes, the regeneration occurs in a descending order from fish to mammals (Fig. 5.5).

In amphibians, after the loss/resection of a part or organ, the stump forms a thin layer of the epidermis, the apical epithelial cap (AEC), beneath which grows the regenerative blastema, a mass of cells produced by dedifferentiation of cells such as chondrocytes, osteoblasts, and fibroblasts, reminiscent of the development of the embryonic limb bud. Nerve axons are necessary for the dedifferentiation and

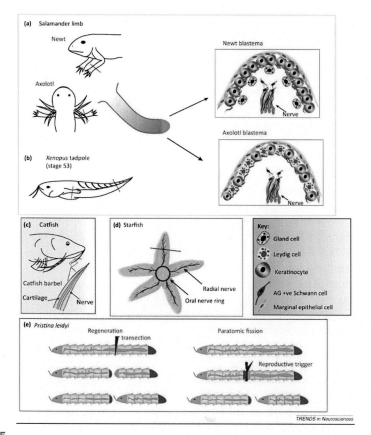

FIG. 5.5

Examples of nerve-dependent regeneration in various phyla. (A) Regeneration of the forelimb in the newt and larval axolotl. After amputation of the limb, regeneration proceeds by the formation of a blastema (*green*). The panel on the right shows diagrammatic illustrations of the blastema. In the newt, anterior gradient (AG) protein is expressed in Schwann cells at the end of the nerve and in gland cells underlying the wound epithelium (Echeverri and Tanaka, 2005). In axolotls, Leydig cells of the wound epithelium, in addition to Schwann cells, express AG. (B) *Xenopus* tadpoles are capable of limb regeneration until metamorphosis at stage 53. Tail amputation results in regeneration of the spinal cord, as well as a full-length tail. (C) The catfish can regenerate its sensory barbels after injury. The barbel is supported by a cartilaginous core, and nerves run along the axis of the tissue to the sensory receptors. (D) In starfish, the oral nerve ring is critical for arm regeneration. Each arm of the starfish is innervated by radial nerves, which radiate from the oral nerve ring. (E) Regeneration of the worm *Pristina leidyi* occurs after transection, as well as after paratomic fission. Favorable environmental conditions trigger fission at the middle of the body. A fission zone containing a presumptive head and tail forms before fission, and scission of the body occurs through the midplane (*red*). After transection of the worm, the anterior segment regenerates a tail, whereas the posterior fragment regenerates a new head.

Kumar, A., Godwin, J.W., Gates, P.B., Garza-Garcia, A.A., Brockes, J.P., 2007. Molecular basis for the nerve dependence of limb regeneration in an adult vertebrate. Science 318, 772–777.

the redifferentiation of the blastema cells to produce the regenerate. No blastema forms in the absence of innervation. After blastema formation is completed, it may undergo morphogenesis even in the absence of innervation, but it will only produce a miniature regenerate (Hall, 1998; Stocum, 2011). Similarly, simultaneous denervation and amputation of the tail of the weakly electric fish, *Sternopygus macrurus*, prevents its regeneration to a large extent (Unguez and Zakon, 2002).

The indispensability of the innervation for blastula development is related to secretion by local axons in the stump of a number of growth factors and other substances, among which the most important seem to be Fgf and Bmp (Fgf8 and Bmp7 in the axolotl, *Ambystoma mexicanum*) (Satoh et al., 2016; Makanae et al., 2016, 2020). The spinal cord is also capable of transforming the wound healing responses into regeneration processes, but, even in its absence, administration of Fgf2 + Fgf8 + Bmp7 induced formation of blastema. Further analysis, however, revealed that blastema cells induced by Fgf2 + Fgf8 + Bmp7 could participate in the regeneration of several tissues but could not organize a patterned tail (Makanae et al., 2016). It is noteworthy that Fgfs and Bmps secreted by the trigeminal nerve seem to control tooth regeneration in *A. mexicanum* (Makanae et al., 2020).

The essential role of the innervation in regeneration is also corroborated by the experimental fact that deviation of the hind limb innervation in the neighboring posterior regions induces the formation of supernumerary limbs in the European newt, Triton (Satoh et al., 2009) (Fig. 5.6).

Regeneration of the eye lens in chickens (Reza and Yasuda, 2004a; Reza and Yasuda, 2004b) is induced by the retina, which is an extension of the forebrain neuroepithelium (Behesti et al., 2009). The lens GRN (gene regulatory network) is neurally activated by signals from the optic vesicle/retina, which is also the source of the secreted protein Pax-6 (Reza and Yasuda, 2004b) in the presumptive lens ectoderm that is essential for lens development.

FIG. 5.6

Axolotl with two left limbs. The additional left limb was induced experimentally by deviation of the brachial nerve to the lateral wound and grafting of a posterior skin graft.

From Satoh, A., James, M.A., Gardiner, D.M., 2009. Axolotl limb regeneration. J. Bone Joint Surg. Am. 91, 90–98.

4 Inherited adaptation to the changing environment

Homeostasis, the ability of animals to maintain their basic functions and structure and internal environment within relatively narrow limits even under widely varying environmental conditions, is inseparable from their developmental plasticity, the ability to adapt the phenotype to the changing conditions of the environment or perfect it under unchanged environmental conditions.

The degree of the phenotypic (physiological, behavioral, morphological, and life history) plasticity is species-specific and inherited. It enables animals to cope with, and compensate for, adverse environmental influences that do not exceed the limits determined by the reaction norm, the animal's capability to adaptively respond.

Animal adaptations may be grouped in four distinct classes:

- reaction norm, comprising only quantitative changes in existing phenotypic traits rather than emergence of new traits, hence may be described as intragenerational plasticity (not to be discussed in this chapter as irrelevant to the subject),
- intergenerational plasticity, implying emergence and transmission of *new or modified traits* from the parent to the offspring of the first generation, but disappearing thereafter in the absence of the triggering agent,
- transgenerational plasticity, dealing with the emergence of new traits that are transmitted beyond the first generation, in two, or many (Rechavi et al., 2011), or even an indefinite (Vastenhouw et al., 2003) number of consecutive generations, and
- evolutionary change, emergence of new or modified traits that are "permanently" inherited in future generations.

Given that adequate evidence on the neural mechanisms of the intergenerational inheritance is presented in Chapter 4, Section 4, in the following we will focus on the neural mechanisms of the transgenerational plasticity, which in all likelihood bears important clues on the nature, origin, and mechanisms of the evolutionary change.

4.1 Induction of transgenerational plasticity

4.1.1 Evidence on the neural involvement in transgenerational plasticity

In response to unfavorable stressful conditions such as low food availability, over-crowding, and high temperature, the tiny nematode worm *Caenorhabditis elegans* arrests the development at the motionless first larval stage (L1), entering the Dauer stage to continue development only after the stress condition disappears. As adults, worms that experienced unfavorable conditions and the Dauer arrest develop smaller body but live longer and transmit the longevity trait for three generations (Webster et al., 2018). The developmental pathway starts from specific olfactory neurons in the head of the worm with the secretion of insulin-like peptides (ILPs), including DAF-2 (an ILP type), which by binding its membrane receptor triggers a signal transduction pathway that suppresses expression of the *daf-16* gene (Lee and Ashrafi,

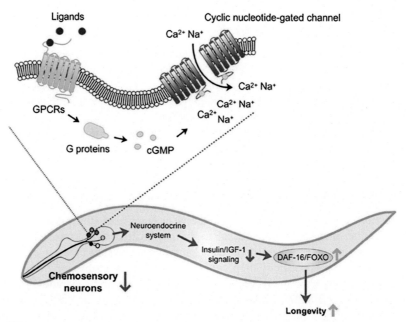

FIG. 5.7

Model of lifespan control by chemosensation and insulin/IGF-1 signaling in *C. elegans*. A subset of ciliated neurons in the head region (for example, in the amphid) perceive chemosensory cues and trigger signal transduction cascades. Upon binding to ligands, chemosensory G protein-coupled receptors (GPCRs) activate G proteins, which lead to cGMP (cyclic guanosine monophosphate) production. This in turn opens cyclic nucleotide-gated channels and increases Ca^{2+} and Na^+ influx. Inhibiting chemosensory neuronal structure or function in specific neurons may alter neuroendocrine signaling and reduces insulin/IGF-1 signaling, which activates the DAF-16/FOXO transcription factor and promotes longevity.

From Jeong, D-E., Artan, M., Seo, K., Lee S-J., 2012. Regulation of lifespan by chemosensory and thermosensory systems: findings in invertebrates and their implications in mammalian aging. Front. Genet. 18 October 2012 | https://doi.org/10.3389/fgene.2012.00218.

2008). Under stressful conditions, the olfactory neurons cease synthesizing DAF-2. This causes accumulation of DAF-16, leading to the Dauer arrest of the development of the worm and longevity that are inherited for at least three generations even under normal environmental conditions (Webster et al., 2018) (Fig. 5.7).

Male attracting pheromone, SRD1, in *C. elegans* is secreted by AWA neurons. Young wild-type *C. elegans* hermaphrodites do not secrete the pheromone, but older worms that cannot produce sperm and younger hermaphrodite worms, cultivated for 10–15 generations at mildly stressful temperature 25 °C, secrete the pheromone and become attractive to males. The trait is transmitted transgenerationally for 3–6 generations reared in normal temperatures. The transgenerational inheritance is induced by heritable neuronal small RNAs and the germline Argonaute HRDE-1 (heritable RNAi deficient 1) (Toker et al., 2020).

RDE-4-dependent neuronal endogenous small RNAs regulate germline endogenous small interfering RNAs (siRNAs), which via Argonaute HRDE-1 regulate gene expression in germline, thus controlling the chemotaxis behavior for at least three generations (Posner et al., 2019). The loss of *rde-4* makes mutants unable to respond to chemical attractants (Tonkin and Bass, 2003).

Another case of transgenerational inheritance of acquired morphological traits is described in one of the smallest crustaceans, *Daphnia cucullata*. For more than a century, it was observed that these organisms display cyclomorphosis: they produce long pointed helmets that reach their largest size in summer to gradually downsize and lose them in the winter. Initially only considered to be a response of the organism to seasonal temperature fluctuations, by the end of the last century, it was demonstrated that it was also a transgenerational defensive morphology that the small cladoceran develops in response to the detection of its predator, the phantom midge *Chaoborus* larvae or their kairomone (Agrawal et al., 1999). The helmet prevents or makes it difficult to be eaten by the predator. It is demonstrated that the kairomone in *Daphnia* is received by chemosensillae of the first antennae, transmitted to the brain, where it is integrated into cholinergic neurons and translated into a specific neuronally stimulated developmental pathway (Barry, 2002; Weiss et al., 2015; Miyakawa et al., 2010).

Phase change in locusts. According to the habitat they live in, two locust species *Schistocerca gregaria* and *Locusta migratoria*, adopt one of two alternative life histories: that of the lonely individual (solitary phase) or they form swarms of migrating locusts (gregarious phase). The change occurs suddenly, and phases display striking differences in morphology (brain size, body size, color, etc.), physiology (neuroendocrinological status), and behavior (tendency to migrate with their crowd, timing of response to moving objects).

Transition from the solitary to the swarm-forming gregarious phase of the motionless green locust larvae under the influence of sensory (olfactory, visual, tactile, and auditory) stimuli is abrupt. They change behavior and abandon the solitary life to join and fly with crowds of conspecific locusts. The gregarious converts display the distinctive traits of the crowd: their body color changes from cryptic green to dark, along with changes in the morphometry, head and brain size, the number of sensilla, etc.

The chemical nature of the species-specific pheromones stimulating transition into the gregarious phase has been unknown, but recently the compound 4-vinylanisole was identified as an aggregation pheromone in *L. migratoria* (Guo et al., 2020).

There are two distinct neural pathways that can jointly or separately (the latter is less efficient) induce transition from the solitary to gregarious phase: the cephalic pathway comprising olfactory (pheromonal) and visual stimuli and the tactile pathway induced by contacts with other locusts (Rogers and Ott, 2015) (Fig. 5.8).

As little as 4 h after being exposed to the combined visual and olfactory stimuli, solitary locusts display the same gregarious behavior as those that have been gregarious for many generations. At the molecular level, the transition is characterized by a

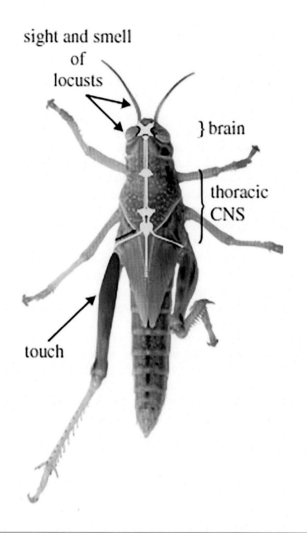

FIG. 5.8

Last larval instar solitarious locust showing the two separate sensory pathways by which behavioral gregarization can be induced: a thoracic pathway (*red*) activated by touch stimuli directed to a hind femur and a cephalic pathway (*blue*) activated by the combined sight and smell of other locusts.

From Rogers, S.M., Ott, S.R., 2015. Differential activation of serotonergic neurons during short- and long-term gregarization of desert locusts. Proc. Biol. Sci. 282 (1800), 20142062. https://doi.org/10.1098/rspb.2014.2062.

transient elevation of the level of the neurotransmitter serotonin in the thoracic ganglia, but not in the brain, which declines sharply after 24 h (Rogers and Ott, 2015).

The brain pathway of induction of the gregarious phase in the desert locust *Schistocerca gregaria* begins with the reception of olfactory stimuli (aggregation pheromones) by olfactory neurons, which in the form of electrical spike trains are transmitted to the interneurons of the frontal antennal lobe in the locust brain and sequentially to the mushroom body and the lateral protocerebrum for further processing (Anton and Hansson, 1996). The pathway stimulates increased production of serotonin in the metathoracic ganglion (Rogers and Ott, 2015), but the fact that allatectomized locusts, which cannot secrete neuropeptide allatotropins and juvenile hormone (JH), do not respond to aggregation pheromones suggests that the neurally induced JH may also be involved in the phase transition (Ignell et al., 2001). A simplified developmental mechanism of the pheromonal induction of gregarization in locusts would look as follows:

Antennal olfactory receptor neurons ➔ *brain interneurons* ➔ *projection neurons* ➔ *neurons of the mushroom calyces* ➔ *neurons of lateral protocerebrum* ➔ *onset of the behavioral and morphological traits of the gregarious phase.*

The tactile (touch) pathway starts with the reception of tactile stimuli via the mechanosensory trichoid sensillae of the outer side of the upper portion of the hind limb that via the metathoracic nerve 5 are transmitted to the thoracic CNS. Gregarization can also be induced by electric stimulation of the nerve 5 (Rogers et al., 2003). Serotonin also has a key role in the tactile pathway of induced gregarization.

Gregarization is inherited transgenerationally. In regard to the number of generations the gregarization is maintained in locusts, Rogers and Ott say that they used in their experiments locusts that "have been in the gregarious phase for many generations" (Rogers and Ott, 2015); hence, it seems plausible that under constant gregarizing conditions, which prevail in nature, the gregarization may be perpetuated, or, at least, there is no visible reason to think otherwise.

The transgenerational mechanism of the gregarization is not known in detail, but now we have a reliable general picture of the process, including the integration and processing of the pheromonal stimuli in the nervous system of the locust (Fig. 5.9).

The phase change involves activation of 532 genes, and 90 of them are differentially methylated in the solitary versus gregarious forms (Ernst et al., 2015). Both locust species transmit their phase to the progeny and revert to the alternative phase under appropriate conditions (Miller et al., 2008; Cullen et al., 2010). Increased sense of smell of conspecific individuals in phase transition implies perception of crowding in the locust brain, which is followed by demonstrated alterations in neurons, especially in synaptic morphology, as well as types and amounts of neurotransmitters (Rogers et al., 2004; Badisco et al., 2011) and neuropeptides (Claeys et al., 2006) secreted in the nervous system.

Experimental injection of the neurotransmitter serotonin alone, or its analogs, induces the transition of solitary locusts to the gregarious phase (Anstey et al., 2009). Several genes (*pale*, *henna*, and *vat1*) involved in the biosynthesis of the neurotransmitter dopamine are also related to transitioning to the gregarious phase, and experimental injection of dopamine also induces the onset of the gregarious behavior in locusts (Ma et al., 2011).

It has been demonstrated that *S. gregaria* possesses enzymes, such as HAT (histone acetyl transferase), HDAC (histone deacetylase), two homologs of DNA methyltransferases, MBD (methyl binding protein), and even "a number of

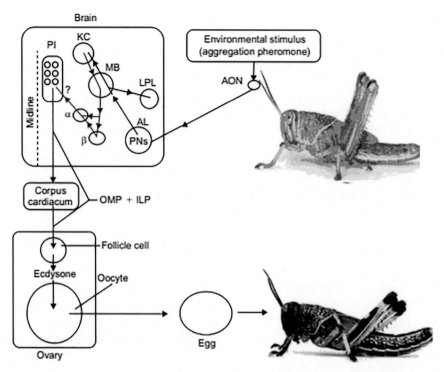

FIG. 5.9

Simplified diagram of generation and the flow of information for transgenerational plasticity (phase transition) in *L. migratoria*. External stimuli received by antennal olfactory neurons are transformed into electric trains and sequentially transmitted to the antennal lobe, mushroom body, Kenyon cells, and pars intercerebralis of the insect brain. Secretory neurons of the pars intercerebralis secrete ovarian maturation parsin (OMP) and insulin-like neuropeptides (ILPs). These inducers, which are initially deposited in the corpus cardiacum, via hemolymph stimulate increased production of ecdysone in the ovary. The increased presence of ecdysone in the egg cells induces the production of the gregarious offspring of solitarious locusts. Abbreviations: α, α-lobe; β, β-lobe; AON, antennal olfactory neuron; AL, antennal lobe; ILP, insulin-like neuropeptides; KC, Kenyon cells; MB, mushroom body; LPL, lateral protocerebral lobe; OMP, ovarian maturation parsin; PI, pars intercerebralis; PNs, projection neurons of the antennal lobe.

From Cabej, N.R., 2019a. Epigenetic Principles of Evolution. second ed. Academic Press. London, San Diego, Cambridge MA, Oxford. p. 364.

components of the epigenetic machinery" (Boerjan et al., 2011). Methylation, acetylation, and phosphorylation are also observed in locust histones. Most histone-modifying enzymes are also found in eggs and testes of male adult locusts and are differentially expressed in the brains of gregarious versus solitary locusts (Guo et al., 2016). However, in distinction from the neural changes that are demonstrated to induce phase transition, it is not known whether these epigenetic changes are simply correlational or causal regarding the phase transition in locusts (Fig. 5.10).

There is evidence that egg foam in locusts contains a gregarizing factor, but the nature and origin of the factor are not determined yet (McCaffery et al., 1998; Ben Hamouda et al., 2009).

Unpredictable maternal stress combined with unpredictable maternal separation (MSUS) causes in the mouse offspring a type of traumatic stress characterized by neuroendocrine disorders and changed behaviors (reduced avoidance and fear, reduced aversive response, and depression-like behavior in forced swimming test). The condition is transmitted to the F1 and F2 male offspring associated with increased synthesis of several miRNAs (microRNAs), such as miR-375-3p and 5p, miR-200b-3p, miR-672-5p, etc., in the brain, serum, and sperm. Injection of MSUS sperm miRNAs into fertilized oocytes of normal mice led to the development of traumatic stress in F2 reared under normal conditions, demonstrating the role of these miRNAs in the transgenerational appearance of the stress condition (Gapp et al., 2014).

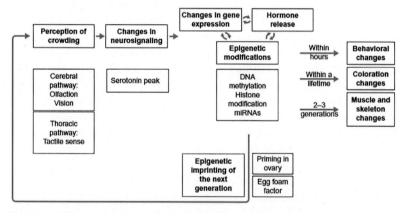

FIG. 5.10

Hypothetical model for epigenetic remodeling in locust phase transitions. Crowding causes profound differences in neuronal and hormonal signaling, gene expression, and epigenetic modifications that eventually lead to significant changes in behavior, physiology, and morphology on different timescales. Hormones, gene expression, and epigenetic modifications influence each other. Epigenetic marks will perpetuate these changes over moltings and egg formation. Eggs may be primed in the ovary and in the egg pod by an egg foam factor. When offspring also experience crowding, epigenetic alterations may accumulate that subsequently lead to morphological changes and the phenotype of long-term gregarious locusts.

From Ernst, U.R., Van Hiel, M.B., Depuydt, G., Boerjan, B., De Loof, A., Schoofs, L., 2015. Epigenetics and locust life phase transitions. J. Exp. Biol. 218, 88–99. https://doi.org/10.1242/jeb.107078.

Mice exposed to MSUS showed upregulation of polyunsaturated fatty acid (PUFA) metabolism and downregulation of aldosterone and its mineralocorticoid receptor (MR). The pharmacological blockade of MR mimics the MSUS effects. It was observed that PPARs (peroxisome proliferation-activated receptors) are upregulated in the sperm of adult MSUS males and intraperitoneal injection of a PPAR receptor agonist induced MSUS condition in normal mice. When control male mice injected with serum collected from MSUS male mice were bred with control females, their offspring had reduced weight and lower blood glucose under stress, similarly to MSUS males. This indicates that some blood factor(s) transferred the MSUS information from the circulating parental blood to the germline. The experiments suggested that altered sperm PPAR, by modifying the expression of zygotic genes, transmits the MSUS phenotype to the offspring (van Steenwyk et al., 2019).

Toxoplasma gondii is a single-celled parasite of humans that may invade the CNS, cause neuropsychiatric disorders, and weaken the immune system. Mice experimentally infested with *T. gondii* also display similar symptoms of dysregulation of neuronal development and transmit these characteristics to healthy offspring of F1 and F2 generations. It was observed that sperm cells of infested mice had altered small RNA profiles. Microinjection of the total sperm small RNAs of infested mice into the egg cell of healthy mice also recapitulated symptoms of the sick mice. It was demonstrated that symptoms of experimental toxoplasmosis in mice are transgenerationally inherited in F1 and F2 in the absence of the parasite and that the trait is inherited transgenerationally via sperm small RNAs (Tyebji et al., 2020) (Fig. 5.11).

Mice conditioned before conception to fear acetophenone (Ace) smell (by associating the smell with a mild electrical shock) produced offspring that display the same fear to Ace in the next two unconditioned generations (F1-Ace and F2-Ace). Investigators found that these mice had a larger number of olfactory neurons and a changed morphology of the olfactory glomerulus M71 in the brain, as well as a higher rate of expression of the olfactory receptor Olfr151 gene in olfactory neurons caused by its hypomethylation (Dias and Ressler, 2014a). The experiment indicates that information from the nervous system was conveyed to the gamete(s) that enabled the unconditioned F1-Ace and F2-Ace generations to develop the parental sensitivity to acetophenone and the increase in the number of neurons and the size of the relevant M71 olfactory glomerulus. As for the mechanism, based on the present knowledge, the investigators proposed that "hypomethylation of Olfr151 in F0 sperm may lead to inheritance of the hypomethylated Olfr151 in F1 MOE (main olfactory epithelium) and F1 sperm, creating an inheritance cascade" but "the epigenetic basis of this inheritance might not be histone based, instead relying on other mechanisms, such as DNA methylation (as reported above) or noncoding RNA" (Dias and Ressler, 2014a). The study was criticized on theoretical-statistical grounds (Francis, 2014) but repudiated by investigators (Dias and Ressler, 2014b).

Endocrine disruptors are a major group of substances that induce pathological changes in the reproductive function of animals, and the pathologies are often inherited in the offspring for two to three generations. Typically, vinclozolin is a fungicide with adverse effects on the neuroendocrine functions of the brain, behavior,

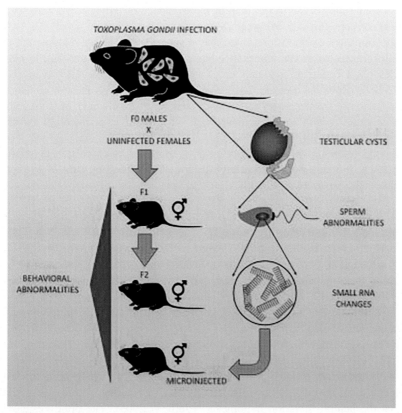

FIG. 5.11

Transgenerational inheritance of behavioral abnormalities in mice seronegative for toxoplasma antibodies.

From Tyebji, S., Hannan, A.J., Tonkin, C.J., 2020. Pathogenic infection in male mice changes sperm small RNA profiles and transgenerationally alters offspring behavior. Cell Rep. 31, 4, 107573.

reproductive functions, etc. Prenatal exposure to vinclozolin decreased the sperm number and increased apoptosis of sperm in the seminiferous tubules (Anway et al., 2005), leading to masculinization of females and feminization of males (Buckley et al., 2006), which are transmitted to the F2 (Anway et al., 2005; Krishnan et al., 2018) and F3 generations (Brieño-Enríquez et al., 2015). The adverse effects have been correlated with alteration in primordial germs cells (PGCs) of several miRNAs, such as miR-23b and miR-21 in three successive generations of males (Brieño-Enríquez et al., 2015) and changes in DNA methylation regions (DMRs), also known as epimutations. The recent evidence showed that the number of DMRs increased from 290 in the F1 to 981 in the F3 generation. Most of these DMRs are found not in promoters of genes but in intergenic regions and distal to genes (Beck et al., 2017; Skinner et al., 2012), and the presence of clusters of overrepresented genes around DMRs led to the idea of the existence of epigenetic control centers (ECRs) consisting of 2–5-Mb long DNA strands.

The empirical evidence presented above suggests that the developmental pathways for transgenerational inheritance begin in the nervous system but, in most of the described cases, no relevant epigenetic modifications are involved (Harris et al., 2012; Robichaud et al., 2012; Weiss, 2018), although these organisms are in possession of the necessary molecular machinery, and changes in DNA methylation in the *Daphnia pulex* genome, e.g., have been experimentally produced (Strepetkaitė et al., 2016).

Having briefly reviewed a representative group of cases of transgenerational inheritance, it is noteworthy to briefly describe the interesting case of eye loss in the cave fish, *Astyanax mexicanus*.

4.1.2 Neural determination of the transgenerational plasticity: The case of eye loss in cave fish

Hypogean forms of the *Astyanax mexicanus* evolved from epigean forms at least four times in the last 10,000 or 20,000–30,000 years (Fumey et al., 2018) as a result of the transition from the surface to the subterranean waters. The transition is conspicuously characterized by the loss of eye structures and sight, along with the reduced size of the optic tectum in the brain, changes in teeth and jaw morphology, loss of pigmentation, etc.

Extensive changes in morphology, physiology, and behavior evolved within an evolutionarily very short period and involved no known changes in relevant genes or genetic information (Klaus et al., 2013). "It appears that eye gene cascades are completely operational in cavefish embryos.... Experiments provide evidence against the neutral mutation hypothesis" (Jeffery, 2005). The optic vesicles and the hypothalamus develop from the forebrain (telencephalon and diencephalon), which in turn develops from the neural plate (Pottin et al., 2011). In the cavefish, the optic cup, neural retina, and lens vesicle begin developing but soon are eliminated via programed cell death (Langecker et al., 1993; Jeffery et al., 2000). Increased Shh expression in the prechordal plate and hypothalamus of the cavefish induces the earlier onset of Fgf8 in the telencephalon, thus impairing the process of eye development (Pottin et al., 2011).

Expanded expression of Shh in the midline of the neural plate and the resulting contraction of the *pax6* expression in the anterior of the neural plate prevent the formation of optic vesicles in the cavefish (Jeffery et al., 2000) (Fig. 5.12).

Injection of Shh mRNA in the surface fish embryos induces lens programmed cell death, leading to the development of the blind adult fish (Yamamoto et al., 2004). The decisive role of the heterotopy of Fgf8 expression preventing eye development in cavefish is corroborated by the fact that administration of SU5402, an inhibitor of Fgf8 receptor, mimics the typical surface fish phenotype (Pottin et al., 2011).

Thus, all the steps determining eye loss in the cave fish, the premature onset of Fgf8, extended expression of *Shh*, and resulting contraction of *pax6* expression field, take place in the neural plate-neural tube-hypothalamus. Interestingly, the surface

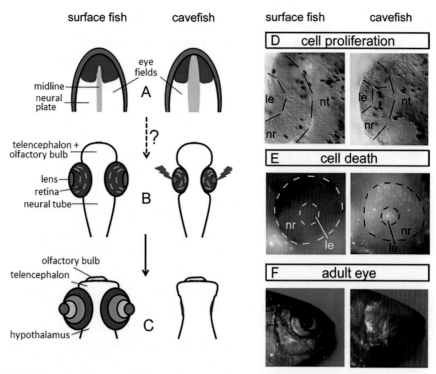

FIG. 5.12

A possible mechanism for eye degeneration in cavefish. (A–C) Successive events of eye development and degeneration. (D–F) Examples of cellular events representing the major steps in this degeneration. (A) Neural plate stage (approx. 10–12 h postfertilization) when Shh midline signaling is enlarged in cavefish, and the subsequent eyefield delineated by *pax6* expression is smaller in cavefish. The lens will differentiate from the placodal field at the border of the neural plate. (B) Lens apoptosis in cavefish, starting at ~24 hpf, while the neural retina develops in a relatively normal fashion, with an attempt to generate retinal layers (*concentric dotted lines*). (C) Adult stage, when the cavefish eyes are regressed, have sunk into the orbits, and are covered by the skin. Abbreviations: le, lens; nr, neural retina; nt, neural tube.

From Rétaux, S., Pottin, K., Alunni, A., 2008. Shh and forebrain evolution in the blind cavefish Astyanax mexicanus. Biol. Cell 100(3), 139–147. https://doi.org/10.1042/BC20070084.

Astyanax embryos raised for 2 years in complete darkness display several metabolic, neurological, and morphological changes necessary for the survival of *A. mexicanus* surface colonizers, such as higher growth hormone, starvation resistance, and cortisol levels as well as lower serotonin and metabolic rates, almost all of them resembling cavefish adaptive phenotypes, although nonadaptive, even maladaptive, effects are also observed. The phenotypic plasticity of *Astyanax* was also associated with changes in expression of the stress-related genes, *hsp90aa1.2*, *dnmt1*, and *dnmt3bb*, caused by the dark environment. The offspring of the surface fish displayed similar adaptations when raised in darkness and investigators add: "According to our results, the next generation may be slightly better equipped for coping with the cave environment by increased plasticity and small adaptive switches in traits" (Bilandzija et al., 2020).

In the long term, natural selection and the Baldwin effect may stabilize and fine-tune the evolution of the surface fish into a cave form.

5 On the source and nature of the information for transgenerational plasticity

When it comes to building a new morphological trait, the most difficult issue is neither the source of the material (biomolecules) nor the energy necessary for its development. The crux of the matter is how to get the information for building the new trait that was not ancestrally possessed. The new trait is a supracellular structure comprising thousands to billions of cells of up to hundreds of different types arranged in strictly determined spatial patterns rather than a pile of randomly assembled cells. Information of some kind is necessary to determine the specific arrangement of cells, often in complex and convoluted forms, in the new trait. The genetic information based on the triplet code, which carries the information for protein biosynthesis, is inapt for determining the spatial arrangement of cells of different types in molding the animal structure, and this is the reason that no scientific hypothesis has ever been presented to show how it could determine that arrangement.

The inquiry into the nature of information for phenotypic adaptation is intimately linked to the origin of the information. The simplest and safest approach to the issue of the source of information for molding new morphological traits is to examine the causal chain or signal cascades inducing inter- and transgenerational plasticity described in Chapter 4 and this chapter. Starting from the proximal causes of the new trait, we can stepwise go upstream the elements of the signal cascade until the ultimate source. We have already seen that in all the considered cases of inter- and transgenerational plasticity the signal cascade begins with a brain electrical or chemical signal, the output of the processing of an external/internal stimulus in the brain. While we know what neural processing does, we are almost clueless about the nature of the processing and computation performed in the brain.

Various environmental conditions/stimuli may disturb the homeostatic equilibrium and affect the physiological and structural integrity of the animal, triggering an adaptive response to restore the normal state. Sometimes the response may be the development of a new/modified trait that is transmitted to one to several generations. The emergence and inheritance of the new/modified trait imply investment of new ancestrally not possessed information, and we need to determine whether the information is provided to the organism externally by the environmental stimuli/cues or it is generated in an adaptive and species-specific response to them.

The concept of information implies a sender and a receiver of the information, and if the environmental stimulus is the sender, it would be expected to produce the same phenotypic result in different animal species, rather than different and sometimes opposite phenotypic results, as often is observed to occur.

5.1 Adaptive phenotypic information is generated in the nervous system

The external environment is in continual change, and these changes are received by the nervous system via the sensory organs. However, the nervous system does not respond adaptively to all these changes indiscriminately. It only responds to changes that it recognizes as stimuli, i.e., that adversely affect the structure and function of the organism or, anticipatively, to changes that presage the approach of such stimuli. Whether the environmental change will be perceived as a stimulus requiring an adaptive response or as a neutral agent requiring no response depends not only on the size of the change per se but on a species-specific set point that has evolutionarily evolved in the CNS as well.

External stimuli are challenges to the organism's homeostasis to which the nervous system responds by inducing adaptive phenotypic changes. These changes are the phenotypic embodiment of information generated via the processing of environmental stimuli in specific neural circuits rather than off-the-shelf inborn responses of the instinct type.

The role of the processing of stimuli in neural circuits as a trigger of a causal chain that leads to the emergence of a new trait implies a teleonomic purpose (Mayr, 1961). The processing assesses the biological adverse or possible favorable effects of the stimulus and activates available mechanisms to respectively counteract the harmful effects of the stimulus or exploit the advantages it may offer. The neural processing, thus, is an adaptation-oriented biological computation intended to generate information for inducing adaptive phenotypic changes in response to the changing circumstances in the environment. Since the information *generated* by neural circuits in response to external and internal stimuli intends to adapt the animal phenotype to the environmental change, it makes sense to call it *adaptive phenotypic information* (API).

In clear distinction from the genetic information that is materially carried by, and stored in, DNA nucleotide sequences, API may be transient information generated ex tempore by the processing of stimuli in neural circuits, but it is not stored in neural circuits, although the latter retain the potential to generate it in response to stimuli. Empirically is demonstrated that API is transmitted physically to the following generations via specific substances (neurohormones, miRNAs, etc.), which activate little known developmental pathways that lead to the activation in the offspring of the same neural circuits that induce the development of the parentally inherited new trait in the absence of the triggering stimulus.

The stimulus per se cannot induce the expression of any gene. It is the nervous system that, based on the phylogenetic experience, via neural processing, generates the adaptive information necessary for activating a developmental mechanism to correct or prevent harmful effects of the stimulus or even fashion a new or improved trait. The adaptive information is transmitted via a chemical (neurohormone, neuro-modulator, neurotransmitter, etc.) that trigger activation of a signal cascade and a signal transduction pathway leading to activation of a specific gene. The release of the chemical in the nervous system serves as a command for activating a signal cascade (Cabej, 2019a) that transfers the neurally derived adaptive information into the cell nucleus to express the gene.

The role of the CNS in inter- and transgenerational adaptive phenotypic changes was adequately dealt with in Chapter 4 and this chapter. Just to refresh the reader's memory, let us remember the case of the intergenerational induction of diapause in the silkworm *Bombyx mori*. The proximate cause for the onset of diapause is dia-pause hormone (DH), secreted by particular neurons of the dorsal protocerebrum (Shimizu et al., 1997), which in turn is induced by the neurotransmitter dopamine released by Lb (labial secretory) neurons of the Brain-SOG (subesophageal gan-glion) (Noguchi and Hayakawa, 2001). The silkworm deposits DH in the eggs. To corroborate the role of neurons in worm's diapause, it was demonstrated that abla-tion of the labial (posterior) neurons of the SOG impairs induction of diapause eggs and experimental elevation of the dopamine level switches to diapause the silkworm larvae that otherwise would not enter diapause. Reduced to its essentials, the signal cascade leading to the onset of silkworm diapause appears as follows:

External stimulus (shortening of photoperiod) → perception of the shortening of the photoperiod in the visual system → processing of the stimulus in other brain cen-ters → activation of the dopaminergic system → release of diapause hormone (DH) by the brain → transport of DH in the eggs via *hemolymph.*

Far from a spontaneous phenomenon, the onset of diapause in the offspring of the silkworms that experienced the shortening of the photoperiod is a strictly predictable phenomenon implying investment of appropriate adaptive information. The informa-tion embodied in the synthesis and secretion of DH in hemolymph is generated by processing in the silkworm's brain of photoperiod change, which presages the approach of the cold winter that is unfavorable for the growth of the worm. The infor-mation is not provided by the shortening of photoperiod, as is proved by the fact that the shortening of the photoperiod does not induce diapause in other closely related species of the genus *Bombyx*.

Among natural cues animals use to adapt their phenotype are night–day cycles, yearly seasonal cycles, moon phase, temperature, humidity, etc., but each of the above and other external stimuli is provided with different adaptive "meaning" in different species. So, e.g., the lengthening of the photoperiod stimulates reproductive activity in cows, but contrarily, sheep take the shortening of the photoperiod as a signal for beginning the reproductive activity; the fall of night stimulates sleep in most animals, but activity in the foraging and preying (mice, owls, bats, etc.) ani-mals. Were these environmental stimuli information (=instructions) to animals, it would be expected all these species to react similarly, not in different and often opposing ways, to the same "instructions."

Activation of the mechanisms to make use of, or counteract, the effect of the stimulus is just another way of saying that the processing of stimuli in the neural circuit(s) generates the adaptive information. The information is coded in the form of spike trains, i.e., spike rates and spike patterns (temporal coding). The exact nature of the computation the neural circuits use in stimulus processing to generate the adaptive information is not known, and we are almost in the dark on what exactly takes place in the neural circuit between the reception of the stimulus and production of the chemical output.

What we know is that by transforming stimulus into a signal that induces expression of a specific gene or GRN (gene regulatory network), the nervous system establishes a causal relationship between the stimuli and the gene that otherwise would not exist (Cabej, 2019b) (the stimulus cannot induce the gene by direct contact). The relationship between the stimulus and the expression of the specific gene it leads to, thus, is indirect. It is the processing of the stimulus in the neural circuit that establishes that specific relationship.

The ability of animals to establish physically not existing causal relationships between various environmental stimuli and specific genes was increased immensely with the evolution of the neuron and the nervous system. This observation seems to be corroborated by the fact that two lower animal groups that have no nervous system, placozoa and sponges, evolved so little during the last ~600 million years they still represent dead ends of evolution, in sharp contrast with the groups that evolved nervous systems.

It is also worth noting that the neural activity can also induce the expression of specific genes without involving extracellular signal cascades via the direct contact of nerve endings with adjacent cells (Hegstrom et al., 1998). This is also the case with the production of several membrane receptor isoforms that are regulated by nerve endings on adjacent cells via the neurally regulated splicing process (Hermey et al., 2017; Li et al., 2007; Razanau and Xie, 2013; Iijima et al., 2016; Suzuki et al., 2017) and so are many nuclear receptors (reviewed by Cabej, 2020d). Neural activity in the synapses is responsible for the expression of microexons in neurons of the human brain (Scheckel and Darnell, 2015), and misregulation of exons in the brain leads to the development of autism spectrum disorder (ASD) (Irimia et al., 2014).

5.2 Transgenerational plasticity vs evolutionary change: How big is the difference?

Unquestionably, transgenerational plasticity and evolutionary change are distinct biological phenomena, but both essentially deal with the emergence and transmission of discrete new or modified traits to the progeny; hence, they are similar in regard to the result they lead to.

Conventionally, however, there is a major distinction between the evolutionary change and transgenerational plasticity; while the former is a long-term, "irreversible" effect transmitted to the "endless" future generations, transgenerational changes are reversible and last from one to many (Rechavi et al., 2011) and even an indefinite (Vastenhouw et al., 2003) number of generations in the absence of the triggering stimulus/environmental conditions. In other words, the evolutionary change is maintained in the absence of the triggering condition/stimulus indefinitely, whereas the transgenerational plasticity requires the presence of the stimulus after a varying number of generations.

A closer examination of the relevant experimental evidence, nevertheless, shows that exemptions that defy this general rule are observed in experiments, blurring the line of separation between them. Let us illustrate this with some experimental evidence. The fish *Astyanax mexicanus* has evolved a cave form that lost eyes and underwent other morphological changes in jaws, teeth, taste buds, skin depigmentation, etc. (Teyke, 1990) within the last ~10,000 or so years. However, exposure of the larvae of the eyeless fish to light for 1 month led to the development of eyes (Romero and Green, 2005) implying that the evolutionary change, loss of eyes is not irreversible.

Drosophila melanogaster flies kept under constant laboratory conditions for hundreds of generations have diverged from the ancestral wild type in several biochemical, physiological, and life-history traits. Among the acquired traits these laboratory strains possess presently is the late life reproduction time, which evolved as an "irreversible" evolutionary change. However, it was observed that reversion of these flies to the ancestral wild conditions led, within 50 generations, to reversion of the ancestral early-life fecundity (Teotónio and Rose, 2000; Teotónio et al., 2002). As for the suspicion that incomplete reversion of some traits may be related to hypothetical undemonstrated mutations, for the sake of argument, we can accept that such an extremely unlikely event of *useful* mutations may happen in one individual fly and even several flies, but it is impossible for an adaptive mutation to affect all the individuals of a whole population of *D. melanogaster* simultaneously. This is accepted by the authors themselves when they state, "There is no data as to the effect (additive or epistatic) of particular novel mutations" (Teotónio and Rose, 2001). The experiments showed that the wild-type traits, in flies kept for hundreds of generations in a changed laboratory environment, are modified into new evolutionary traits, but within 50 generations of rearing in the "wild" conditions, they revert or evolve back to the wild type. These experiments demonstrate how blurred the distinction between the "irreversible" evolutionary traits and the reversible transgenerational plasticity is. A more recent study on turtle ants (genus *Cephalotes*) shows that morphotype and head size of the soldier caste "are extensively reversible, repeatable, and decoupled within soldiers and between soldier and queen castes" (Powell et al., 2020).

Reversion of ancestral phenotypic traits in response to reappearance of ancestral conditions is a widespread phenomenon in the evolution of animals (reviewed by Cabej, 2019g). Reversion of the ancestral morphology, in contradiction with Dollo's law, occurred frequently among many animal groups in the course of evolution. This occurred with reevolution of sexual reproduction in mites of the Crotoniidae family, reversion of eyes in podocopid ostracods (Domes et al., 2007), reversion of eyes in a group of species of the class Ostracoda (Dingle, 2003), shell coiling in gastropods (Gould and Robinson, 1994), reversion of wings in insects (Whiting et al., 2003), reevolution in cyclostome fish of the cartilaginous skeleton of primitive chordates from the bony skeleton of their ancestral ostracoderms (Carter, 1967), etc.

From an evolutionary standpoint, the existence of inter- and transgenerational plasticity is favored by natural selection, and wasting it would be a luxury evolution could not afford. These forms of developmental plasticity are flexible mechanisms of animal adaptation to the fluctuating, ever-changing, but reversible, conditions in the environment, to which irreversible evolutionary changes would be maladaptive. From such a viewpoint, transgenerational inheritance may be considered both the optimal adaptation under the conditions of periodically alternating conditions of the environment and a reservoir of preadaptations. It is tempting to imagine that when environmental changes become permanent, they may serve as a raw material of evolutionary changes.

The obscure line of separation between the evolutionary change and transgenerational plasticity strongly suggests that the knowledge on the mechanisms of emergence of the transgenerational plasticity may be cautiously extrapolated into the mechanisms of the evolutionary changes that occurred in the past and are difficult to be reproduced experimentally.

6 Inheritance of acquired traits in animals

During the last 2 centuries, biologists were divided on whether acquired traits were inherited in successive generations. The Weissman experiments, rediscovery of the gene by the beginning of the 20th century, development of the modern synthesis in the 1930s, and the recognition of the chemical carriers of genes and nature of gene mutations by the middle of the last century, appeared to have swung the pendulum on the direction of the rejection of the possibility. However, the accumulating new evidence presented in this book and other relevant evidence demonstrate that acquired traits may be inherited, and this tempts us to inquire into the mechanisms of the inheritance of acquired traits.

6.1 A brief historical sketch of the concept and evidence on the inheritance of acquired traits

By the beginning of the 19th century, Jean-Baptiste Lamarck (1744–1829) raised the ancient observation on graduality and similarity of living forms to the level of a hypothesis that was commonsensical, plausible, and thought-provoking. Succinctly, his doctrine posits that living organisms evolve according to a physiological principle of the use-and-disuse and an inherent property to adapt their morphology to the changing environment.

In his attempt to prove the gradual evolution of life forms by natural selection, Charles Darwin was confronted with a "crowd of difficulties" (Darwin, 1859a). He believed that preeminent among these difficulties were the discontinuous and abrupt appearance of Cambrian fossils, the frequently observed absence of intermediary forms, the evolution of complex organs such as vertebrate eyes, and the evolution and inheritance of instincts. These difficulties compelled Darwin to adopt the Lamarckian principle of the use-and-disuse of parts as a mechanism of evolution of new or changed organs in animals. Elaborating along these lines, he developed the "provisional hypothesis" of pangenesis, according to which all parts of the organism release particular elements that contribute to gametes information on the structure of the respective parts or organs in the form of gemmules (from Latin gemmule, "little bud") thrown off to gametes as "minute granules or atoms, which circulate freely throughout the system, and subsequently becoming developed into cells like those from which they were derived" (Darwin, 1868). It is noteworthy that Darwin himself believed in the inheritance of the acquired traits (Darwin, 1859a, b, c, d, e; Darwin, 1872).

By the end of the 19th century, German biologist August Weismann (1834–1914), based on negative results obtained from his experiments on the inheritance of acquired traits, rejected the idea of evolution via the use-and-disuse of organs; while supporting the idea of natural selection, his hypothesis of the *Keimplasma* (German, germplasm) denied any role of the soma in evolution. He believed that evolution was determined by changes that occur in the germ cells, which are isolated from any influence of the soma. However, he admitted that the environment can modify heredity by acting only on the Keimplasma: "It appears that such influences may induce various small changes in the molecular structure of the germplasm, which according to our assumption are transmitted over generations, hence are hereditary" (Weissman, 1892).

The hypothesis put forward by American psychologist and philosopher James M. Baldwin (1861–1934), which posits that animals, responding to changes in the environment, adaptively change their behavior, represents an important development in the history of ideas on the inheritance of acquired traits; in the long run, if the environmental change persists, spontaneous variation under the action of natural selection may heritably fix the new behavior (Baldwin, 1896, 1897). This form of genetic assimilation came to be known as "Baldwin effect" (not discussed herein).

By the end of the 19th century, Georges J. Romanes (1848–1894) defended and promoted the Darwinian theory of natural selection and coined the term neoDarwinism. Owing to the work of a generation of biologists (R. Fisher, J.B.S. Haldane, T.G. Dobzhansky, et al.), a new variant of neoDarwinism emerged in the 1930s of the last century. It combined Darwinian theory with genetics and population biology and is known as the modern synthesis, a term invented by Huxley (1942). According to the new paradigm, evolution of living forms is the result of natural selection acting on random gene mutations in germ cells, recombination, and genetic drift that under conditions of reproductive isolation leads to the formation of new species and higher taxa. Modern synthesis developed rapidly into a theory larger than the initial form to include additions and modifications, which did not affect its deepest core, the concept of evolution as a result of random mutations in genes of the germ cells. Accordingly, evolution is mainly the result of natural selection acting on randomly occurring changes in the DNA and gene drift. For a long time, the neoDarwinian paradigm dominated the biological thought, while the inheritance of acquired traits by nongenetic means was regarded simply as a fallacy.

Interestingly, some of the best experimental evidence against the neoDarwinian viewpoint on the inheritance of acquired traits came from within, in the early 1940s, by Conrad Waddington (1905–1975), one of the leading geneticists of the time and now known as the "father of epigenetics" (Noble, 2015). He heat-shocked *Drosophila* in the laboratory and selected cross-veinless individuals treated by heat shock over several generations, finally obtaining flies that expressed the cross-veinless phenotype even in the absence of the heat treatment. Given the extremely low frequency of mutations and the exceedingly low probability of the emergence of the new phenotype in the whole group, Waddington concluded that the phenomenon was not related to changes in genes, but instead to a specific change in the relationship between those same genes and the environment. He coined the term *epigenetics* to describe specifically this interaction between genes and environmental temperature in determining the cross-veinless phenotype (Tronick and Hunter, 2016).

Another milestone in the development of the idea of inheritance of acquired traits represents the experimental work of E.J. Steele (1948–) and others by the late 1970s on the inheritance of immune tolerance to foreign antigens and formulation of the somatic selection hypothesis. After injecting lymphoid cells of mouse AJ strain into neonatal mice of CBA strain, he observed that mating of the F1 CBA males with naïve AJ females led to the production of a generation that was tolerant to H-2 antigens. This was the first experimental demonstration of transgenerational inheritance of an identified molecular phenotype and that a somatic modification can be passed on to the offspring. Based on theoretical premises of Burnet's theory of clonal selection of immunity and Temin's protovirus and provirus hypotheses, he suggested that endogenous viruses may capture an RNA sequence of a gene copy and via reverse transcriptase integrate it into the germline DNA and inherit through generations (Steele, 1979; Gorczynski and Steele, 1980).

In the publication in 1975 of two papers by Riggs (1975) and Holliday and Pugh (1975), investigators suggested that inactivation of one of the two X chromosomes of the female embryonic cells was the result of DNA methylation rather than any gene mutations. This opened a new avenue in the field of epigenetic studies. Soon it was understood that DNA methylation was induced by specific methylases, DNA methyltransferases, that add methyl groups to cytosines of the CpG groups. DNA methylation represses gene expression by blocking the binding sites of transcription factors. How DNA methyltransferases find the appropriate DNA sites to induce methylation is not known, but in multiple described cases, this is determined by the neural activity, which is a mechanism of specific induction of methylation of DNA and histones (Cabej, 2020a).

Over the past 4 decades, ever-increasing evidence on epigenetic marks in DNA and histones (chromatin remodeling) shows that inherited phenotypic changes may occur involving no known changes in genetic information. Later, it was discovered that changes in the expression patterns of small noncoding RNAs (sncRNAs or miR-NAs) may also affect the phenotype without changes in the DNA base sequence. Moreover, nongenetic mechanisms of alternative splicing, by selectively choosing and integrating the segments of genes to be transcribed, can produce numerous protein types from one gene and, thus, influence the phenotype without changing the genetic information. And, finally, it was repeatedly proved that neural signals induce gene expression (Cabej, 2020d), regulate miRNA expression patterns (Cabej, 2020b), alternative splicing (Cabej, 2020c), and the intergenerational and transgenerational inheritance (Cabej, 2019c).

All of the above, along with the fact that no correlation has been found to exist between the number of genes and the complexity of an organism (Srivastava et al., 2010), suggest that nongenetic factors may be involved in the inheritance of acquired traits.

Almost unconsidered for over a century after the publication of *The Origin of Species*, the hypothesis of inheritance of acquired traits rose like a phoenix from the ashes by the later part of the 20th century, stimulated by the discovery of epigenetic modifications of DNA and histones as proximate determinants of phenotypic characters in living organisms and the increasing number of described cases of transgenerational developmental plasticity. All of this and difficulties biologists faced in explaining increasing numbers of newly discovered phenomena within the theoretical framework of the modern synthesis led several authors (Eldredge and Gould, 1971; West-Eberhard, 1986; Jablonka and Lamb, 1989; Schlichting and Pigliucci, 1998; Newman and Müller, 2000; Gilbert, 2001, Cabej, 2008, 2012, 2019a, b, c, d, e, f, g; Gilbert and Epel, 2009; Jablonka and Raz, 2009) to the development of new alternative ideas, hypotheses, and theories to explain the biological evolution or its particular aspects. At various degrees, implicitly or explicitly they tended to favor the long-abandoned, but evergreen, idea of the inheritance of acquired traits.

The discovery of the role of miRNAs in the transmission of the parental changed phenotypes to the progeny over the last two decades represents the most direct and irrefutable evidence of the inheritance of acquired traits in animals. A succinct review of that work and the main contributors is presented in the following subsection of the evidence supporting the neural hypothesis of inheritance of acquired traits.

The term acquired trait herein will be used to describe any adaptive phenotypic novelties involving no change in the genetic information that develops in animals or their offspring and is transmitted to the subsequent generation(s) even in the absence of the initial triggering stimulus. The term is defined broadly to include cases when the new trait develops in the parent(s) or appears first only in the offspring of the individual that perceived the stimulus.

Herein I put forward a hypothesis on a neural mechanism of transgenerational inheritance of acquired traits. The empirical evidence I present to substantiate the hypothesis is adequate. For the sake of simplicity and clarity, I will focus only on some representative cases of transmission to the progeny of acquired behavioral, life-history, and morphological traits.

Given the strong intellectual influence Weissman experiments and Keimplasmatheorie played in thwarting studies on the inheritance of acquired traits, a brief review of the decline of the Weissman barrier is noteworthy.

6.2 The fall of the Weissman barrier

The core of the Weissman concept of inheritance was that the hereditary material of multicellular organisms resides in the germ cells, which, segregated and isolated, are not influenced by somatic cells. The main theoretical evolutionary corollary of the theory was the nonheritance of acquired traits. In modern biological terms, it means that no experiences and information can be transmitted from the somatic cells to the germ cells; the flow of the biological information is strictly unidirectional, from germ cells to somatic cells, which takes place during the development. Accordingly, discrete phenotypic (morphological, physiological, behavioral, and life history) traits that may emerge in somatic cells during the life of animals cannot be transmitted to future generations. Inheritance of acquired traits requires as a *conditio* sine qua non, an appropriate change in the hereditary material of gametes.

For our purpose, an acquired trait will be considered any new discrete trait that emerges in animals and is transmitted to future generations without involving relevant changes in genes. According to this working definition, inheritance of acquired traits is almost synonymous with evolutionary change. As shown earlier, it is difficult to determine the number of generations that inherited the newly emerging trait to be considered evolutionary change.

In our time, vast empirical evidence contradicting the Weismann barrier and its theoretical ramifications is accumulated, rendering the concept obsolete and making appealing the idea of a reconsideration of the mostly ignored idea of the inheritance of acquired traits. The evidence against the Weissman barrier derives mainly from two fields of biological research that involve transmission to the offspring of the discrete traits: intergenerational and transgenerational plasticity.

The inheritance of acquired traits is obviously incompatible with the modern synthesis, hence intellectually perplexing and emotionally averse. However, the Lamarckian doctrine of the inheritance of acquired traits by use-and-disuse was also embraced, although contradictorily, by Darwin, in his attempt to overcome difficulties, especially in view of Fleeming Jenkin's (1833–1885) remark that the rare frequency of useful evolutionary changes would prevent their spread in the population and eventually be lost. Thus, Darwin admitted that "[t]here can be little doubt that use in our domestic animals strengthens and enlarges certain parts, and disuse diminishes them; and that such modifications are inherited" (Darwin, 1859c).

Both forms of plasticity discussed in this chapter, the intergenerational plasticity discussed in Chapter 4 and transgenerational inheritance in this chapter, involve no changes in genes or genetic information in general; both defy the Weissman barrier by involving somatic influence on germ cells and the related inheritance of new emerging traits.

6.3 Neurally mediated inheritance of acquired traits

6.3.1 Neural control of inheritance of acquired traits and the epigenetic marks

Despite the scarcity of evidence on the involvement of epigenetic marks in DNA and histones in the inheritance of acquired traits, these marks are highly specific; they occur "in the right place at the right time" rather than randomly. This obviously indicates that information is invested for them to occur.

The logical and most parsimonious approach to the search for the source of information for epigenetic modifications in DNA and histones is to start from immediate inducers, the proximate causes. Inducers of the DNA methylation are DNA methyltransferases (DNMT1, DNMT3A, and DNMT3B), and histone methylation is induced by histone methyltransferases (HMTs). Adequate experimental evidence shows the synthesis of these enzymes under hormonal control (Yamagata et al., 2009). In mice, changes in the level of steroid hormones induce changes in methylation of the promoters of vasopressin and estrogen-α receptor genes in the brain of adult male rats (Auger et al., 2011). The hormone Tr3 (triiodothyronine) recruits corepressor complexes containing histone deacetylases (Shi, 2013) and histone methyltransferases (Matsuura et al., 2012). Hormones, such as JH (juvenile hormone) and ecdysteroids, are involved in the process of "chromatin untagging" (removal of methyl groups from histones) in both the germinal vesicle and the male pronucleus (De Loof et al., 2013), and so on.

Interestingly, removal of the pituitary, which controls the hormone production by the endocrine glands, leads to a drastic loss in the methyltransferase activity (Burke et al., 1983). In this context, it is important to remember that the synthesis of all hormones in endocrine glands, including the pituitary, is determined by signal cascades that start in the brain (the hypothalamus in vertebrates).

In many cases, production of methylases is under the direct control of neuronal activity (Ricq et al., 2016). So, e.g., the experimental elevation of neuronal activity by administration of pilocarpine (an activator of cholinergic receptors of the muscarinic type) induces production in the mouse hippocampus of many methylases and demethylases, KTMs (lysine methyltransferases), and KDMs (lysine demethylases) (Wijayatunge et al., 2014).

Neuronal activity also induces production of the H3K27 histone demethylase Kdm6b (Jmjd3) (Wijayatunge, 2012). Neural activity determines histone acetylation and gene expression (Tang et al., 2009; Méjat et al., 2005), while denervation leads to histone deacetylation by accumulation of HDACs (histone deacetylases) (Cohen et al., 2009; Kawano et al., 2015). It also determines methylation both globally and at specific sites of DNA and histones. For instance, neuronal activation in mice leads to the induction of DNA methylation at more than 30,000 specific sites of their genome (Kawano et al., 2015; Guo et al., 2011). Alternatively, neuronal activity induces DNA demethylation leading to transcription of affected genes (Martinowich et al., 2003; Nelson et al., 2008; Ma et al., 2009).

Maltreatment and psychological traumas in children also neurally alter their global DNA methylation patterns (Matosin et al., 2017) and sperm DNA methylation (Roberts et al., 2018), while environmental stressors, via the limbic system, induce hypothalamic neurons to produce CRF (corticotrophin-releasing factor), which stimulates the pituitary to secrete ACTH (adrenocorticotrophic hormone), leading to secretion of methylated or demethylated glucocorticoids by adrenals (Bakusic et al., 2017). Even fear or the perceived (real or imaginary) danger leads to changes in DNA methylation and gene transcription in mammals (Zovkic and Sweatt, 2013). Psychosocial stress related to PTSD (posttraumatic stress disorder) or child abuse in humans may alter global and gene-specific DNA methylation patterns (Smith et al., 2011).

Neural activity induces histone acetylation and regulates transcription of activity-inducible genes (Chen et al., 2019), including histone acetylation of the sestrin 2 promoter (Soriano et al., 2009), while experimental enhancement of the activity of proprioceptive dorsal root ganglion (DRG) leads to Creb-binding protein (Cbp)-mediated histone acetylation (Hutson et al., 2019). Neuronal activation recruits to the promoter of the neurexin gene nrxn1α, the histone methyltransferase Ash1L, which methylates the histone H3 (H3K36me2) (Zhu et al., 2016).

6.3.2 Neural control of inheritance of acquired traits and miRNA expression patterns

miRNAs are involved in the inheritance of acquired traits (see later), but there is sufficient evidence that neuronal activity is a major regulator of miRNA expression (Park and Tang, 2009; Nudelman et al., 2010; Sim et al., 2014; Mathew et al., 2016; Sambandan et al., 2017). In *Drosophila*, neuronal activity regulates the expression of miR-134 (Fiore et al., 2009; Eacker et al., 2011), and in the mouse, miRNA expression in cultures of hippocampal neurons (Wayman et al., 2008). Perception of different light intensities induces changes in levels of miR-183/96/182 cluster, miR-204, and miR-211 in the retina (Krol et al., 2010). Experimental depression in rats upregulates the expression of miR-27a in the hippocampus (Cui and Xu, 2018). Mice exposed to artificial light at night show increased expression of miRNAs, miR-140-5p, 185-5p, 326-5p, and 328-5p, which, by repressing the expression of Rev-erba nuclear receptors in liver cells, cause fat accumulation in the mouse liver (Borck et al., 2018). It is interesting to observe that muscle denervation leads to a 10-fold increase in the expression of miR-206. Similarly, a threefold increase in miR-206 expression (Williams et al., 2009) is observed during amyotrophic lateral sclerosis, characterized by degeneration of motor neurons, indicating that innervation is an inhibitor of miR-206 synthesis.

Often, innervation upregulates miRNAs. So, e.g., a month after denervation, downregulation of miR-1 and miR-133a occurs in rat muscle, but after reinnervation and 4 months after denervation, these same miRs increase by about threefold (Jeng et al., 2009).

The above partial evidence indicates that neural activity determines the spatio-temporal patterns of miRNA expression.

6.3.3 The nervous system conveys parental information to gametes
Selective deployment of cytoplasmic factors in eggs

The oocyte is maternally supplied not only with nutritive substances, hormones, secreted proteins, neurotransmitters/neuromodulators, miRNAs, etc., but also, most importantly, with mRNAs, which are translated during the early development, before the activation of the embryonic genome but also later until, and even after, the phylotypic stage. Being transcriptionally and translationally inactive (Zalokar, 1965), throughout most of oogenesis, the insect oocyte receives vitellogenin first from surrounding follicle cells via gap junctions formed under neural control both via neural activity-dependent (Shimizu and Stopfer, 2013) and activity-independent modes (Koulakoff et al., 2008). The squeezing of the nurse cells' content supplies the oocyte, besides various mRNA types, with numerous miRNAs, which may also play the role of carriers of the maternal acquired traits. Among the identified sncRNAs provided by nurse cells are piRNAs (Tóth et al., 2016), as well as let-7 miRNA, miR-30, and miR-34 (Tang et al., 2007; Soni et al., 2013).

Recently it is reported that the experimentally introduced double-stranded RNA (dsRNA) into the *C. elegans* circulation, via endocytosis, enters the oocyte, inducing specific intergenerational silencing of matching genes in the embryo (Devanapally et al., 2015; Marré et al., 2016) (Fig. 5.13).

In mammals, neurotransmitters serotonin and dopamine are also involved in the regulation of receptor-mediated endocytosis (Raote et al., 2013), and dopamine through its D2 receptors induces endocytosis of VEGF receptors (Basu et al., 2001).

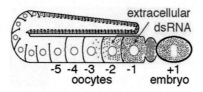

FIG. 5.13

Extracellular dsRNA can accumulate without a cytosolic entry in proximal oocytes and subsequently within embryos. Model illustrating that extracellular dsRNA can be transported through oocytes to progeny with or without entry into the cytosol.

From Marré, J., Traver, E.C., Jose, A.M., 2016. Extracellular RNA is transported from one generation to the next in Caenorhabditis elegans. Proc. Natl. Acad. Sci. U. S. A. 113 (44), 12496–12501. https://doi.org/10.1073/pnas.1608959113.

Zebra finches deposit more yolk carotenoid when paired to a low genetic quality male and more testosterone when paired to a low parental quality male (Bolund et al., 2008). This is in contrast with the female mallard (*Anas platyrhynchos*) that lays larger eggs after copulating with preferred males and smaller eggs after copulating with less preferred males (Cunningham and Russell, 2000). Such examples suggest a role of the animal perception in selective deposition of maternal factors in the oocyte.

Reproductive organs in *Drosophila* are innervated by OA/TA (octopamine/tyramine) neurons, which have "a strong impact on muscle activity" of the reproductive organs (Pauls et al., 2018) and may be necessary to squeeze nurse cells during the transport of mRNAs into the oocyte. Cytoplasmic factors (parentally expressed genes) do not remain where they are transported initially but are relocated to specific sites in the oocyte. Their transport to destinations takes place along microtubules and is implemented by adaptively changing microtubule length and orientation. This implies that some kind of information is provided to regulate "at will" microtubule length. Where does the information come from?

The only agents capable of determining "at will" the length of microtubules that we know of are neuromodulators epinephrine and acetylcholine released by local innervation (Hanlon et al., 1990; Mäthger et al., 2004; Wardill et al., 2012; Gonzalez-Bellido et al., 2014; Gur et al., 2015) that instantly and adaptively regulate the distance between microtubules separating guanine platelets to produce the camouflage or cryptic coloration in fish (Kasukawa et al., 1987; Mäthger et al., 2003; Yoshioka et al., 2011).

The abovementioned empirical evidence on the regulatory function of innervation on microtubules makes it reasonable to extrapolate a similar role of the innervation of the ovariole peritoneal sheath (Middleton et al., 2006) in determining the length of microtubules and, consequently, the sites of the placement of mRNAs in the oocyte.

Selective deployment of miRNAs into spermatozoa

Whereas it has long been known that the oocyte stores mRNAs and miRNAs for later use, it now appears that the sperm also accumulates miRNAs that are critical in normal development. In the last two decades, miRNAs have emerged as transmitters of the parental modified phenotypes to the progeny. Now, selective expression and secretion of miRNAs by cells is a well-known phenomenon (Pigati et al., 2010; Collino et al., 2010; Alvarez-Erviti et al., 2011; Guduric-Fuchs et al., 2012).

The involvement of the nervous system in the inheritance of acquired traits is demonstrated even in one of the smallest known worms, *C. elegans*. The worm that has only 302 neurons and no brain learns to avoid the pathogenic bacteria *Pseudomonas aeruginosa* and transmits the avoidance behavior to the progeny. The inheritance of the avoidance behavior requires the expression of TGF-β (transforming growth factor β) in the sensory neurons and activation of PIWI Argonaute small RNA (Moore et al., 2019).

As mentioned earlier, under deteriorating environmental conditions, *C. elegans* enters the Dauer stage, and when the conditions are restored, it reproduces normally.

It was observed, however, that in the absence of the metabolic regulator, AMP-activated protein kinase (AMPK) after recovering from the Dauer stage, these worms are reproductively sterile. Investigators found that during the Dauer stage, neuronal AMPK serves to ensure germline quiescence and to maintain its integrity via endogenous small RNA pathways (Wong and Roy, 2020).

Spermatozoa are also carriers of parental epigenetic marks and miRNAs in mammals. This was first shown in 2006 when Rassoulzadegan and colleagues demonstrated that microinjection of one or two miRNAs (miR-221 or miR-222) into wild-type fertilized mouse oocytes transmitted to the progeny the paramutated phenotype of white patches in feet and tails. It is noteworthy that the same result they obtained by microinjection of brain miRNAs into the wild-type fertilized oocytes (Rassoulzadegan et al., 2006). The experiment caused a snowball effect in studies on the role of small noncoding regulatory RNAs (Lee et al., 2009) in the inheritance of acquired traits in animals. Grandjean et al. demonstrated that microinjection of a mouse miRNA, miR-124, in one-celled mouse embryos produced a paramutant generation that was 30% heavier than normal mice and the trait was inherited to the third generation (Grandjean et al., 2009). Later, members of the same group showed that a high-sugar diet (Western-like diet) in male mice caused obesity and metabolic disorders and upregulation of miR-19b in sperm. Microinjection of miR-19b into naïve one-celled embryos recapitulated the modified parental phenotype (Grandjean et al., 2016).

One of the earliest identified small RNA classes, the "mature-sperm-enriched tRNA-derived small RNAs" (mse-tsRNAs), is extremely enriched in mature mouse sperm (Peng et al., 2012). Another small noncoding RNA class is altered in male mice exposed to a high-fat diet (HFD). These mice experience a metabolic disorder and display an altered profile of tsRNAs (transfer RNA-derived small RNAs) in sperm. Microinjection of sperm tsRNAs from HFD males produced the same metabolic disorder in F1 progeny (Chen et al., 2016), but this effect and the relevant alterations in miRNAs are abolished if tRNA methyltransferase (DNMT2) is deleted (Zhang et al., 2018).

Microinjection of sperm tsRNAs of the progeny of HFD dams into normal mouse zygotes reproduces a maternal obese phenotype and addictive-like behaviors that persist for up to three generations (Sarker et al., 2019).

Cocaine-addicted rats display cocaine-seeking behavior and transmit it to F1 and F2 generations, whereas changes in DNA methylation are inherited only in F1 (Le et al., 2017). It is interesting to note that specific alterations in the sperm miRNA profile of humans on a high-sugar diet occur very rapidly within a week (Nätt et al., 2019). The stress-dependent transcription factor 7 (ATF7) binds to promoters of ~2300 genes and regulates their expression in testicular germ cells. In mice exposed to a low-protein diet, ATF7 is phosphorylated and downregulates H3K9me2 on its target genes while inducing the expression of some tRNA fragments in spermatozoa. These changes are maintained in F1 (Yoshida et al., 2020; Zhang and Chen, 2020). A total of 164 male mice exposed to a low-protein diet showed decreased let-7 levels and increased tRFs (tRNA fragments) levels as well as repression of MERVL target genes. Injection of tRFs from sperms obtained from cauda epididymis similarly suppressed expression of these genes. It was found that tRFs are trafficked to sperm in epididymosomes (special vesicles released by the epididymal epithelial cells), and investigators concluded that "the paternal diet can influence offspring phenotype via information in sperm" (Sharma et al., 2016).

The fact that 60% of the 350 miRNAs detected in the mouse epididymis are present in the epididymosomes and that epididymal miRNAs are transferred to sperms suggests that epididymosomes may perform transfer of miRNAs to sperm cells (Gapp and Bohacek, 2018), while the relative underrepresentation of miRNAs in epididymosomes suggests that they are selectively incorporated in these secretory vesicles (Martone et al., 2019). On their surface, epididymosomes contain milk fat globule-EGF 8 (MFGE8), which binds to the postacrosomal domain of the sperm head by using mechanoenzyme DNM1 (dynamin1) to form a transient fusion that delivers the epididymosome cargo to the sperm (Zhou et al., 2019). During the epididymal transit, epididymosomes provide sperms with groups of miRNAs sequentially secreted by each epididymal segment (Belleannée, 2015; Chu et al., 2015). It is proposed that environmental experience coded in the form of miRNAs is transmitted by sperm cells to the offspring via epididymosomal vesicles (Eaton et al., 2015).

The majority (213) of miRNAs are common in both the epididymal epithelial cells and spermatozoa. The modification of the sperm miRNA repertoire in the process of sperm maturation includes the loss of 113 miRNAs and the acquisition of 115 miRNA types (Nixon et al., 2015).

How the segmental release of miRNAs by epididymis is regulated in the course of the sperm transit within its lumen is not known, but it is experimentally demonstrated that epididymal innervation plays an indispensable role in providing sperms with epididymal miRNAs, as indicated by the fact that the sympathetic denervation of this organ inhibits the embryo development (Ricker et al., 1997; Ricker, 1998). The successful application of ICSI (intracytoplasmic sperm injection) with sperms obtained from the human vas efferens seems to contradict these results. However, the fact that ICSI offspring shows a drastic decrease in the total sperm count (~48%) and sperm motility (~32%) (Belva et al., 2016) and that the experimental acceleration in sperm transit time of the rat epididymis lowers the fertility rate (Dal Bianco Fernandez et al., 2007) corroborates the role of the epididymal miRNAs in the process. During their travel along the epididymal lumen (caput, corpus, and cauda epididymis), spermatozoa increase the number and modify the repertoire of their miRNAs.

6.3.4 Transport of parental factors from the nervous system to the gametes via body fluids

Injection in the nematode *C. elegans* of a double-stranded RNA (dsRNA) that targets the gene *ceh13* induces the development of a small dumpy phenotype that is inherited with 30% penetrance in an indefinite number of generations (Vastenhouw et al., 2003).

In response to external unfavorable conditions, *Daphnia pulex* brain secretes alla-totropins, which stimulate the formation of hormone methyl farnesoate (MF), the crustacean juvenile hormone (LeBlanc and Medlock, 2015). MF binds its receptor MfR, which with methoprene tolerant (MET) and steroid receptor coactivator (SRC) forms a complex that serves as a transcription factor for expression of male sex genes *dsx* and *8960* that are deposited in eggs.

In response to high temperatures of >25 °C for a relatively long period of time, a dopaminergic circuit stimulates the labial secretory neurons of the subesophageal ganglion (SOG) of the silkworm *Bombyx mori* to secrete diapause hormone (DH), which targets oocytes (Kitagawa et al., 2005) and induces diapause in the progeny that otherwise would not diapause. Ablation of the labial neurons impairs the onset of diapause (Noguchi and Hayakawa, 2001).

In *Drosophila*, a brain signal, a "cephalic event" (Handler and Postlethwait, 1977), controls the vitellogenin uptake by the oocyte from the hemolymph. Generally, the receptor-mediated endocytosis of oocytes in *Drosophila* (Raikhel and Lea, 1985; Richard et al., 2001) and in the red cotton stainer *Dysdercus koenigii* (Venugopal and Kumar, 2000) is neurohormonally regulated.

Neurotransmitters released by the central nervous system are transported to the oocyte to suppress/reduce the probability of entering the diapause in the offspring of the flesh fly *Sarcophaga bullata* (Henrich and Denlinger, 1982; Webb and Denlinger, 1998), and injection into the thoracic ganglia of the neurotransmitter sero-tonin alone, or its analogs, induces in locusts the transgenerational phase transition from the solitary to the gregarious phase, displaying all the associating morphological, physiological, and behavioral changes, whereas blocking the action of serotonin prevents transition to the gregarious phase (Anstey et al., 2009).

The most impressive experimental evidence on the role of the brain in the deployment of miRNAs in sperm and passing them on patrilineally to embryos was reported in 2020. It was demonstrated that injection of human pre-MIR94–1 in the mouse brain induces the appearance of human pre-MIR94–1 in the contralateral side of the brain, cerebellum, and lymph nodes as well as in the vas deferens and epididymis, but not in the blood or liver. Mating the brain-injected mice with control female mice led to the reappearance of human pre-MIR94–1 in the same organs in one-third of the offspring of F1 generation (O'Brien et al., 2020). The experiment indicates that the brain, apparently via lymph and VE, selectively transported pre-MIR94–1 into sperm, thus leading to the transgenerational reappearance of pre-MIR94–1 in the same organs in the F1 generation.

And now to summarize, we are faced with two firmly established facts:

First, *Drosophila* nurse cells provide the oocyte with some vital transcription factors although the oocyte synthesizes many of the mRNAs it needs to translate before the maternal-to-zygotic transition, whereas in vertebrates, development of the oogonium into an oocyte and mature ovum is function of the reproductive system under brain control, primarily via the HPA (hypothalamus-pituitary-ovarian) axis.

Second, in mammalians, the epididymal epithelial cells selectively secrete and deposit paternal miRNAs in epididymal vesicles, which fuse into sperms as they pass through the epididymal vesicles in a process that requires the epididymal innervation.

The common sense says that neither nurse cells nor epididymal epithelial cells know what ova and sperms need to have to perform their reproductive functions. Nor do gametes themselves know what cytoplasmic factors they need to synthesize for performing their reproductive function; metazoan reproduction is a systemic function involving integration and coordination of activities of multiple organs at the organismal level rather than a monopoly of particular specialized cells.

7 Why the nervous system?

All the examples of transgenerational plasticity presented in this chapter, as well as those on intergenerational plasticity described in Chapter 4, share the fact that new traits emerge as predictable deterministic outcomes, implying that information is used to make them happen. Tracking back the flow of information along the causal chain that leads to the onset of the acquired trait shows that the causal chain originates in the nervous system.

Why in the nervous system? Why does not another organ system, organ, or cell act as originator of the information for the emergence of new traits or even simply to maintain its relative steady state? Theoretically, it may be argued that the functioning of the enormously and continually changing complex biological system requires an organ system, organ, or a hypothetical "omniscient" cell that could figure what the organism needs at every moment.

To do this, it must be in possession of.

1. Information on the homeostatic levels of thousands of phisico-chemical parameters and the species-specific morphology,
2. A pervasive presence throughout the animal body to monitor the state of the system,
3. A control system to compare the state of the system with the homeostatic state, and
4. Ability to generate and transmit to affected parts information or "instructions" to restore the normal state and replace dead cells in the unavoidably degrading structures.

A system that meets all the above criteria is an integrated control system.

There is no other organ system or organ, besides the nervous system, capable of carrying out all the functions of an integrated control system: monitor the state of the system, detect deviations from the norm throughout the animal body, and correct them by regulating gene expression. Via afferent nerve pathways, it constantly receives billions of bits data on the state of the system and compares them with the norm as is the case, e.g., with specific brain set points regulating body temperature, water content, and levels of pituitary and endocrine hormones, to mention only a few of the numerous homeostatic physicochemical parameters in body fluids.

It also sends signals to the target cells and organs to restore any lost structure as it is plainly seen, among other things, in the process of the regeneration of lost parts of the animal body.

The appearance and inheritance of acquired traits involving no known changes in genetic information remain a great enigma in modern theoretical biology. Given that the acquired trait implies involvement of information, the question arises, What is the nature and source of the information animals use to mold new traits?

As pointed out earlier, external stimuli represent data about the changing environment instead of any kind of information or instructions for what to do. Data and stimuli are taken as "problems to solve" (Cabej, 2019d) in the course of neural processing. The nature of the processing of external and internal stimuli in the nervous system still represents a black box. We know "what" it generates, but we do not know "how" it does it. We assume the information generated in the black box is electrically coded in the form of the frequency (rate-based code) and timing of spikes (Ponulak and Kasiński, 2011).

In general terms, the output of the processing of the stimuli in the neural circuits is a chemical agent that by triggering a specific signal cascade and transduction pathway in the target cell turns on a tissue-specific gene, increasing from 0 to 1 the probability that the stimulus expresses the specific gene(s) (Cabej, 2019e). By inducing the expression of a gene that the stimulus per se cannot induce, the neural processing turns an improbable event into a certain result. In doing so, the neural processing establishes an otherwise nonexisting relationship between the external signal and the gene, according to the following generalized schema:

External/internal stimulus → Sensory organs → CNS processing → Output of the processing → Activation of a signal cascade → Activation of a signal transduction pathway → Expression of a tissue-specific gene.

Hence, it is the nervous system rather than the particular stimulus that determines which of the thousands of genes in the genome will be induced in response to the stimulus (Cabej, 2019f). The signal cascades that lead to the emergence and inheritance of acquired traits leave no doubt that the nervous system is the source of that information, but the question grows extremely complex when it comes to determining the nature, storage, and generation of the information involved in the process (Cabej, 2019d).

From the above viewpoint, external/internal stimuli represent only cues, not information, as suggested, among other things, by the fact that, in response to the same stimuli, different organisms respond in different ways, by activating different signal cascades and producing different phenotypic results. It is not the stimulus per se, but its processing in the nervous system that generates the information necessary to express/suppress the specific gene. There is abundant and solid evidence that the nervous system, by processing internal/external stimuli, transforms the latter into specific electrical spike trains and chemical signals, determined by both the nature of the stimulus and the connectivity of neurons in neural circuits. By processing this electrical information in neural circuits, according to its "best guess" (Katz and

Shatz, 1996), the brain modifies its connections and neural circuits (Katz and Shatz, 1996; Penn et al., 1998; Penn, 2001; Segal et al., 2003) and generates a chemical output (neuropeptide, neurotransmitter, or neuromodulator) (Shimizu et al., 1989; Shimizu et al., 1997; Yamanaka et al., 2000; Ichikawa, 2003; Rogers et al., 2004; Claeys et al., 2006), or an electrical signal (Morrison, 2004; Thorens, 2011), which activates a specific signal cascade or gene regulatory network resulting in the development of a phenotypic trait. Besides the role in determining patterns of miRNA expression considered earlier in the chapter, neural activity is demonstrated to induce specific changes in epigenetic marks (methylation, acetylation, etc.) in DNA (Guo et al., 2011; Ma et al., 2009; Oliveira et al., 2012; Bayraktar and Kreutz, 2018) and histones (Guan et al., 2002; West and Greenberg, 2011; Santoro and Dulac, 2012; Ding et al., 2017).

8 A tentative mechanism of emergence and inheritance of acquired traits

The identification of the nervous system as the source of information for the emergence of acquired traits does not tell us anything about the mechanisms of their transmission to the offspring, and this is a crucial point in the present discourse on the inheritance of acquired traits.

Virtually, there is no other way to transmit acquired traits to the progeny, but via the gametes, ova, and sperms. But how does this transmission occur? Until recently, this question represented an impenetrable enigma. The Weissman barrier stated that the germline is stringently segregated from somatic cells so that somatic cells, including neurons, cannot influence gametes, ova, and sperm. In recent years, however, evidence is accumulating that the "barrier" was a temporary illusion. Somatic inducers penetrate into gametes, and epigenetic signatures of the parentally acquired traits selectively escape the global epigenetic erasure in gametes.

It appears that the evidence at hand warrants an attempt to piece together the links in the chain of causation of the acquired traits. As argued and demonstrated earlier in this chapter, the cascade of events leading to the transmission of the acquired trait from parent(s) to the offspring reveals the nervous system/CNS as the source of the heritable information. By processing the sensory (visual, auditive, olfactory, thermosensory, chemosensory, etc.) input, the nervous system generates a neural output in the form of a chemical signal that starts a specific neuroendocrine cascade, resulting in production and deposition in gametes of hormones, neuromodulators, epigenetic marks, miRNAs, etc. Acquired traits appear in parent(s) that experienced specific stimuli that transmitted the adaptive trait to the offspring, or they appear first in the offspring of these parent(s). In both cases, the parental nervous system is the source of the information for the development of the new trait.

In this chapter, I have presented adequate empirical evidence, demonstrating that the nervous system/CNS:

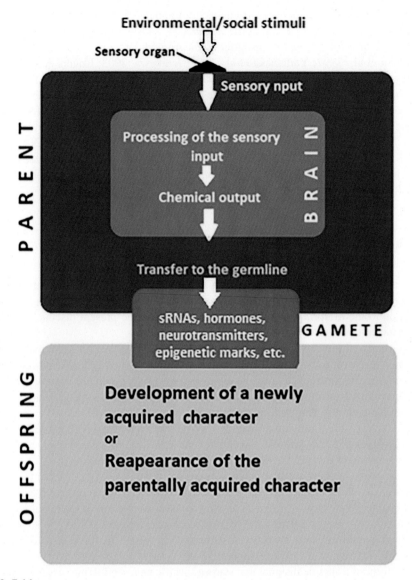

FIG. 5.14

A generalized diagram of mechanisms of the emergence and transmission of acquired traits to the offspring. The input of environmental stimuli is processed in the central nervous system, leading to the release of chemical output that activates a particular neuroendocrine cascade. The cascade continues with deposition in the gametes of relevant hormones, miRNAs, and other inducers as well as specific epigenetic tags that convert the parentally provided information into the acquired trait.

1. Starts signal cascades that lead to the deposition in the germline of inducers of the acquired trait,
2. Is involved in the process of production of specific mRNAs, hormones, neuromodulators, and other inducers in gamete/zygote (miRNAs), enabling the onset and transmission of the acquired trait in the offspring, and
3. Provides information for determining the exact placement of epigenetic tags on DNA and histones and induces adaptive changes in miRNA expression patterns.

These empirically demonstrated facts warrant the conclusion that the CNS/nervous system is the ultimate source of the information necessary for the development and inheritance of acquired traits in eumetazoans, as is diagrammatically represented in Fig. 5.14.

Validation of this neural hypothesis of inheritance of acquired traits may have theoretical consequences for the theory of evolution.

References

Adams, C. S., Korytko, A. I., & Blank, J. L. (2001). A novel mechanism of body mass regulation. *J. Exp. Biol.*, *204*, 1729–1734.

Agrawal, A. A., Laforsch, C., & Tollrian, R. (1999). Transgenerational induction of defences in animals and plants. *Nature*, *401*, 60–63.

Alvarez-Erviti, L., Seow, Y., Yin, H., Betts, C., Lakhal, S., Wood, M. J., et al. (2011). Delivery of siRNA to the mouse brain by systemic injection of targeted exosomes. *Nat. Biotechnol.*, *29*(4), 341–345. https:/doi.org/10.1038/nbt.1807.

Anstey, M. L., Rogers, S. M., Ott, S. R., Burrows, M., & Simpson, S. J. (2009). Serotonin mediates behavioral gregarization underlying swarm formation in desert locusts. *Science*, *323*(5914), 627–630. https://doi.org/10.1126/science.1165939.

Anton, S., & Hansson, B. S. (1996). Antennal lobe interneurons in the desert locust *Schistocerca gregaria* (Forskal): processing of aggregation pheromones in adult males and females. *J. Comp. Neurol.*, *370*, 85–96.

Anway, M. D., Cupp, A. S., Uzumcu, M., & Skinner, M. K. (2005). Epigenetic transgenerational actions of endocrine disruptors and male fertility. *Science*, *308*(5727), 1466–1469. https://doi.org/10.1126/science.1108190.

Auger, C. J., Coss, D., Auger, A. P., & Forbes-Lorman, R. M. (2011). Epigenetic control of vasopressin expression is maintained by steroid hormones in the adult male rat brain. *Proc. Natl. Acad. Sci. U. S. A.*, *108*(10), 4242–4247. https://doi.org/10.1073/pnas.1100314108.

Badisco, L., Huybrechts, J., Simonet, G., Verlinden, H., Marchal, E., Huybrechts, R., et al. (2011). Transcriptome analysis of the desert locust central nervous system: production and annotation of a *Schistocerca gregaria* EST database. *PLoS One*, *6*(3). https://doi.org/10.1371/journal.pone.0017274, e17274.

Bakusic, J., Schaufeli, W., Claes, S., & Godderis, L. (2017). Stress, burnout and depression: a systematic review on DNA methylation mechanisms. *J. Psychosom. Res.*, *92*, 34–44. https://doi.org/10.1016/j.jpsychores.2016.11.005.

Baldwin, J. M. (1896). A new factor in evolution. *Am. Nat.*, *30*, 441–451.

Baldwin, J. M. (1897). Organic selection. *Science*, *5*(121), 634–636. https://doi.org/10.1126/science.5.121.634.

Barry, M. J. (2002). Progress toward understanding the neurophysiological basis of predator induced morphology in *Daphnia pulex*. *Physiol. Biochem. Zool.*, *75*(2), 179–186. https://doi.org/10.1086/339389.

Basu, S., Nagy, J. A., Pal, S., Vasile, E., Eckelhoefer, I. A., Bliss, V. S., et al. (2001). The neurotransmitter dopamine inhibits angiogenesis induced by vascular permeability factor/vascular endothelial growth factor. *Nat. Med.*, *7*, 569–574. https://doi.org/10.1038/87895.

Bayraktar, G., & Kreutz, M. R. (2018). Neuronal DNA methyltransferases: epigenetic mediators between synaptic activity and gene expression? *Neuroscientist*, *24*(2), 171–185. https://doi.org/10.1177/1073858417707457.

Beck, D., Sadler-Riggleman, I., & Skinner, M. K. (2017). Generational comparisons (F1 versus F3) of vinclozolin induced epigenetic transgenerational inheritance of sperm differential DNA methylation regions (epimutations) using MeDIP-Seq. *Environ. Epigenet.*, *3*(3). https://doi.org/10.1093/eep/dvx016.

Behesti, H., Papaioannou, V. E., & Sowden, J. C. (2009). Loss of Tbx2 delays optic vesicle invagination leading to small optic cups. *Dev. Biol.*, *333*, 360–372.

Belleannée, C. (2015). Extracellular microRNAs from the epididymis as potential mediators of cell-to-cell communication. *Asian J. Androl.*, *17*, 730–736.

Belva, F., Bonduelle, M., Roelants, M., Michielsen, D., Van Steirteghem, A., Verheyen, G., et al. (2016). Semen quality of young adult ICSI offspring: the first results. *Hum. Reprod.*, *31*(12), 2811–2820. https://doi.org/10.1093/humrep/dew245.

Ben Hamouda, A., Ammar, M., Ben Hamouda, M. H., & Bouain, A. (2009). The role of egg pod foam and rearing conditions of the phase state of the Asian migratory locust Locusta migratoria migratoria (Orthoptera, Acrididae). *J. Insect Physiol.*, *55*(7), 617–623. https://doi.org/10.1016/j.jinsphys.2009.03.008.

Bilandzija, H., Hollifield, B., Steck, M., Meng, G., Ng, M., Koch, A. D., et al. (2020). Phenotypic plasticity as a mechanism of cave colonization and adaptation. *eLife*, *9*. https://doi.org/10.7554/eLife.51830, e51830.

Boerjan, B., Sas, F., Ernst, U. R., Tobback, J., Lemière, F., Vandegehuchte, M. B., et al. (2011). Locust phase polyphenism: does epigenetic precede endocrine regulation? *Gen. Comp. Endocrinol.*, *173*(1), 120–128. https://doi.org/10.1016/j.ygcen.2011.05.003.

Bolund, E., Schielzeth, H., & Forstmeier, W. (2008). Compensatory investment in zebra finches: females lay larger eggs when paired to sexually unattractive males. *Proc. Biol. Sci.*, *276*(1657), 707–715. https://doi.org/10.1098/rspb.2008.1251.

Borck, P. C., Batista, T. M., Vettorazzi, J. F., Soares, G. M., Lubaczeuski, C., Guan, D., et al. (2018). Nighttime light exposure enhances Rev-erbα-targeting microRNAs and contributes to hepatic steatosis. *Metabolism*, *85*, P250–P258. https://doi.org/10.1016/j.metabol.2018.05.002.

Bourque, C. W. (2008). Central mechanisms of osmosensation and systemic osmoregulation. *Nat. Rev. Neurosci.*, *9*, 519–531.

Brieño-Enríquez, M. A., García-López, J., Cárdenas, D. B., Guibert, S., Cleroux, E., & Děd, L. (2015). Exposure to endocrine disruptor induces transgenerational epigenetic deregulation of microRNAs in primordial germ cells. *PLoS One*, *10*(4). https://doi.org/10.1371/journal.pone.0124296, e0124296.

Brownell, I., Guevara, E., Bai, C. B., Loomis, C. A., & Joyner, A. L. (2011). Nerve-derived sonic hedgehog defines a niche for hair follicle stem cells capable of becoming epidermal stem cells. *Cell Stem Cell*, *8*(5), 552–565. https://doi.org/10.1016/j.stem.2011.02.021.

Buckley, J., Willingham, E., Agras, K., & Baskin, L. S. (2006). Embryonic exposure to the fungicide vinclozolin causes virilization of females and alteration of progesterone receptor expression in vivo: an experimental study in mice. *Environ. Health*, *5*, 4. https://doi.org/10.1186/1476-069X-5-4.

Burke, W. J., Davis, J. W., & Joh, T. H. (1983). The role of the compartmentalization of epinephrine in the regulation of phenylphenolamine N-methyltransferase synthesis in rat adrenal medulla. *Endocrinology*, *113*, 1102–1110. https://doi.org/10.1210/endo-113-3-1102.

Cabej, N. R. (2008). *Epigenetic Principles of Evolution*. Dumont, NJ: Albanet.

Cabej, N. R. (2012). *Epigenetic Principles of Evolution*. Amsterdam–Boston: Elsevier Publishing.

Cabej, N. R. (2019a). *Epigenetic Principles of Evolution* (2nd ed., p. 364). London, San Diego, Cambridge MA, Oxford: Academic Press.

Cabej, N. R. (2019b). *Ibid* (p. XXXVIII).

Cabej, N. R. (2019c). *Ibid* (pp. 337–379).

Cabej, N. R. (2019d). *Ibid* (pp. 241–244).

Cabej, N. R. (2019e). *Ibid* (p. 368).

Cabej, N. R. (2019f). *Ibid* (pp. 54–59). 247–251, 368–369.

Cabej, N. R. (2019g). *Ibid* (pp. 535–561).

Cabej, N. R. (2020a). *Epigenetic Mechanisms of the Cambrian Explosion* (pp. 83–95). London, San Diego, Cambridge MA, Oxford UK: Academic Press.

Cabej, N. R. (2020b). *Epigenetic Mechanisms of the Cambrian Explosion* (pp. 100–104). London, San Diego, Cambridge MA, Oxford UK: Academic Press.

Cabej, N. R. (2020c). *Ibid* (pp. 104–108).

Cabej, N. R. (2020d). A neural mechanism of nuclear receptor expression and regionalization. *Dev. Dyn.*, *249*(10), 1172–1181.

Carter, G. S. (1967). *Structure and Habit in Vertebrate Evolution* (p. 68). Seattle: University of Washington Press.

Chen, Q., Yan, M., Cao, Z., Li, X., Zhang, Y., Shi, J., et al. (2016). Sperm tsRNAs contribute to intergenerational inheritance of an acquired metabolic disorder. *Science*, *351*(6271), 397–400. https://doi.org/10.1126/science.aad7977.

Chen, L.-F., Lin, Y. T., Gallegos, D. A., Hazlett, M. F., Gómez-Schiavon, M., Yang, M. G., et al. (2019). Enhancer histone acetylation modulates transcriptional bursting dynamics of neuronal activity-inducible genes. *Cell Rep.*, *26*(5), 1174–1188. https://doi.org/10.1016/j.celrep.2019.01.032.

Chong, T., Collins, J., Brubacher, J., Zarkower, D., & Newmark, P. A. (2013). A sex-specific transcription factor controls male identity in a simultaneous hermaphrodite. *Nat. Commun.*, *4*, 1814. https://doi.org/10.1038/ncomms2811.

Chu, C., Zheng, G., Hu, S., Zhang, J., Xie, S., Ma, W., et al. (2015). Epididymal region-specific miRNA expression and DNA methylation and their roles in controlling gene expression in rats. *PLoS One*, *10*(4). https://doi.org/10.1371/journal.pone.0124450, e0124450.

Claeys, I., Breugelmans, B., Simonet, G., Van Soest, S., Sas, F., De Loof, A., et al. (2006). Neuroparsin transcripts as molecular markers in the process of desert locust (*Schistocerca gregaria*) phase transition. *Biochem. Biophys. Res. Commun.*, *341*(2), 599–606. https://doi.org/10.1016/j.bbrc.2006.01.011. 48.

Coble, J. P., Cassell, M. D., Davis, D. R., Grobe, J. L., & Sigmund, C. D. (2014). Activation of the renin-angiotensin system, specifically in the subfornical organ is sufficient to induce fluid intake. *Am. J. Physiol. Regul. Integr. Comp. Physiol.*, *307*(4), R376–R386. https://doi.org/10.1152/ajpregu.00216.2014.

Cohen, T. J., Barrientos, T., Hartman, Z. C., Garvey, S. M., Cox, G. A., & Yao, T.-P. (2009). The deacetylase HDAC4 controls myocyte enhancing factor-2-dependent structural gene expression in response to neural activity. *FASEB J.*, *23*(1), 99–106. https://doi.org/10.1096/fj.08-115931.

Collino, F., Deregibus, M. C., Bruno, S., Sterpone, L., Aghemo, G., Viltono, L., et al. (2010). Microvesicles derived from adult human bone marrow and tissue specific mesenchymal stem cells shuttle selected pattern of miRNAs. *PLoS One*, *5*(7). https://doi.org/10.1371/journal.pone.0011803, e11803.

Collins, J. J., III, Hou, X., Romanova, E. V., Lambrus, B. G., Miller, C. M., Saberi, A., et al. (2010). Genome-wide analyses reveal a role for peptide hormones in planarian germline development. *PLoS Biol.*, *8*(10), e1000509. 2010 Oct.

Cui, D., & Xu, X. (2018). DNA methyltransferases, DNA methylation, and age-associated cognitive function. *Int. J. Mol. Sci.*, *19*(5), 1315. https://doi.org/10.3390/ijms19051315.

Cullen, D. A., Sword, G. A., Dodgson, T., & Simpson, S. J. (2010). Behavioural phase change in the Australian plague locust, *Chortoicetes terminifera*, is triggered by tactile stimulation of the antennae. *J. Insect Physiol.*, *56*(8), 937–942. https://doi.org/10.1016/j.jinsphys.2010.04.023.

Cunningham, E. J., & Russell, A. F. (2000). Egg investment is influenced by male attractiveness in the mallard. *Nature*, *404*(6773), 74–77. https://doi.org/10.1038/35003565.

Dal Bianco Fernandez, C., Porto, E. M., Arena, A. C., & De Grava Kempinas, W. (2007). Effects of altered epididymal sperm transit time on sperm quality. *Int. J. Androl.*, *31*(4), 427–437. https://doi.org/10.1111/j.1365-2605.2007.00788.x.

Darwin, C. R. (1859). *The Origin of Species by Means of Natural Selection* (1st ed., p. 171). London: John Murray.

Darwin, C. R. (1859). *The Origin of Species by Means of Natural Selection* (1st ed., p. 6). London: John Murray.

Darwin, C. R. (1859). *Ibid* (p. 134).

Darwin, C. R. (1859). *Ibid* (p. 454).

Darwin, C. R. (1859). *Ibid* (p. 479).

Darwin, C. R. (1868). *The Variation of Animals and Plants Under Domestication. vol. II* (p. 374). London: John Muray.

Darwin, C. R. (1872). *The Origin of Species by Means of Natural Selection*. sixth ed. with additions and corrections (p. 131). London: John Murray.

De Loof, A., Boerjan, B., Ernst, U. R., & Schoofs, L. (2013). The mode of action of juvenile hormone and ecdysone: towards an epi-endocrinological paradigm? *Gen. Comp. Endocrinol.*, *188*(1), 35–45. https://doi.org/10.1016/j.ygcen.2013.02.004.

Devanapally, S., Ravikumar, S., & Jose, A. M. (2015). Transport of RNA from neurons to the germline. *Proc. Natl. Acad. Sci.*, *112*(7), 2133–2138. https://doi.org/10.1073/pnas.1423333112.

Dias, B. G., & Ressler, K. J. (2014a). Parental olfactory experience influences behavior and neural structure in subsequent generations. *Nat. Neurosci.*, *17*(1), 89–96. https://doi.org/10.1038/nn.3594.

Dias, B. G., & Ressler, K. J. (2014b). Reply to Gregory Francis. *Genetics*, *198*(2), 453. https://doi.org/10.1534/genetics.114.169904.

Ding, X., Liu, S., Tian, M., Zhu, T., Li, D., Wu, J., et al. (2017). Activity-induced histone modifications govern Neurexin-1 mRNA splicing and memory preservation. *Nat. Neurosci.*, *20* (5), 690–699. https://doi.org/10.1038/nn.4536.

Dingle, R. V. (2003). Some paleontological implications of putative, long-term, gene reactivation. *J. Geol. Soc. London*, *160*, 815–818.

Domes, K., Norton, R. A., Maraun, M., & Scheu, S. (2007). Reevolution of sexuality breaks Dollo's law. *Proc. Natl. Acad. Sci. U. S. A.*, *104*, 7139–7144.

Eacker, S. M., Keuss, M. J., Berezikov, E., Dawson, V. L., & Dawson, T. M. (2011). Neuronal activity regulates hippocampal miRNA expression. *PLoS One*, *6*(10). https://doi.org/10.1371/journal.pone.0025068, e25068.

Eaton, S. A., Jayasooriah, N., Buckland, M. E., Martin, D. I., Cropley, J. E., & Suter, C. M. (2015). Roll over Weismann: extracellular vesicles in the transgenerational transmission of environmental effects. *Epigenomics*, *7*(7), 1165–1171. https://doi.org/10.2217/epi.15.58.

Echeverri, K., & Tanaka, M. (2005). Proximodistal patterning during limb regeneration. *Dev. Biol.*, *79*(2), 391–401.

Eldredge, N., & Gould, S. J. (1971). *Punctuated Equilibria: An Alternative to Phyletic Gradualism*. San Francisco: Freeman, Cooper & Co.

Eliasson, B., Rawshani, A., Axelsen, M., Hammarstedt, A., & Smith, U. (2017). Cephalic phase of insulin secretion in response to a meal is unrelated to family history of type 2 diabetes. *PLoS One*, *12*(3). https://doi.org/10.1371/journal.pone.0173654, e0173654.

Ernst, U. R., Van Hiel, M. B., Depuydt, G., Boerjan, B., De Loof, A., & Schoofs, L. (2015). Epigenetics and locust life phase transitions. *J. Exp. Biol.*, *218*, 88–99. https://doi.org/10.1242/jeb.107078.

Fan, S. M., Chang, Y. T., Chen, C. L., Wang, W.-H., Pan, M.-K., Chen, W.-P., et al. (2018). External light activates hair follicle stem cells through eyes via an ipRGC-SCN-sympathetic neural pathway [published correction appears in Proc. Natl. Acad. Sci. U S A. 115(51), E12121]. *Proc. Natl. Acad. Sci. U. S. A.*, *115*(29), E6880–E6889. https://doi.org/10.1073/pnas.1719548115.

Fiore, R., Khudayberdiev, S., Christensen, M., Siegel, G., Flavell, S. W., Kim, T.-. K., et al. (2009). Mef2-mediated transcription of the miR379-410 cluster regulates activity-dependent dendritogenesis by fine-tuning Pumilio2 protein levels. *EMBO J.*, *28*(6), 697–710. https://doi.org/10.1186/s12862-018-1156-7.

Francis, G. (2014). Too much success for recent groundbreaking epigenetic experiments. *Genetics*, *198*(2), 449–451. https://doi.org/10.1534/genetics.114.163998.

Fujita, S., Bohland, M. A., Sanchez-Watts, G., Watts, A. G., & Donovan, C. M. (2007). Hypoglycemic detection at the portal vein is mediated by capsaicin-sensitive primary sensory neurons. *Am. J. Physiol. Endocrinol. Metab.*, *293*, E96–E101.

Fumey, J., Hinaux, H., Noirot, C., Thermes, C., Rétaux, S., & Casane, D. (2018). Evidence for late Pleistocene origin of *Astyanax mexicanus* cavefish. *BMC Evol. Biol.*, *18*(1), 43. https://doi.org/10.1186/s12862-018-1156-7.

Gapp, K., & Bohacek, J. (2018). Epigenetic germline inheritance in mammals: looking to the past to understand the future. *Genes Brain Behav.*, *17*(3), 1–12. e12407 https://doi.org/10.1111/gbb.12407.

Gapp, K., Jawaid, A., Sarkies, P., Bohacek, J., Pelczar, P., Prados, J., et al. (2014). Implication of sperm RNAs in transgenerational inheritance of the effects of early trauma in mice. *Nat. Neurosci.*, *17*(5), 667–669. https://doi.org/10.1038/nn.3695.

Gilbert, S. F. (2001). Ecological developmental biology: developmental biology meets the real world. *Dev. Biol.*, *233*, 1–12. https://doi.org/10.1006/dbio.2001.0210.

Gilbert, S. F., & Epel, D. (2009). *Ecological Developmental Biology: Integrating Epigenetics, Medicine, and Evolution*. Sunderland MA: Sinauer Associates Incorporated. https://works.swarthmore.edu/fac-biology/141.

Gonzalez-Bellido, P. T., Wardill, T. J., Buresch, K. C., Ulmer, K. M., & Hanlon, R. T. (2014). Expression of squid iridescence depends on environmental luminance and peripheral 2 ganglion control. *J. Exp. Biol.*, *217*(Pt. 6), 850–858. https://doi.org/10.1242/jeb.091884.

Gorbman, A., & Davey, K. (1991). Endocrines. In C. L. Prosser (Ed.), *Neural and Integrative Animal Physiology* (4th ed., pp. 693–754). New York: Wiley-Liss.

Gorczynski, R. M., & Steele, E. J. (1980). Inheritance of acquired immunologic tolerance to foreign histocompatibility antigens in mice. *Proc. Natl. Acad. Sci. U. S. A.*, *77*, 2871–2875. https://doi.org/10.1073/pnas.77.5.2871.

Gould, S. J., & Robinson, B. A. (1994). The promotion and prevention of recoiling in a maximally snaillike vermetid gastropod: a case study for the centenary of Dollo's law. *Palaeobiology*, *20*, 368–390.

Grandjean, V., Gounon, P., Wagner, N., Martin, L., Wagner, K. D., Bernex, F., et al. (2009). The miR-124-Sox9 paramutation: RNA-mediated epigenetic control of embryonic and adult growth. *Development*, *136*(21), 3647–3655. https://doi.org/10.1242/dev.041061.

Grandjean, V., Fourré, S., De Abreu, D., Derieppe, M. A., Remy, J. J., & Rassoulzadegan, M. (2016). RNA-mediated paternal heredity of diet-induced obesity and metabolic disorders. *Sci. Rep.*, *5*, 18193. https://doi.org/10.1038/srep18193.

Guan, Z., Giustetto, M., Lomvardas, S., Kim, J.-H., Miniaci, M. C., Schwartz, J. H., et al. (2002). Integration of long-term-memory-related synaptic plasticity involves bidirectional regulation of gene expression and chromatin structure. *Cell*, *111*(4), 483–493. https://doi.org/10.1016/s0092-8674(02)01074-7.

Guduric-Fuchs, J., O'Connor, A., Camp, B., O'Neill, C. L., Medina, R. J., & Simpson, D. A. (2012). Selective extracellular vesicle-mediated export of an overlapping set of microRNAs from multiple cell types. *BMC Genomics*, *13*, 357. https://doi.org/10.1186/1471-2164-13-357.

Guo, S., Jiang, F., Yang, P., Liu, Q., Wang, X., & Kang, L. (2016). Characteristics and expression patterns of histone-modifying enzyme systems in the migratory locust. *Insect Biochem. Mol.*, *76*, 18–28.

Guo, J. U., Ma, D. K., Mo, H., Ball, M. P., Jang, M.-H., Bonaguidi, M. A., et al. (2011). Neuronal activity modifies DNA methylation landscape in the adult brain. Neuronal activity modifies DNA methylation landscape in the adult brain. *Nat. Neurosci.*, *14*(10), 1345–1351. https://doi.org/10.1038/nn.2900.

Guo, X., Yu, Q., Chen, D., Weil, J., Yang, P., Yu, J., et al. (2020). 4-Vinylanisole is an aggregation pheromone in locusts. *Nature*, *584*(7822), 584–588. https://doi.org/10.1038/s41586-020-2610-4.

Gur, D., Palmer, B. A., Leshem, B., Oron, D., Fratzl, P., Weiner, S., et al. (2015). The mechanism of color change in the neon tetra fish: A light-induced tunable photonic crystal array. *Angew. Chem. Int. Ed.*, *54*(42), 12426–12430. https://doi.org/10.1002/anie.201502268.

Hall, B. K. (1998). *Evolutionary Developmental Biology* (2nd ed., p. 339). London: Chapman & Hall.

Hammel, H. T., Jackson, D. C., Stolwijk, J. A. J., Hardy, J. D., & Stroeme, S. B. (1963). Temperature regulation by hypothalamic proportional control with an adjustable set point. *J. Appl. Physiol.*, *18*, 1146–1154.

Handler, A. M., & Postlethwait, J. H. (1977). Endocrine control of vitellogenesis in *Drosophila melanogaster*: effects of the brain and corpus allatum. *J. Exp. Zool., 202*(3), 389–402. https://doi.org/10.1002/jez.1402020309.

Hanlon, R. T., Cooper, K. M., & Budelmann, B. U. (1990). Physiological color change in squid iridophores. *Cell Tissue Res., 259*(1), 3–14. https://doi.org/10.1007/BF00571424.

Harris, K. D. M., Bartlett, N. J., & Lloyd, V. K. (2012). *Daphnia* as an emerging epigenetic model organism. *Genet Res. Int., 2012*. https://doi.org/10.1155/2012/147892, 147892.

Hegstrom, C. D., Riddiford, L. M., & Truman, J. W. (1998). Steroid and neuronal regulation of ecdysone receptor expression during metamorphosis of muscle in the moth, *Manduca sexta*. *J. Neurosci., 18*, 1786–1794.

Henrich, V. C., & Denlinger, D. L. (1982). A maternal effect that eliminates pupal diapause in progeny of the flesh fly, *Sarcophaga bullata*. *J. Insect Physiol., 28*(10), 881–884. https://doi.org/10.1016/0022-1910(82)90102-0.

Hermey, G., Blüthgen, N., & Kuhl, D. (2017). Neuronal activity-regulated alternative mRNA splicing. *Int. J. Biochem. Cell. Biol., 91*(Pt. B), 184–193. https://doi.org/10.1016/j.biocel.2017.06.002.

Hevener, A. L., Bergman, R. N., & Donovan, C. M. (1997). Novel glucosensor for hypoglycemic detection localized to the portal vein. *Diabetes, 46*, 1521–1525.

Hevener, A. L., Bergman, R. N., & Donovan, C. M. (2001). Hypoglycemic detection does not occur in the hepatic artery or liver: findings consistent with a portal vein glucosensor locus. *Diabetes, 40*, 399–403.

Holliday, R., & Pugh, J. E. (1975). DNA modification mechanisms and gene activity during development. *Science, 187*(4173), 226–232. https://doi.org/10.1126/science.187.4173.226.

Hutson, T. H., Kathe, C., Palmisano, I., Bartholdi, K., Hervera, A., De Virgiliis, F., et al. (2019). Cbp-dependent histone acetylation mediates axon regeneration induced by environmental enrichment in rodent spinal cord injury models. *Sci. Transl. Med., 487*(11). https://doi.org/10.1126/scitranslmed.aaw2064, eaaw2064.

Huxley, J. (1942). *Evolution—The Modern Synthesis*. London: G. Allen & Unwin Ltd.

Ichikawa, T. (2003). Firing activities of neurosecretory cells producing diapause hormone and its related peptides in the female silkmoth, *Bombyx mori*. I. Labial cells. *Zool. Sci., 20*(8), 971–978. https://doi.org/10.2108/zsj.20.971.

Ignell, R., Couillaud, F., & Anton, S. (2001). Juvenile hormone-mediated plasticity of aggregation behavior and olfactory processing in adult desert locusts. *J. Exp. Biol., 204*, 249–259.

Iijima, T., Hidaka, C., & Iijima, Y. (2016). Spatio-temporal regulations and functions of neuronal alternative RNA splicing in developing and adult brains. *Neurosci. Res., 109*, 1–8. https://doi.org/10.1016/j.neures.2016.01.010.

Ikeya, T., Galic, M., Belawat, P., Nairz, K., & Hafen, E. (2002). Nutrient-dependent expression of insulin-like peptides from neuroendocrine cells in the CNS contributes to growth regulation in *Drosophila. Curr. Biol., 12*, 1293–1300.

Irimia, M., Weatheritt, R. J., Ellis, J., et al. (2014). A highly conserved program of neuronal microexons is misregulated in autistic brains. *Cell, 159*(7), 1511–1523. https://doi.org/10.1016/j.cell.2014.11.035.

Jablonka, E., & Lamb, M. (1989). The inheritance of acquired epigenetic variations. *J. Theor. Biol., 139*(1), 69–83. https://doi.org/10.1016/S0022-5193(89)80058-X.

Jablonka, E., & Raz, G. (2009). Transgenerational epigenetic inheritance: prevalence, mechanisms, and implications for the study of heredity and evolution. *Q. Rev. Biol., 84*(2), 131–176. https://doi.org/10.1086/598822.

Jansson, J.-O., Palsdottir, V., Hägg, D. A., Schéle, E., Dickson, S. L., Anesten, F., et al. (2018). Body weight homeostat that regulates fat mass. *Proc. Natl. Acad. Sci. U. S. A.*, *115*(2), 427–432. https://doi.org/10.1073/pnas.1715687114.

Jeffery, W. R. (2005). Adaptive evolution of eye degeneration in the Mexican blind cavefish. *J. Hered.*, *96*(3), 185–196. https://doi.org/10.1093/jhered/esi028.

Jeffery, W. R., Strickler, A. G., Guiney, S., Heyser, D., & Tomarev, S. I. (2000). Prox1 in eye degeneration and sensory organ compensation during development and evolution of the cavefish *Astyanax*. *Dev. Genes Evol.*, *210*(5), 223–230. https://doi.org/10.1007/s004270050308.

Jeng, S. F., Rau, C. S., Liliang, P. C., Wu, C.-J., Lu, T.-H., Chen, Y.-C., et al. (2009). Profiling muscle-specific microRNA expression after peripheral denervation and reinnervation in a rat model. *J. Neurotrauma*, *26*(12), 2345–2353.

Kasukawa, H., Oshima, N., & Fujii, R. (1987). Mechanism of light-reflection in blue damselfish motile iridophore. *Zool. Sci.*, *4*(2), 243–257.

Katz, L. C., & Shatz, C. J. (1996). Synaptic activity and the construction of cortical circuits. *Science*, *274*(5290), 1133–1138. https://doi.org/10.1126/science.274.5290.1133.

Kawano, F., Nimura, K., Ishino, S., Nakai, N., Nakata, K., & Ohira, Y. (2015). Differences in histone modifications between slow- and fast-twitch muscle of adult rats and following overload, denervation, or valproic acid administration. *J. Appl. Physiol.*, *119*(10), 1042–1052. https://doi.org/10.1152/japplphysiol.00289.2015.

Kitagawa, N., Shiomi, K., Imai, K., Niimi, T., Yaginuma, T., & Yamashita, O. (2005). Establishment of a sandwich ELISA system to detect diapause hormone, and developmental profile of hormone levels in egg and subesophageal ganglion of the silkworm, *Bombyx mori*. *Zool. Sci.*, *22*, 213–221.

Klaus, S., Mendoza, J. C. E., Liew, J. H., Plath, M., Meier, R., & Yeo, D. C. J. (2013). Rapid evolution of troglomorphic characters suggests selection rather than neutral mutation as a driver of eye reduction in cave crabs. *Biol. Lett.*, *9*(2), 20121098. https://doi.org/10.1098/rsbl.2012.1098.

Koulakoff, A., Ezan, P., & Giaume, C. (2008). Neurons control the expression of connexin 30 and connexin 43 in mouse cortical astrocytes. *Glia*, *56*(2), 1299–1311. https://doi.org/10.1002/glia.20698.

Krishnan, K., Mittal, N., Thompson, L. M., Rodriguez-Santiago, M., Duvauchelle, C. L., Crews, D., et al. (2018). Effects of the endocrine-disrupting chemicals, vinclozolin and polychlorinated biphenyls, on physiological and sociosexual phenotypes in F2 generation Sprague-Dawley rats. *Environ. Health Perspect.*, *126*(9), 97005. https://doi.org/10.1289/EHP3550.

Krol, J., Busskamp, V., Markiewicz, I., Stadler, M. B., Ribi, S., Richter, J., et al. (2010). Characterizing light-regulated retinal microRNAs reveals rapid turnover as a common property of neuronal microRNAs. *Cell*, *141*(4), 618–631. https://doi.org/10.1016/j.cell.2010.03.039.

Langecker, T. G., Schmale, H., & Wilkens, H. (1993). Transcription of the opsin gene in degenerate eyes of cave dwelling *Astyanax fasciatus* (Teleostei, Characidae) and its conspecific ancestor during early ontogeny. *Cell Tissue Res.*, *273*(1), 183–192. https://doi.org/10.1007/BF00304625.

Le, Q., Yan, B., Yu, X., Li, Y., Song, H., Zhu, H., et al. (2017). Drug-seeking motivation level in male rats determines offspring susceptibility or resistance to cocaine-seeking behaviour. *Nat. Commun.*, *8*, 15527. https://doi.org/10.1038/ncomms15527.

LeBlanc, G. A., & Medlock, E. K. (2015). Males on demand: The environmental–neuro-endocrine control of male sex determination in daphnids. *FEBS J.*, *282*(21), 4080–4093.

Lee, B. H., & Ashrafi, K. (2008). A TRPV channel modulates, *C. elegans* neurosecretion, larval starvation survival, and adult lifespan. *PLoS Genet.*, *4*. https://doi.org/10.1371/journal.pgen.1000213.

Lee, Y. S., Shibata, Y., Malhotra, A., & Dutta, A. (2009). A novel class of small RNAs: tRNA-derived RNA fragments (tRFs). *Genes Dev.*, *23*(22), 2639–2649. https://doi.org/10.1101/gad.1837609.

Li, Q., Lee, J. A., & Black, D. L. (2007). Neuronal regulation of alternative pre-mRNA splicing. *Nat. Rev. Neurosci.*, *8*(11), 819–831. https://doi.org/10.1038/nrn2237.

Ma, D. K., Jang, M.-H., Guo, J. U., Kitabatake, Y., Chang, M.-L., Pow-Anpongkul, N., et al. (2009). Neuronal activity–induced Gadd45b promotes epigenetic DNA demethylation and adult neurogenesis. *Science*, *323*(5917), 1074–1077. https://doi.org/10.1126/science.1166859.

Ma, Z., Guo, W., Guo, X., Wang, X., & Kang, L. (2011). Modulation of behavioral phase changes of the migratory locust by the catecholamine metabolic pathway. *Proc. Natl. Acad. Sci. U. S. A.*, *108*(10), 3882–3887. https://doi.org/10.1073/pnas.1015098108.

Makanae, A., Mitogawa, K., & Satoh, A. (2016). Cooperative inputs of Bmp and Fgf signaling induce tail regeneration in urodele amphibians. *Dev. Biol.*, *410*, 45–55.

Makanae, A., Tajika, Y., Nishimura, K., Saito, N., Tanaka, J.-I., & Satoh, A. (2020). Neural regulation in tooth regeneration of *Ambystoma mexicanum*. *Sci. Rep.*, *10*, 9323. https://doi.org/10.1038/s41598-020-66142-2.

Marré, J., Traver, E. C., & Jose, A. M. (2016). Extracellular RNA is transported from one generation to the next in *Caenorhabditis elegans*. *Proc. Natl. Acad. Sci. U. S. A.*, *113*(44), 12496–12501. https://doi.org/10.1073/pnas.1608959113.

Martinowich, K., Hattori, D., Wu, H., Fouse, S., He, F., Hu, Y., et al. (2003). DNA methylation related chromatin remodeling in activity-dependent BDNF gene regulation. *Science*, *302*(5646), 890–893. https://doi.org/10.1126/science.1090842.

Martone, J., Marian, D., Desideri, F., & Ballarino, M. (2019). Non-coding RNAs shaping muscle. *Front. Cell Dev. Biol.*, *7*, 394. https://doi.org/10.3389/fcell.2019.00394.

Mathew, R. S., Tatarakis, A., Rudenko, A., Johnson-Venkatesh, E. M., Yang, Y. J., Murphy, E. A., et al. (2016). A microRNA negative feedback loop downregulates vesicle transport and inhibits fear memory. *eLife*, *5*. https://doi.org/10.7554/eLife.22467, e22467.

Mäthger, L. M., Land, M. F., Siebeck, U. E., & Marshall, N. J. (2003). Rapid colour changes in multilayer reflecting stripes in the paradise whiptail, *Pentapodus paradiseus*. *J. Exp. Biol.*, *206*(Pt20), 3607–3613. https://doi.org/10.1242/jeb.00599.

Mäthger, L. M., Collins, T. F. T., & Lima, P. A. (2004). The role of muscarinic receptors and intracellular Ca^{2+} in the spectral reflectivity changes of squid iridophores. *J. Exp. Biol.*, *207*(Pt11), 1759–1769. https://doi.org/10.1242/jeb.00955.

Matosin, N., Cruceanu, C., & Binder, E. B. (2017). Preclinical and clinical evidence of DNA methylation changes in response to trauma and chronic stress. *Chronic Stress (Thousand Oaks)*, *1*. https://doi.org/10.1177/2470547017710764. 2017; Feb 1. 2470547017710764.

Matsuura, K., Fujimoto, K., Das, B., Fu, L., Lu, C. D., & Shi, Y.-B. (2012). Histone H3K79 methyltransferase Dot1L is directly activated by thyroid hormone receptor during *Xenopus* metamorphosis. *Cell Biosci.*, *2*, 25. https://doi.org/10.1186/2045-3701-2-25.

Mayr, E. (1961). Cause and effect in biology. *Science*, *134*, 1501–1506.

McCaffery, A. R., Simpson, S. J., Islam, M. S., & Roessingh, P. (1998). A gregarizing factor present in the egg pod foam of the desert locust *Schistocerca gregaria*. *J. Exp. Biol.*, *201* (3), 347–363. 9427669.

McLeod, C. J., Wang, L., Wong, C., & Jones, D. L. (2010). Stem cell dynamics in response to nutrient availability. *Curr. Biol.*, *20*(23), 2100–2105.

Méjat, A., Ramond, F., Bassel-Duby, R., Khochbin, S., Olson, E. N., & Schaeffer, L. (2005). Histone deacetylase 9 couples neuronal activity to muscle chromatin acetylation and gene expression. *Nat. Neurosci.*, *8*(3), 313–321. PMID: 15711539 (web archive link).

Middleton, C. A., Nongthomba, U., Parry, K., Sweeney, S. T., Sparrow, J. C., & Elliott, C. J. H. (2006). Neuromuscular organization and aminergic modulation of contractions in the *Drosophila* ovary. *BMC Biol.*, *4*, 17. https://doi.org/10.1186/1741-7007-4-17.

Miller, G. A., Islam, M. S., Claridge, T. D. W., Dodgson, T., & Simpson, S. J. (2008). Swarm formation in the desert locust *Schistocerca gregaria*: isolation and NMR analysis of the primary maternal gregarizing agent. *J. Exp. Biol.*, *211*(3), 370–376. https://doi.org/10.1242/jeb.013458.

Miyakawa, H., Imai, M., Sugimoto, N., Ishikawa, Y., Ishikawa, A., Ishigaki, H., et al. (2010). Gene up-regulation in response to predator kairomones in the water flea, *Daphnia pulex*. *BMC Dev. Biol.*, *10*, 45. https://doi.org/10.1186/1471-213X-10-45.

Moore, R. S., Kaletsky, R., & Murphy, C. T. (2019). Piwi/PRG-1 Argonaute and TGF-β mediate transgenerational learned pathogenic avoidance. *Cell*, *177*(7), 1827–1841.e12. https://doi.org/10.1016/j.cell.2019.05.024.

Morrison, S. F. (2004). Central pathways controlling brown adipose tissue thermogenesis. *News Physiol. Sci.*, *19*(2), 67–74. https://doi.org/10.1152/nips.01502.2003.

Morrison, S. F. (2016). Central control of body temperature. *F1000Res.*, *5*. https://doi.org/10.12688/f1000research.7958.1. F1000 Faculty Rev-880. Published 2016 May 12.

Moullé, V. S., Tremblay, C., Castell, A. L., Vivot, K., Ethier, M., Fergusson, G., et al. (2019). The autonomic nervous system regulates pancreatic β-cell proliferation in adult male rats. *Am. J. Physiol. Endocrinol. Metab.*, *317*(2), E234–E243. https://doi.org/10.1152/ajpendo.00385.2018.

Nätt, D., Kugelberg, U., Casas, E., Nedstrand, E., Zalavary, S., Henriksson, P., et al. (2019). Human sperm displays rapid responses to diet. *PLoS Biol.*, *17*(12). https://doi.org/10.1371/journal.pbio.3000559, e3000559.

Nebigil, C. G., & Maroteaux, L. (2001). A novel role for serotonin in heart. *Trends Cardiovasc. Med.*, *11*, 329–335.

Nebigil, C. G., Choi, D.-S., Dierich, A., Hickel, P., Le Meur, M., Messaddeq, N., et al. (2000). Serotonin 32B receptor is required for heart development. *Proc. Natl. Acad. Sci. U. S. A.*, *97*, 9508–9513. https://doi.org/10.1073/pnas.97.17.9508.

Nelson, E. D., Kavalali, E. T., & Monteggia, L. M. (2008). Activity-dependent suppression of miniature neurotransmission through the regulation of DNA methylation. *J. Neurosci.*, *28* (2), 395–406. https://doi.org/10.1523/JNEUROSCI.3796-07.2008.

Newman, S. A., & Müller, G. B. (2000). Epigenetic mechanisms of character origination. *J. Exp. Zool. B Mol. Dev. Evol.*, *288*(4), 304–317. https://doi.org/10.1002/1097-010x(20001215)288:4<304::aid-jez3>3.0.co;2-g.

Nixon, B., Stanger, S. J., Mihalas, B. P., Reilly, J. N., Anderson, A. L., Tyagi, S., et al. (2015). The microRNA signature of mouse spermatozoa is substantially modified during epididymal maturation. *Biol. Reprod.*, *93*(4), 91. https://doi.org/10.1095/biolreprod.115.132209.

Noble, D. (2015). Conrad Waddington and the origin of epigenetics. *J. Exp. Biol.*, *218*, 816–818. https://doi.org/10.1242/jeb.120071.

Noguchi, H., & Hayakawa, Y. (2001). Dopamine is a key factor for the induction of egg diapause of the silkworm, *Bombyx mori*. *Eur. J. Biochem.*, *268*, 774–780.

Nudelman, A. S., DiRocco, D. P., Lambert, T. J., Garelick, M. G., Le, J., Nathanson, N. M., et al. (2010). Neuronal activity rapidly induces transcription of the CREB-regulated microRNA-132, in vivo. *Hippocampus*, *20*(4), 492–498. https://doi.org/10.1002/hipo.20646.

O'Brien, E. A., Ensbey, K. S., Day, B. W., Baldock, P. A., & Barry, G. (2020). Direct evidence for transport of RNA from the mouse brain to the germline and offspring. *BMC Biol.*, *18*(1), 45. https://doi.org/10.1186/s12915-020-00780-w.

Oliveira, A. M., Hemstedt, T. J., & Bading, H. (2012). Rescue of aging-associated decline in Dnmt3a2 expression restores cognitive abilities. *Nat. Neurosci.*, *15*(8), 1111–1113. https://doi.org/10.1038/nn.3151.

Park, C. S., & Tang, S. J. (2009). Regulation of microRNA expression by induction of bidirectional synaptic plasticity. *J. Mol. Neurosci.*, *38*(1), 50–56. https://doi.org/10.1007/s12031-008-9158-3.

Pauls, D., Blechschmidt, C., Frantzmann, F., El Jundi, B., & Selcho, M. (2018). A comprehensive anatomical map of the peripheral octopaminergic/tyraminergic system of *Drosophila melanogaster*. *Sci. Rep.*, *8*, 15314. https://doi.org/10.1038/s41598-018-33686-3.

Peng, H., Shi, J., Zhang, Y., Zhang, H., Liao, S., Li, W., et al. (2012). A novel class of tRNA-derived small RNAs extremely enriched in mature mouse sperm. *Cell Res.*, *22*(11), 1609–1612. https://doi.org/10.1038/cr.2012.141.

Penn, A. A. (2001). Early brain wiring: activity-dependent processes. *Schizophr. Bull.*, *27*(3), 337–347. https://doi.org/10.1093/oxfordjournals.schbul.a006880.

Penn, A. A., Riquelme, P. A., Feller, M. B., & Shatz, C. J. (1998). Competition in retinogeniculate pattern driven by spontaneous activity. *Science*, *279*(5359), 2108–2112. https://doi.org/10.1126/science.279.5359.2108.

Pigati, L., Yaddanapudi, S. C., Iyengar, R., Kim, D. J., Hearn, S. A., Danforth, D., et al. (2010). Selective release of microRNA species from normal and malignant mammary epithelial cells. *PLoS One*, *5*(10). https://doi.org/10.1371/journal.pone.0013515150, e13515.

Ponulak, F., & Kasiński, A. (2011). Introduction to spiking neural networks: information processing, learning and applications. *Acta Neurobiol. Exp.*, *71*(4), 409–433. 22237491.

Posner, R., Toker, I. A., Antonova, O., Star, E., Anava, S., Azmon, E., et al. (2019). Neuronal small RNAs control behavior transgenerationally. *Cell*, *177*, 1814–1826.e15. https://doi.org/10.1016/j.cell.2019.04.029.

Pottin, K., Hinaux, H., & Rétaux, S. (2011). Restoring eye size in *Astyanax mexicanus* blind cavefish embryos through modulation of the Shh and Fgf8 forebrain organising centres. *Development*, *138*(12), 2467–2476. https://doi.org/10.1242/dev.054106.

Powell, S., Price, S. L., & Kronauer, D. J. C. (2020). Trait evolution is reversible, repeatable, and decoupled in the soldier caste of turtle ants. *Proc. Natl. Acad. Sci. U. S. A.*, *117*(12), 6608–6615. https://doi.org/10.1073/pnas.1913750117.

Raikhel, A. S., & Lea, A. O. (1985). Hormone-mediated formation of the endocentric complex in mosquito oocytes. *Gen. Comp. Endocrinol.*, *57*(3), 422–433. https://doi.org/10.1016/0016-6480(85)90224-2.

Raote, I., Bhattacharyya, S., & Panicker, M. M. (2013). Functional selectivity in serotonin receptor 2A (5-HT2A) endocytosis, recycling, and phosphorylation. *Mol. Pharmacol.*, *83*(1), 42–50. https://doi.org/10.1124/mol.112.078626.

Rassoulzadegan, M., Grandjean, V., Gounon, P., Vincent, S., Gillot, I., & Cuzin, F. (2006). RNA-mediated non-mendelian inheritance of an epigenetic change in the mouse. *Nature*, *441*(7092), 469–474. https://doi.org/10.1038/nature04674.

Razanau, A., & Xie, J. (2013). Emerging mechanisms and consequences of calcium regulation of alternative splicing in neurons and endocrine cells. *Cell. Mol. Life Sci.*, *70*(23), 4527–4536. https://doi.org/10.1007/s00018-013-1390-5.

Rechavi, O., Minevich, G., & Hobert, O. (2011). Transgenerational inheritance of an acquired small RNA-based antiviral response in C. elegans. *Cell*, *147*(6), 1248–1256. https://doi.org/10.1016/j.cell.2011.10.042.

Reza, H. M., & Yasuda, K. (2004a). Lens differentiation and crystallin regulation: a chick model. *Int. J. Dev. Biol.*, *48*, 805–817.

Reza, H. M., & Yasuda, K. (2004b). The involvement of neural retina Pax6 in lens fiber differentiation. *Dev. Neurosci.*, *26*, 318–327.

Richard, D. S., Jones, J. M., Barbarito, M. R., Cerula, S., Detweiler, J. P., Fisher, S. J., et al. (2001). Vitellogenesis in diapausing and mutant *Drosophila melanogaster*: further evidence for the relative roles of ecdysteroids and juvenile hormones. *J. Insect Physiol.*, *47*(8), 905–913. https://doi.org/10.1016/S0022-1910(01)00063-4.

Ricker, D. D. (1998). The autonomic innervation of the epididymis: its effects on epididymal function and fertility. *J. Androl.*, *19*(1), 1–4. 9537285.

Ricker, D. D., Crone, J. K., Chamness, S. L., Strader, L. F., Ferrell, J., Goldman, J. M., et al. (1997). Partial sympathetic denervation of the rat epididymis permits fertilization but inhibits embryo development. *J. Androl.*, *18*(2), 131–138. https://doi.org/10.1002/j.1939-4640.1997.tb01893.x.

Ricq, E. L., Hooker, J. M., & Haggarty, S. J. (2016). Activity-dependent regulation of histone lysine demethylase KDM1A by a putative thiol/disulfide switch. *J. Biol. Chem.*, *291*(47), 24756–24767. https://doi.org/10.1074/jbc.M116.734426.

Riggs, A. D. (1975). X inactivation, differentiation and DNA methylation. *Cytogenet. Cell Genet.*, *14*(1), 9–25. https://doi.org/10.1159/000130315.

Roberts, A. L., Gladish, N., Gatev, E., Jones, M. J., Chen, Y., MacIsaac, J. L., et al. (2018). Exposure to childhood abuse is associated with human sperm DNA methylation. *Transl. Psychiatry*, *8*(1), 194. https://doi.org/10.1038/s41398-018-0252-1.

Robichaud, N. F., Sassine, J., Beaton, M. J., & Lloyd, V. K. (2012). The epigenetic repertoire of *Daphnia magna* includes modified histones. *Genet. Res. Int.*, *2012*. https://doi.org/10.1155/2012/174860, 174860.

Rogers, S. M., & Ott, S. R. (2015). Differential activation of serotonergic neurons during short- and long-term gregarization of desert locusts. *Proc. Biol. Sci.*, *282*(1800), 20142062. https://doi.org/10.1098/rspb.2014.2062.

Rogers, S. M., Matheson, T., Despland, E., Dodgson, T., Burrows, M., & Simpson, S. J. (2003). Mechanosensory-induced behavioural gregarization in the desert locust *Schistocerca gregaria*. *J. Exp. Biol.*, *206*, 3991–4002.

Rogers, S. M., Matheson, T., Sasaki, K., Kendrick, K., Simpson, S. J., & Burrows, M. (2004). Substantial changes in central nervous system neurotransmitters and neuromodulators accompany phase change in the locust. *J. Exp. Biol.*, *207*(20), 3603–3617. https://doi.org/10.1242/jeb.01183.

Romero, A., & Green, S. M. (2005). The end of regressive evolution: examining and interpreting the evidence from cave fishes. *J. Fish Biol.*, *67*, 3–32.

Rulifson, E. J., Kim, S. K., & Nusse, R. (2002). Ablation of insulin-producing neurons in flies: growth and diabetic phenotypes. *Science*, *296*, 1118–1120.

Sambandan, S., Akbalik, G., Kochen, L., Rinne, J., Kahlstatt, J., Glock, C., et al. (2017). Activity-dependent spatially localized miRNA maturation in neuronal dendrites. *Science*, *355*(6325), 634–637. https://doi.org/10.1126/science.aaf8995.

Santoro, S. W., & Dulac, C. (2012). The activity-dependent histone variant H2BE modulates the life span of olfactory neurons. *eLife*, *1*, e00070. Published 2012 Dec 13 https://doi.org/10.7554/eLife.00070.

Sarker, G., Sun, W., Rosenkranz, D., Pelczar, P., Opitz, L., Efthymiou, V., et al. (2019). Maternal overnutrition programs hedonic and metabolic phenotypes across generations through sperm tsRNAs. *Proc. Natl. Acad. Sci. U. S. A.*, *116*(21), 10547–10556. https://doi.org/10.1073/pnas.1820810116.

Satoh, A., James, M. A., & Gardiner, D. M. (2009). Axolotl limb regeneration. *J. Bone Joint Surg. Am.*, *91*, 90–98.

Satoh, A., Makanae, A., Nishimoto, Y., & Mitogawa, K. (2016). FGF and BMP derived from dorsal root ganglia regulate blastema induction in limb regeneration in *Ambystoma mexicanum*. *Dev. Biol.*, *417*, 114–125.

Scheckel, C., & Darnell, R. B. (2015). Microexons—tiny but mighty. *EMBO J.*, *34*(3), 273–274. https://doi.org/10.15252/embj.201490651.

Schlichting, C. D., & Pigliucci, M. (1998). *Phenotypic Evolution: A Reaction Norm Perspective*. Sunderland, MA: Sinauer Associates.

Schwartz, Y., Gonzalez-Celeiro, M., Chen, C.-L., Pasolli, H. A., Sheu, S.-H., Fan, S. M.-Y., et al. (2020). Cell types promoting goosebumps form a niche to regulate hair follicle stem cells. *Cell*, *182*(3), 578–593.

Segal, M., Greenberger, V., & Korkotian, E. (2003). Formation of dendritis spines in cultured striatal neurons depends on excitatory afferent activity. *Eur. J. Neurosci.*, *17*(12), 2573–2585. https://doi.org/10.1046/j.1460-9568.2003.02696.x.

Sharma, U., Conine, C. C., Shea, J. M., Boskovic, A., Derr, A. G., Bing, X. Y., et al. (2016). Biogenesis and function of tRNA fragments during sperm maturation and fertilization in mammals. *Science*, *351*(6271), 391–396. https://doi.org/10.1126/science.aad6780.

Shi, Y.-B. (2013). Unliganded thyroid hormone receptor regulates metamorphic timing via the recruitment of histone deacetylase complexes. *Curr. Top. Dev. Biol.*, *105*, 275–297. https://doi.org/10.1016/B978-0-12-396968-2.00010-5.

Shimizu, K., & Stopfer, M. (2013). Gap junctions. *Cell*, *23*(23), R1026–R1031. https://doi.org/10.1016/j.cub.2013.10.067.

Shimizu, I., Matsui, T., & Hasegawa, K. (1989). Possible involvement of GABAergic neurons in regulation of diapause hormone secretion in the silkworm, *Bombyx mori*. *Zool. Sci.*, *6*, 809–819.

Shimizu, I., Aoki, S., & Ichikawa, T. (1997). Neuroendocrine control of diapause hormone secretion in the silkworm, *Bombyx mori*. *J. Insect Physiol.*, *43*, 1101–1109.

Sim, S.-E., Bakes, J., & Kaang, B.-K. (2014). Neuronal activity-dependent regulation of microRNAs. *Mol. Cells*, *37*(7), 511–517. https://doi.org/10.14348/molcells.2014.0132.

Skinner, M. K., Manikkam, M., Haque, M. M., Zhang, B., & Savenkova, M. I. (2012). Epigenetic transgenerational inheritance of somatic transcriptomes and epigenetic control regions. *Genome Biol.*, *13*, R91. https://doi.org/10.1186/gb-2012-13-10-r91.

Smith, A. K., Conneely, K. N., Kilaru, V., Mercer, K. B., Weiss, T. E., Bradley, B., et al. (2011). Differential immune system DNA methylation and cytokine regulation in post-traumatic stress disorder. *Am. J. Med. Genet. B Neuropsychiatr. Genet.*, *156*(6), 700–708. https://doi.org/10.1002/ajmg.b.31212.

Soni, K., Choudhary, A., Patowary, A., Singh, A. R., Bhatia, S., Sivasubbu, S., et al. (2013). miR-34 is maternally inherited in *Drosophila melanogaster* and *Danio rerio*. *Nucleic Acids Res.*, *41*(8), 4470–4480. https://doi.org/10.1093/nar/gkt139.

Soriano, F. X., Papadia, S., Bell, K. F. S., & Hardingham, G. E. (2009). Role of histone acetylation in the activity-dependent regulation of sulfiredoxin and sestrin 2. *Epigenetics*, *4*(3), 152–158. https://doi.org/10.4161/epi.4.3.8753.

Srivastava, M., Simakov, O., Chapman, J., Fahey, B., Gauthier, M. E. A., Mitros, T., et al. (2010). The *Amphimedon queenslandica* genome and the evolution of animal complexity. *Nature*, *466*, 720–726. https://doi.org/10.1038/nature09201.

Steele, E. J. (1979). *Somatic Selection and Adaptive Evolution: On the Inheritance of Acquired Characters* (1st ed.). Toronto: Williams-Wallace.

Stocum, D. L. (2011). The role of peripheral nerves in urodele limb regeneration. *Eur. J. Neurosci.*, *34*(6), 908–916. https://doi.org/10.1111/j.1460-9568.2011.07827.x.

Strepetkaitė, D., Alzbutas, G., Astromskas, E., Lagunavičius, A., Sabaliauskaitė, R., & Arbačiauskas, K. (2016). Analysis of DNA methylation and hydroxymethylation in the genome of crustacean *Daphnia pulex*. *Genes (Basel)*, *7*(1), 1. https://doi.org/10.3390/genes7010001.

Suzuki, S., Ayukawa, N., Okada, C., et al. (2017). Spatio-temporal and dynamic regulation of neurofascin alternative splicing in mouse cerebellar neurons. *Sci. Rep.*, *7*(1), 11405. https://doi.org/10.1038/s41598-017-11319-5.

Takeda, H., Nishimura, K., & Agata, K. (2009). Planarians maintain a constant ratio of different cell types during changes in body size by using the stem cell system. *Zool. Sci.*, *26*, 805–813.

Tang, F., Kaneda, M., O'Carroll, D., Hajkova, P., Barton, S. C., Sun, Y. A., et al. (2007). Maternal microRNAs are essential for mouse zygotic development. *Genes Dev.*, *21*(6), 644–648. https://doi.org/10.1101/gad.418707.

Tang, H., Macpherson, P., Marvin, M., Meadows, E., Klein, W. H., Yang, X.-J., et al. (2009). A histone deacetylase 4/myogenin positive feedback loop coordinates denervation dependent gene induction and suppression. *Mol. Biol. Cell*, *20*(4), 1120–1131. PMCID: PMC2642751.

Teotónio, H., & Rose, M. R. (2000). Variation in the reversibility of evolution. *Nature*, *408*, 463–466.

Teotónio, H., & Rose, M. R. (2001). Perspective: reverse evolution. *Evolution*, *55*, 653–660.

Teotónio, H., Matos, M., & Rose, M. R. (2002). Reverse evolution of fitness in *Drosophila melanogaster*. *J. Evol. Biol.*, *15*(4), 608–617.

Teyke, T. (1990). Morphological differences in neuromasts of the blind cave fish *Astyanax hubbsi* and the sighted fish *Astyanax mexicanus*. *Brain Behav. Evol.*, *35*, 23–30.

Thorens, B. (2011). Brain glucose sensing and neural regulation of insulin and glucagon secretion. *Diabetes Obes. Metab.*, *13*, 82–88. https://doi.org/10.1111/j.1463-1326.2011.01453.x.

Thorens, B. (2014). Neural regulation of pancreatic islet cell mass and function. *Diabetes Obes. Metab.*, *16*(Suppl 1), 87–95.

Toker, I. A., Lev, I., Mor, Y., Gurevich, Y., Fisher, D., Houri-Zeevi, L., et al. (2020). Transgenerational regulation of sexual attractiveness in *C. elegans* nematodes. *bioRxiv*. https://doi.org/10.1101/2020.11.18.389387, 2020.11.18.389387.

Tonkin, L. A., & Bass, B. L. (2003). Mutations in RNAi rescue aberrant chemotaxis of ADAR mutants. *Science*, *302*, 1725. https://doi.org/10.1126/science.1091340.

Tóth, K. F., Pezic, D., Stuwe, E., & Webster, A. (2016). The piRNA pathway guards the germline genome against transposable elements. *Adv. Exp. Med. Biol.*, *886*, 51–77. https://doi.org/10.1007/978-94-017-7417-8_4.

Tronick, E., & Hunter, R. G. (2016). Waddington, dynamic systems, and epigenetics. *Front. Behav. Neurosci.*, *10*, 107. https://doi.org/10.3389/fnbeh.2016.00107.

Tyebji, S., Hannan, A. J., & Tonkin, C. J. (2020). Pathogenic infection in male mice changes sperm small RNA profiles and transgenerationally alters offspring behavior. *Cell Rep.*, *31*(4), 107573. APRIL 28, 2020.

Tzschenke, B., & Basta, D. (2002). Early development of neuronal hypothalamic thermosensitivity in birds: influence of epigenetic temperature adaptation. *Comp. Biochem. Physiol. A*, *131*, 825–832.

Tzschenke, B., & Nichelmann, M. (1997). Influence of prenatal and postnatal acclimation on nervous and peripheral thermoregulation. *Ann. N.Y. Acad. Sci.*, *813*, 87–94 (Abstract).

Ueishi, S., Shimizu, H., & Inoue, Y. H. (2009). Male germline stem cell division and spermatocyte growth require insulin signaling in *Drosophila*. *Cell Struct. Funct.*, *34*(1), 61–69.

Unguez, G. A., & Zakon, H. H. (2002). Skeletal muscle transformation into electric organ in *S. macrurus* depends on innervation. *J. Neurobiol.*, *53*, 391–402.

Vallet, P. G., & Baertschi, A. J. (1982). Spinal afferents for peripheral osmoreceptors in the rat. *Brain Res.*, *239*, 271–274.

van Steenwyk, G., Gapp, K., Jawaid, A., Germain, P.-L., Manuella, F., Tanwar, D. K., et al. (2019). A novel mode of communication between blood and the germline for the inheritance of paternal experiences. *bioRxiv*. https://doi.org/10.1101/653865.

Vastenhouw, N. L., Fischer, S. E., Robert, V. J., Thijssen, K. L., Fraser, A. G., Kamath, R. S., et al. (2003). A genome-wide screen identifies 27 genes involved in transposon silencing in *C. elegans*. *Curr. Biol.*, *13*, 1311–1316. https://doi.org/10.1016/S0960-9822(03)00539-6.

Venugopal, K. J., & Kumar, D. (2000). Role of juvenile hormone in the synthesis and sequestration of vitellogenins in the red cotton stainer, *Dysdercus koenigi* (Heteroptera: Pyrrhocoridae). *Comp. Biochem. Physiol. C*, *127*(2), 153–163. https://doi.org/10.1016/s0742-8413(00)00143-2.

Wang, Y., Zayas, R. M., Guo, T., & Newmark, P. A. (2007). Nanos function is essential for development and regeneration of planarian germ cells. *Proc. Natl. Acad. Sci. U. S. A.*, *104*, 5901–5906.

Wardill, T. J., Gonzalez-Bellido, P. T., Crook, R. J., & Hanlon, R. T. (2012). Neural control of tuneable skin iridescence in squid. *Proc. Biol. Sci.*, *279*(1745), 4243–4252. https://doi.org/10.1098/rspb.2012.1374.

Wayman, G. A., Davare, M., Ando, H., Fortin, D., Varlamova, O., Cheng, H.-Y. M., et al. (2008). An activity-regulated microRNA controls dendritic plasticity by down-regulating p250GAP. *Proc. Natl. Acad. Sci. U. S. A.*, *105*(26), 9093–9098. https://doi.org/10.1073/pnas.0803072105.

Webb, M.-L., & Denlinger, D. L. (1998). GABA and picrotoxin alter expression of a maternal effect that influences pupal diapause in the flesh fly, *Sarcophaga bullata*. *Physiol. Entomol.*, *23*(2), 184–191. https://doi.org/10.1046/j.1365-3032.1998.232073.x.

Webster, A. K., Jordan, J. M., Hibshman, J. D., Chitrakar, R., & Baugh, L. R. (2018). Transgenerational effects of extended Dauer diapause on starvation survival and gene expression plasticity in *Caenorhabditis elegans*. *Genetics*, *210*(1), 263–274. https://doi.org/10.1534/genetics.118.301250.

Weiss, L. C. (2018). Sensory ecology of predator-induced phenotypic plasticity. *Front. Behav. Neurosci.*, *12*, 330. https://doi.org/10.3389/fnbeh.2018.00330.

Weiss, L. C., Leese, F., Laforsch, C., & Tollrian, R. (2015). Dopamine is a key regulator in the signalling pathway underlying predator-induced defences in *Daphnia*. *Proc. Biol. Sci.*, *282* (1816), 20151440. https://doi.org/10.1098/rspb.2015.1440.

Weissman, A. (1892). *Aufsätze über Vererbung und verwandte biologische Fragen* (p. 329). Jena: Fischer. "Es erscheint denkbar, dass solche Einflüsse auch verschiedenartige kleine Abänderungen in der molekularen Structur des Keimplasmas hervorrufen könnten. Da das Keimplasma—unserer Annahme gemäss—sich von einer generation auf die andere überträgt, so müssten also solche Veränderungen erbliche sein".

West, A. E., & Greenberg, M. E. (2011). Neuronal activity-regulated gene transcription in synapse development and cognitive function. *Cold Spring Harb. Perspect. Biol.*, *3*(6), a005744. Published 2011 Jun 1 https://doi.org/10.1101/cshperspect.a005744.

West-Eberhard, M. J. (1986). Alternative adaptations, speciation, and phylogeny. *Proc. Natl. Acad. Sci. U. S. A.*, *83*(5), 1388–1392. PMID 16578790.

Whiting, M. F., Bradler, S., & Maxwell, T. (2003). Loss and recovery of wings in stick insects. *Nature*, *421*, 264–267.

Wijayatunge, R. (2012). *The H3K27 Histone Demethylase Kdm6b (Jmjd3) is Induced by Neuronal Activity and Contributes to Neuronal Survival and Differentiation. Dissertation.* Duke University. https://dukespace.lib.duke.edu/dspace/bitstream/handle/10161/5577/WIJAYATUNGE_duke_0066D_11393.pdf?sequence=1. Accessed on Aug. 10. 2020.

Wijayatunge, R., Chen, L.-F., Cha, Y. M., Zannas, A. S., Frank, C. L., & West, A. E. (2014). The histone lysine demethylase Kdm6b is required for activity-dependent preconditioning of hippocampal neuronal survival. *Mol. Cell. Neurosci.*, *61*, 187–200. https://doi.org/10.1016/j.mcn.2014.06.008.

Williams, A. H., Valdez, G., Moresi, V., Qi, X., McAnally, J., Elliott, J. L., et al. (2009). MicroRNA-206 delays ALS progression and promotes regeneration of neuromuscular synapses in mice. *Science*, *326*(5959), 1549–1554. https://doi.org/10.1126/science.1181046.

Woltereck, R. (1909). Weitere experimentelle Untersuchungen uber Artveränderung, speziell über das Wesen quantitativer Artunterschiede bei Daphniden. *Verh. Deut. Z.*, *19*, 110–172.

Wong, C., & Roy, R. (2020). AMPK regulates developmental plasticity through an endogenous small RNA pathway in *Caenorhabditis elegans*. *Int. J. Mol. Sci.*, *21*(6). https://doi.org/10.3390/ijms21062238. 23 Mar 2020.

Yamagata, Y., Asada, H., Tamura, I., Lee, L., Maekawa, R., Taniguchi, K., et al. (2009). DNA methyltransferase expression in the human endometrium: down-regulation by progesterone and estrogen. *Hum. Reprod.*, *24*(5), 1126–1132. https://doi.org/10.1093/humrep/dep015.

Yamamoto, Y., Stock, D. W., & Jeffery, W. R. (2004). Hedgehog signaling controls eye degeneration in blind cavefish. *Nature*, *431*(7919), 844–847. https://doi.org/10.1038/nature02864.

Yamanaka, A., Tsurumaki, J., & Katsuhiko, E. (2000). Neuroendocrine regulation of seasonal morph development in a bivoltine race (Daizo) of the silkmoth, *Bombyx mori* L. *J. Insect Physiol.*, *46*(5), 803–808. https://doi.org/10.1016/S0022-1910(99)00169-9.

Yoshida, K., Maekawa, T., Ly, N. H., Fujita, S. I., Muratani, M., Ando, M., et al. (2020). ATF7-dependent epigenetic changes are required for the intergenerational effect of a paternal low-protein diet. *Mol. Cell*, *78*(3), 445–458.e6. https://doi.org/10.1016/j.molcel.2020.02.028.

Yoshioka, S., Matsuhana, B., Tanaka, S., Inouye, Y., Oshima, N., & Kinoshita, S. (2011). Mechanism of variable structural colour in the neon tetra: quantitative evaluation of the Venetian blind model. *J. R. Soc. Interface*, *8*(54), 56–66.

Zalokar, M. (1965). Etudes de la formation de l'acide ribonucléique et des protéines chez les insectes. *Rev. Suisse Zool.*, *72*, 241–262.

Zhang, X., & Chen, Q. A. (2020). Twist between ROS and sperm-mediated intergenerational epigenetic inheritance. *Mol. Cell*, *78*(3), 371–373. https://doi.org/10.1016/j.molcel.2020.04.003.

Zhang, Y., Zhang, X., Shi, J., Tuorto, F., Li, X., Liu, Y., et al. (2018). Dnmt2 mediates inter-generational transmission of paternally acquired metabolic disorders through sperm small non-coding RNAs. *Nat. Cell Biol.*, *20*(5), 535–540. https://doi.org/10.1038/s41556-018-0087-2.

Zhao, Z. D., Yang, W. Z., Gao, C., Fu, X., Zhang, W., Zhou, Q., et al. (2017). A hypothalamic circuit that controls body temperature [published correction appears in proc. Natl. Acad. Sci. USA 114(9), E1755]. *Proc. Natl. Acad. Sci. U. S. A.*, *114*(8), 2042–2047. https://doi.org/10.1073/pnas.1616255114.

Zhou, W., Stanger, S. J., Anderson, A. L., Bernstein, I. R., De Iuliis, G. N., McCluskey, A., et al. (2019). Mechanisms of tethering and cargo transfer during epididymosome-sperm interactions. *BMC Biol.*, *17*, 35. https://doi.org/10.1186/s12915-019-0653-5.

Zhu, T., Liang, C., Li, D., Tian, M., Liu, S., Gao, G., et al. (2016). Histone methyltransferase Ash1L mediates activity-dependent repression of neurexin-1α. *Sci. Rep.*, *6*, 26597 (2016) https://doi.org/10.1038/srep26597.

Zovkic, I. B., & Sweatt, J. D. (2013). Epigenetic mechanisms in learned fear: implications for PTSD. *Neuropsychopharmacology*, *38*(1), 77–93. https://doi.org/10.1038/npp.2012.79.

Index

Note: Page numbers followed by *f* indicate figures.

Printed in the United States
by Baker & Taylor Publisher Services